Ecology: Achievement and Challenge

The 41st Symposium of the British Ecological Society jointly sponsored by the Ecological Society of America held at Orlando, Florida, USA 10–13 April 2000

EDITED BY

MALCOLM C. PRESS
Department of Animal and Plant Sciences
University of Sheffield, UK

NANCY J. HUNTLY
Center for Ecological Research and Education
Idaho State University, USA

AND

SIMON LEVIN
Department of Ecology and Evolutionary Biology
Princeton University, USA

Blackwell
Science

© 2001 by
Blackwell Science Ltd
Editorial Offices:
Osney Mead, Oxford OX2 0EL
25 John Street, London WC1N 2BS
23 Ainslie Place, Edinburgh EH3 6AJ
350 Main Street, Malden
 MA 02148-5018, USA
54 University Street, Carlton
 Victoria 3053, Australia
10, rue Casimir Delavigne
 75006 Paris, France

Other Editorial Offices:
Blackwell Wissenschafts-Verlag
GmbH
Kurfürstendamm 57
10707 Berlin, Germany

Blackwell Science KK
MG Kodenmacho Building
7–10 Kodenmacho Nihombashi
Chuo-ku, Tokyo 104, Japan

Iowa State University Press
A Blackwell Science Company
2121 S. State Avenue
Ames, Iowa 50014-8300, USA

First published 2001

Set by Best-set Typesetter Ltd., Hong Kong
Printed and bound in Great Britain at MPG Books Ltd, Bodmin, Cornwall

The Blackwell Science logo is a trade mark of Blackwell Science Ltd, registered at the United Kingdom Trade Marks Registry

The right of the Author to be identified as the Author of this Work has been asserted in accordance with the Copyright, Designs and Patents Act 1988.

All rights reserved. No part of this publication may be reproduced, stored in a retrieval system, or transmitted, in any form or by any means, electronic, mechanical, photocopying, recording or otherwise, except as permitted by the UK Copyright, Designs and Patents Act 1988, without the prior permission of the copyright owner.

A catalogue record for this title is available from the British Library

ISBN 0-632-05879-X Paperback
 0-632-05878-1 Hardback

Library of Congress
Cataloging-in-Publication Data

British Ecological Society.
 Symposium (41st : 2000 :
 Orlando, Fla.)
  Ecology: achievement and challenge: the 41st Symposium of the British Ecological Society jointly sponsored by the Ecological Society of America, held at Orlando, Florida, USA 10–13 April 2000 / edited by Malcolm C. Press, Nancy J. Huntly, and Simon Levin.
   p. cm.
  Includes bibliographical references.
  ISBN 0-632-05878-1 — ISBN 0-632-05879-X (pbk.)
  1. Ecology—Congresses.
 I. Press, Malcolm C. II. Huntly, Nancy J. III Levin, Simon A.
 IV. Title.

QH540.B75 2000
577—dc21
                    2001025043

DISTRIBUTORS

Marston Book Services Ltd
PO Box 269
Abingdon, Oxon OX14 4YN
(*Orders*: Tel: 01235 465500
        Fax: 01235 465555)

The Americas
Blackwell Publishing
c/o AIDC
P.O. Box 20
50 Winter Sport Lane
Williston, VT 05495-0020
(*Orders*: Tel: 800 216 2522
        Fax: 802 864 7626)

Australia
Blackwell Science Pty Ltd
54 University Street
Carlton, Victoria 3053
(*Orders*: Tel: 3 9347 0300
        Fax: 3 9347 5001)

For further information on Blackwell Science, visit our website:
www.blackwell-science.com

Contents

List of Contributors, vii

History of the British Ecological Society, ix

Preface, xi

Part 1: Evolution and Population Biology

1 Genetics and ecology, 3
  L. Partridge

2 Testing Antonovics' five tenets of ecological genetics: experiments with bacteria at the interface of ecology and genetics, 25
  R.E. Lenski

3 Sociality and population dynamics, 47
  T.H. Clutton-Brock

4 Studies of the reproduction, longevity and movements of individual animals, 67
  I. Newton

Part 2: Functional and Community Ecology

5 Specificity, links and networks in the control of diversity in plant and microbial communities, 95
  A.H. Fitter

6 Global change and the linkages between physiological ecology and ecosystem ecology, 115
  J.R. Ehleringer, T.E. Cerling and L.B. Flanagan

7 Biodiversity, ecosystem processes and climate change, 139
  J.H. Lawton

8 Plant functional types, communities and ecosystems, 161
  J.P. Grime

9 Effects of diversity and composition on grassland stability and productivity, 183
  D. Tilman

CONTENTS

Part 3: Ecology of Changing Environments

10 Climate change and steady state in temperate hardwood forests, 211
    M.B. Davis

11 Experimental plant ecology: some lessons from global change research, 227
    Ch. Körner

12 Keeping track of carbon flows between biosphere and atmosphere, 249
    J. Grace, P. Meir and Y. Malhi

13 Climate and plants: past, present and future interactions, 271
    F.I. Woodward

Part 4: Ecosystems, Management and Human Impacts

14 Lost linkages and lotic ecology: rediscovering small streams, 295
    J.L. Meyer and J.B. Wallace

15 Plant–mammal interactions: lessons for our understanding of nature and implications for biodiversity conservation, 319
    R. Dirzo

16 Ecological economic theory for managing ecosystem services, 337
    J. Roughgarden and P.R. Armsworth

17 Alternate states of ecosystems: evidence and some implications, 357
    S.R. Carpenter

Part 5: Concluding Remarks

18 Concluding remarks, 387
    J.H. Brown

Index, 397

List of Contributors

P.R. Armsworth
Department of Biological Sciences, Stanford University, Stanford, CA 94305, USA

J.H. Brown
Department of Biology, University of New Mexico, Albuquerque, NM 87131, USA

S.R. Carpenter
Center for Limnology, University of Wisconsin, Madison, WI 53706, USA

T.E. Cerling
Department of Geology and Geophysics, University of Utah, Salt Lake City, UT 84112, USA

T.H. Clutton-Brock
Department of Zoology, University of Cambridge, Downing Street, Cambridge CB2 3EJ, UK

M.B. Davis
Department of Ecology, Evolution and Behavior, University of Minnesota, 1987 Upper Buford Circle, St. Paul, MN 55108, USA

R. Dirzo
Departamento de Ecología Evolutiva, Instituto de Ecología, UNAM, Ap. Post. 70-275, Mexico 04510 DF

J.R. Ehleringer
Department of Biology, University of Utah, 257 South 1400 East, Salt Lake City, UT 84112, USA

A.H. Fitter
Department of Biology, University of York, York YO10 5YW, UK

L.B. Flanagan
Department of Biological Sciences, University of Lethbridge, 4401 University Drive, Lethbridge, Alberta T1K 3M4, Canada

J. Grace
Institute of Ecology and Resource Management, University of Edinburgh, Darwin Building, King's Buildings, Edinburgh EH9 3JU, UK

J.P. Grime
Unit of Comparative Plant Ecology, Department of Animal and Plant Sciences, University of Sheffield, Western Bank, Sheffield S10 2TN, UK

Ch. Körner
Institute of Botany, University of Basel, Schönbeinstrasse 6, CH-4056, Switzerland

J.H. Lawton
NERC Centre for Population Biology, Imperial College, Silwood Park, Ascot SL5 7PY, UK; and Natural Environment Research Council, Polaris House, Swindon, SN2 1EU, UK

LIST OF CONTRIBUTORS

R.E. Lenski
Center for Microbial Ecology, Michigan State University, East Lansing, MI 48824, USA

Y. Malhi
Institute of Ecology and Resource Management, University of Edinburgh, Darwin Building, King's Buildings, Edinburgh EH9 3JU, UK

P. Meir
Institute of Ecology and Resource Management, University of Edinburgh, Darwin Building, King's Buildings, Edinburgh EH9 3JU, UK

J.L. Meyer
Institute of Ecology, University of Georgia, Athens, GA 30602, USA

I. Newton
Centre for Ecology and Hydrology, Monks Wood Research Station, Abbots Ripton, Huntingdon, Cambridgeshire PE28 2LS, UK

L. Partridge
Department of Biology, University College London, Wolfson House, 4 Stephenson Way, London NW1 2HE, UK

J. Roughgarden
Department of Biological Sciences, Stanford University, Stanford, CA 94305, USA

D. Tilman
Department of Ecology, Evolution and Behavior, University of Minnesota, 1987 Upper Buford Circle, St. Paul, MN 55108, USA

J.B. Wallace
Department of Entomology, University of Georgia, Athens, GA 30602, USA

F.I. Woodward
Department of Animal and Plant Sciences, University of Sheffield, Western Bank, Sheffield S10 2TN, UK

History of the British Ecological Society

The British Ecological Society is a learned society, a registered charity and a company limited by guarantee. Established in 1913 by academics to promote and foster the study of ecology in its widest sense, the Society currently has around 5000 members spread around the world. Members include research scientists, environmental consultants, teachers, local authority ecologists, conservationists and many others with an active interest in natural history and the environment. The core activities are the publication of the results of research in ecology, the development of scientific meetings and the promotion of ecological awareness through education. The Society's mission is:

To advance and support the science of ecology and publicize the outcome of research, in order to advance knowledge, education and its application.

The Society publishes four internationally renowned journals and organizes at least two major conferences each year plus a large number of smaller meetings. It also initiates a diverse range of activities to promote awareness of ecology at the public and policy maker level in addition to developing ecology in the education system, and it provides financial support for approved ecological projects. The Society is an independent organization that receives little outside funding.

British Ecological Society
26 Blades Court
Deodar Road, Putney
London SW15 2NU
United Kingdom
Tel.: +44 (0)20 8871 9797
Fax: +44 (0)20 8871 9779
E-mail: general@ecology.demon.co.uk
ULR: http://www.demon.co.uk/bes

The British Ecological Society is a limited company, registered in England No. 15228997 and a Registered Charity No. 281213.

Preface

This volume contains 18 chapters that formed the basis of the first joint meeting between the British Ecological Society and the Ecological Society of America. The 'millennium symposium' meeting was held in Orlando, Florida in April 2000, with the aim of celebrating achievements in ecology in the past and identifying challenges for the future. Major topics are addressed under four broad themes: evolution and population biology; functional and community ecology; the ecology of changing environments; and ecosystems, management and human impacts. The volume gives a state-of-the-art account of key aspects of contemporary ecology and of its interfaces with related disciplines, such as physiology, genetics and economics. The international team of authors includes some of the most eminent ecologists of our time. The volume discusses ecological processes over an enormous range of scales, from predator–prey relations between a bacterium and a phage to global carbon cycling, and draws examples from the atmospheric, terrestrial, fresh-water and marine environments. Many of the chapters take the achievements of others who have made fundamental contributions to ecological science as their starting point. From here, they review key important questions that have been resolved and identify those that are outstanding, drawing widely from studies by contemporary ecologists. The result is 17 review chapters (plus a synthesis chapter) that will be invaluable to both advanced students and researchers in ecology. In addition to the specific challenges identified by the authors, at least two generic issues emerged from the meeting: identifying the correct questions to ask and the importance of communicating ecological science to a wide audience including policy makers. Finally, we would like to thank the many members of both Societies as well as their staff members, who contributed so much towards the planning and execution of the stimulating meeting that formed the basis of this book.

M.C. Press,
N.J. Huntly
S. Levin

Part 1
Evolution and population biology

Chapter 1
Genetics and ecology

L. Partridge

Introduction

This chapter outlines the use of an experimental, genetical approach to the evolution of ageing and thermal evolution of body size of ectotherms. These two topics have in common the necessity to identify both ecological and genetic inputs to understand the distributions of traits in nature. In addition, they both involve important trade-offs between different fitness-related traits, and understanding the mechanisms underlying these will be highly informative about fundamental constraints on the phenotypes that can be achieved. They thus motivate an experimental, analytical approach.

Because this paper was prepared for the first joint meeting of the British Ecological Society and the Ecology Society of America, it seemed an opportune moment to gauge the extent to which ecologists in general, and the American and British ecological communities in particular, use experimental or genetical approach to their research. The aim of the survey was to gauge the extent to which the two communities regard experiments or genetics as of central importance in ecology, and the main pure ecology journals were therefore examined. Many other, more specialized, journals exist, both within ecology and on its interfaces with other subjects, and experimental and genetical studies will be represented in these. But the aim here is to determine the extent to which these approaches are regarded as mainstream in the areas that ecologists themselves define as making up the current core of the subject.

A study can be classed as experimental if a variable is manipulated by the experimenter, and the behaviour of a response variable compared to that of a control or other experimental treatments. Studies that report measurements or case–control comparisons would therefore not qualify as experimental, whereas a common garden rearing experiment would. Theoretical papers can be similarly classified according to the type of data they are designed to analyse. Inevitably there will be a few grey areas in such a classification, and independent reporters might not be perfectly concordant. However, most papers were straightforward to classify, and it is likely that the procedure is a robust one.

Department of Biology, University College London, Wolfson House, 4 Stephenson Way, London NW1 2HE UK

The survey

The papers published in *Ecology*, *Journal of Animal Ecology* and *Journal of Ecology* in the years 1986–88 showed a marked difference in the proportion of experimental studies between the American journal (*Ecology*) and the two British ones. Nearly 60% of the papers in *Ecology* contained experimental work, whereas only 41% and 34% of the papers in *Journal of Animal Ecology* and *Journal of Ecology*, respectively, did so (Partridge 1989; Table 1.1a). The proportions were very similar in the years 1992–93 (Table 1.1b). By 1998–99, when *Ecological Monographs* (USA) and *Functional Ecology* (UK) were included in the analysis (Table 1.1c), the proportion of experimental papers published in *Ecology*, but not in the two British journals, had shown a significant decline from those in the previous two surveys. The difference between the American journals and *Journal of Animal Ecology* persisted, but was not significant for *Journal of Ecology* and *Functional Ecology*. Experimental work may therefore be more valued in the USA.

An alternative interpretation is that American and British journals have become specialized on a different mix of scientific outputs, and that the two communities of ecologists do not differ in the proportion of experimental studies that they undertake. To address this issue, the papers in the 1998–99 survey were additionally classified by the national affiliation of their first author. The data for those papers where the first author was from an address in the USA or the UK are shown in Table 1.2. Only 16 papers originating from the UK were published in the two American journals over the survey period, 12% of the total output from the UK published in these five journals. In contrast, 28% of the total output from the USA was published in the three British journals, a statistically significant difference. When Americans published in *Journal of Animal Ecology*, the paper was significantly more likely to be observational than when it was placed in an American journal. In contrast, the British were no more likely to report an experimental study in an American journal than in a British one. In the total output from the USA there was a significantly higher proportion of experimental work (47.2%) than in the British (37.6%).

An experimental approach to scientific questions allows causation of events to be established, with the effects of confounding variables controlled. In ecology an experimental approach is not always feasible. Issues such as ecological history must, of necessity, be studied by a process of observation and inference, because history cannot be rerun. Feasibility of experiments is also limited by spatial scale and by expense. In addition, observational studies are often needed before an informed experimental approach can be taken. Overall, the data suggest that ecologists do routinely use an experimental approach to their work, that this type of approach is held in higher regard in the USA than in the UK, but that this regard may be on the decline.

In contrast, genetical approaches are rarely used in the studies reported in these mainstream ecology journals (Table 1.3). For the 1992–93 and 1998–99 surveys, the proportion of genetical papers never exceeded 10%, and was usually under 5%. The highest figure, 8.1% for *Ecology* in 1998–99, was inflated by two special issues,

Table 1.1 Numbers and proportions of papers published in mainstream ecology journals, classified as observational or experimental (see text for details). The 1986–88 census period ran from July 1986 to June 1988 inclusive (Partridge 1989), while the 1992–93 and 1998–99 census periods were the two full calendar years. For 1986–88 and 1992–93, the proportion of experimental papers in the two British journals did not differ significantly (1986–88 chi squared 1.386, $P = 0.2391$, 1992–93 chi squared 2.147, $P = 0.1428$), while both figures were significantly lower than those for *Ecology* (for *Journal of Animal Ecology* 1986–88 chi squared 11.047 $P = 0.0009$, 1992–93 chi squared 24.4, $P = 0.0001$; for *Journal of Ecology* 1986–88 chi squared 23.212, $P < 0.0001$, 1992–93 chi squared 7.521, $P = 0.0061$). For 1998–99, the proportion of experimental papers published in *Ecology*, but not in the two British journals, showed a significant decline (1986–88 vs. 1998–99 chi squared 6.583, $P = 0.0103$; 1992–93 vs. 1998–99 chi squared 7.645, $P = 0.0057$). The proportion of experimental papers was significantly higher in *Ecology* than in *Journal of Animal Ecology* (chi squared 14.614, $P < 0.0001$), and higher in *Ecological Monographs* than in *Journal of Animal Ecology* (chi squared 7.017, $P = 0.0081$), but did not differ significantly between the two American journals and either *Journal of Ecology* or *Functional Ecology*. The proportion of experimental work reported in *Journal of Animal Ecology* was significantly lower than in the other two British journals.

(a) 1986–88

Journal	Observational	Experimental	% Experimental
Ecology	150	199	57
Journal of Animal Ecology	85	58	41
Journal of Ecology	103	53	34

(b) 1992–93

Journal	Observational	Experimental	% Experimental
Ecology	166	224	59.7
Journal of Animal Ecology	94	47	37.7
Journal of Ecology	61	45	44.4

(c) 1998–99

Journal	Observational	Experimental	% Experimental
Ecology	229	210	47.8
Ecological Monographs	22	24	55.8
Journal of Animal Ecology	113	50	30.7
Journal of Ecology	85	62	42.2
Functional Ecology	113	88	43.8

Table 1.2 Observational and experimental papers by geographical location of senior author, 1998–99. The total proportion of experimental papers produced by American authors was significantly higher than that produced by British authors (chi squared = 3.918, $P = 0.0478$).

Journal	Senior author	Observational	Experimental	% Experimental
Ecology	USA	150	146	49.3
	UK	10	5	33.3
Ecological Monographs	USA	15	17	53.1
	UK	0	1	100
Journal of Animal Ecology	USA	24	6	20.0
	UK	35	12	25.5
Journal of Ecology	USA	23	27	54.0
	UK	10	10	50
Functional Ecology	USA	28	19	40.4
	UK	28	22	44.0
Total	USA	240	215	47.2
	UK	83	50	37.6

Table 1.3 Numbers and proportions of papers that did or did not contain a genetical component in the 1992–93 and 1998–99 surveys. See text for further details.

(a) 1992–93

Journal	Non-genetical	Genetical	% Genetical
Ecology	312	15	4.6
Journal of Animal Ecology	114	8	6.6
Journal of Ecology	93	6	6.1

(b) 1998–99

Journal	Non-genetical	Genetical	% Genetical
Ecology	439	21	8.1
Ecological Monographs	46	1	2.1
Journal of Animal Ecology	163	4	2.4
Journal of Ecology	147	4	2.6
Functional Ecology	201	1	0.49

on hybridization and on the use of molecular markers in ecology. In those papers that used a genetical approach, the aim was frequently to estimate ecological parameters such as population structure, selfing rate in plants, and fungal community structure. Only a small minority of papers was concerned with the qualitative differences between individuals produced by differences in genotype, in traits such as foraging behaviour, resistance to parasitism, life history and adaptation to climate.

Experimental analysis and genetics in ecology
Ecology is the study of the distribution and abundance of organisms. Many areas of ecology demand an analytical, experimental or a genetical approach. Life histories evolve in response to demographic forces imposed by the environment, and they in turn have a bearing on the ways that population dynamics respond to environmental change. To understand life-history diversity, we need to understand the mechanistic constraints on what organisms can achieve. Many studies have revealed the existence of trade-offs, for instance between reproduction and survival, but we do not understand why this trade-off occurs in any organism. Adaptation to climate can play a role in determining geographical range and patterns of gene flow, and also raises issues of the relative roles of phenotypic plasticity and genetic change in producing local adaptation. But why cannot organisms function equally well at a wide range of temperatures? Experimental work and genetic analysis will be needed to solve these problems. If we understand these the mechanisms underlying these constraints, and the trade-offs that organisms are forced to make, then we shall have a much better understanding of why we see only the restricted range of phenotypes that in fact occur in nature. These points are here illustrated by work on the evolution of ageing and on thermal evolution of body size of ectotherms.

The evolution of ageing

Background
Ageing is a drop in survival probability and fertility with advancing adult age. Several long-term field studies, particularly of birds and mammals, have shown that ageing is apparent in natural populations (e.g. Loison *et al.* 1999; Chapter 4, this volume). Despite its deleterious consequences, ageing can evolve because the force of natural selection declines at older ages. Extrinsic causes of mortality, such as disease, predation and accidents cause death even in the absence of ageing. Carriers of mutations with deleterious effects later in the lifespan are therefore more likely to have died of other causes by the age when the effects of the mutation become apparent. Ageing can then evolve as a result of greater accumulation under mutation–selection balance of mutations with deleterious effects later in life (Medawar 1946, 1952; Hamilton 1966; Charlesworth 1990, 1994) and through pleiotropy (trade-offs) between fitness early and late in life (Williams 1957, 1966; Kirkwood & Rose 1991; Partridge & Barton 1993; Charlesworth 1994). These two theories are not mutually exclusive, and they both predict that the intrinsic rate of ageing will evolve in response to the level of extrinsic hazard experienced (Ricklefs 1998; Keller & Genoud 1997).

Evaluating the relative roles of mutation accumulation and pleiotropy in the evolution of ageing is important for both medicine and population biology. To the extent that different parts of the adult period are free to evolve independently, it should be possible to achieve ameliorations of the ageing process without deleterious side-effects. If, on the other hand, different parts of the life history are mutually constrained by pleiotropy, then there may be a tendency for interventions that ameliorate ageing to have deleterious effects at other ages. To the extent that ageing is a consequence of trade-offs with traits that are beneficial earlier in life, it is part of the suite of life history traits that trade-off with one another in the evolution of life histories generally (Stearns 1992), rather than an independent phenomenon.

Testing for mutation accumulation and pleiotropy
Several approaches have been used to test the mutation-accumulation theory, nearly all using experiments with *Drosophila*. Under mutation accumulation, additive genetic variation for mortality rate should increase with age (Hughes & Charlesworth 1994). Initial work suggested that such an increase occurs (Engström *et al.* 1992; Hughes & Charlesworth 1994), but later experiments and improved data analysis have shown that genetic variance for age-specific mortality rates goes down at advanced ages (Promislow *et al.* 1995; Shaw *et al.* 1999), an effect that may be related to the deceleration of mortality rate at later ages often observed in large cohorts (e.g. Pletcher *et al.* 1998; Pletcher & Curtsinger 2000). The mutation-accumulation theory also predicts that inbreeding depression should increase for later age classes, and evidence in support of this idea has been produced (Charlesworth & Hughes 1996). However, as these authors pointed out, old individuals may be more susceptible to the effects of inbreeding depression, and the results of this kind of test are therefore ambiguous. The most direct evidence has come from experiments in which new mutations are allowed to accumulate sheltered from the effects of selection, and their age-specific effects examined. While some evidence suggests that mutations only affect the mortality rates of particular age classes (Pletcher *et al.* 1998), that evidence is stronger for the younger than for the older age classes and, as mutations accumulate, so does the tendency for their combined effects to span several age classes (Pletcher *et al.* 1999).

These experiments have not produced strong evidence for mutational effects on mortality rates specific to late ages, although making an unambiguous, direct test for such effects has proved difficult. It would be useful, and easier, to make equivalent measurements for effects on age-specific fertility. The results of one artificial selection experiment suggested that reversal of the effects of mutation accumulation was unimportant in producing the response to selection for increased lifespan (Sgrò & Partridge 1999). Finally, if mutation accumulation of highly age-specific mutations were important in producing ageing, then we might expect to see a wall of death at the age where selection first becomes ineffective at countering mutation pressure (Partridge & Barton 1993; Charlesworth & Partridge 1997; Pletcher & Curtsinger 1998). Such a wall of death is not in general observed, suggesting that mutational effects at late ages in general have correlated effects earlier in life.

The pleiotropy (trade-off) theory has been tested mainly by artificial selection by age at breeding, again in *Drosophila* (Wattiaux 1968; Rose & Charlesworth

1981). 'Young' lines, in which only young adults reproduce, are contrasted with 'old' lines, in which adults reproduce for some time before the eggs that will produce the next generation are collected (e.g. Luckinbill et al. 1984; Rose 1984). More death occurs before the eggs are collected in the 'old' lines, and they are therefore more strongly selected for adult survival. A direct response to selection is usually observed, with adult survival increasing in the 'old' lines (e.g. Luckinbill et al. 1984; Rose 1984; Partridge & Fowler 1992; Roper et al. 1993; Zwaan et al. 1995; Partridge et al. 1999). If the pleiotropy theory of ageing applies in practice, then the direct responses to selection in 'old' selection lines should have deleterious consequences earlier in life. The most repeatable correlated response has been lower fecundity early in life in 'old' lines (e.g. Luckinbill et al. 1984; Rose 1984; Zwaan et al. 1995; Partridge et al. 1999).

The correlated response of early fertility to the increase in longevity in 'old' selection lines is evidence for a cost of reproduction (Williams 1966), but does not in itself constitute evidence for the pleiotropy theory of the evolution of ageing. The pleiotropy theory is one of delayed gene effects, with the beneficial early effect and the late deleterious one occurring sequentially rather than simultaneously. A time lag between the early deleterious effect (low fertility) and the later beneficial one (increased survival at late ages) must be established as effects of the same genetic variants. Otherwise, the results could be explained by a combination of risk (Partridge & Andrews 1985) and mutation accumulation, if reproduction is harmful to simultaneous but not subsequent survival, and some reversal of the effects of mutation accumulation for survival at later ages has also occurred in 'old' selection lines.

In two sets of selection lines, it was demonstrated that mortality rates increased more rapidly with age in the 'young' lines (Service et al. 1998; Sgrò & Partridge 1999) (Fig. 1.1). This difference in mortality trajectories was associated with lowered early fertility in the 'old' lines and not, apparently, in any other life-history trait (Partridge et al. 1999). The difference in mortality trajectories between the 'young' and 'old' selection line females was completely abolished by sterilization, either by X-irradiation or by a female-sterile mutant (Sgrò & Partridge 1999). This result indicates that the difference in death rates between the 'young' and 'old' lines was entirely attributable to the difference between them in early fertility. It was therefore possible to deduce the timing of the effect on mortality of the difference in early fertility by direct comparison of fertile 'young' and 'old' line females (Fig. 1.2). There was a substantial time lag between the difference in early fertility and the difference in mortality that it caused, providing support for the pleiotropy theory of the evolution of ageing. This delayed cost of reproduction could provide quite a general mechanism for the kind of delayed pleiotropic genetic effect that will lead to the evolution of ageing.

Future challenges
Evidence for the importance of pleiotropy in causing ageing to evolve is stronger than that for a role for mutation accumulation. However, nearly all of that evidence

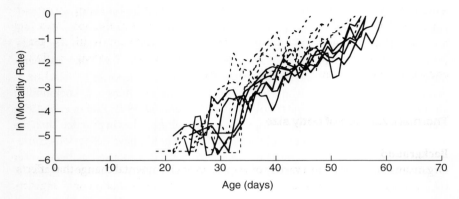

Figure 1.1 Age-specific mortality rates of female *Drosophila melanogaster* (ln proportion of those entering each timed interval that died during it). Dashed lines, five replicate 'young' lines, selected by breeding from young adults; solid lines, five replicate 'old' lines, selected by breeding from 'old' adults. The lines are further described in the text and Partridge et al. (1999) and Sgrò and Partridge (1999).

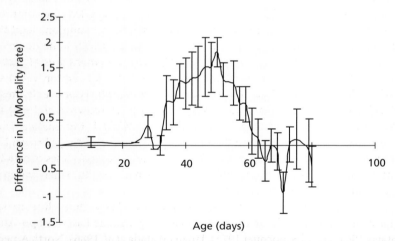

Figure 1.2 Difference in age-specific mortality rate (with 95% CL) between 'young' and 'old' line females. The difference in fertility between the lines is over by day 28, and all females are sterile by day 40. (Reproduced with permission from Partridge and Sgrò 1999.)

has come from work with a single organism, *Drosophila melanogaster*, and it is important that the generality of the findings be established. We understand very little of the mechanisms underlying the delayed effect of early fertility upon subsequent mortality rates or, indeed, of the mechanisms underlying life-history trade-offs generally. Reproduction may cause damage directly, or it may cause withdrawal of resources from repair. Reproduction itself may not be costly, and the apparent cost of reproduction may be attributable instead to enabling metabolic processes that would not be generally recognized as part of reproduction. These are issues that require a mechanistic, experimental analysis, and identification of genes producing both the response to artificial selection for the rate of ageing and differences between species and populations for this trait.

Thermal evolution of body size

Background

Organisms can respond in a variety of ways to an environmental change that affects their fitness. In the extreme, extinction may occur. Partial extinction or changes in patterns of dispersal may result in a change in geographical range. Phenotypic plasticity, as in developmental or acclimatory responses within one or a few generations, can also ameliorate environmental stresses. On a longer time scale genetic, evolutionary change can occur. The interplay between these responses will determine geographical and habitat range and patterns of dispersal and hence gene flow.

Several species of ectotherm show geographical, latitudinal clines in body size with larger individuals at higher latitudes. These clines are revealed when representatives of different populations are reared under standard conditions, and they have been found in the house fly *Musca domestica* (Bryant 1977), the honey bee *Apis melifera* (Alpatov 1929a), an ant lion *Myrmeleon immaculatus* (Arnett & Gotelli 1999), a copepod *Scottolana canadensis* (Lonsdale & Levinton 1985) and several species of fish (the Atlantic silverside *Menidia menidia* (Conover & Present 1990; Billerbeck *et al.* 2000; striped bass *Morone saxatilis* (Brown *et al.* 1998) and the Mummichog *Fundulus heteroclitus* (Schultz *et al.* 1996)). Latitudinal size clines have also been found in several species of *Drosophila* (see Partridge & French 1996 for review). The selective agents responsible for these latitudinal clines may not be the same across different taxa, and counter-examples occur when voltinism changes with latitude (Mousseau 1997).

Drosophila melanogaster has a global distribution, and latitudinal clines in body size have been found in all the main continents (e.g. Middle East–Europe–Africa (Tantawy 1961; David & Bocquet 1975), Japan (Watada *et al.* 1986), North America (Coyne & Beecham 1987; Long & Singh 1995), Eastern Europe–Central Asia (Imasheva *et al.* 1994), Australia (James *et al.* 1995, 1997) and South America (Van't Land *et al.* 1999). These latitudinal clines can evolve rapidly. For instance a cline appeared in *D. subobscura* between 10 and 20 years after it was introduced to North America (Huey *et al.* 2000). Phenotypic clines in body size are evident in flies from different populations collected directly from the field, as well as when they are reared in standard conditions in the laboratory (James *et al.* 1997; Van't Land *et al.* 1995), with the clines steeper in nature (Fig. 1.3). A similar suite of changes has been found in fish, with counter-gradient variation in growth rate, northerly populations growing more rapidly in common garden experiments, but less rapidly in

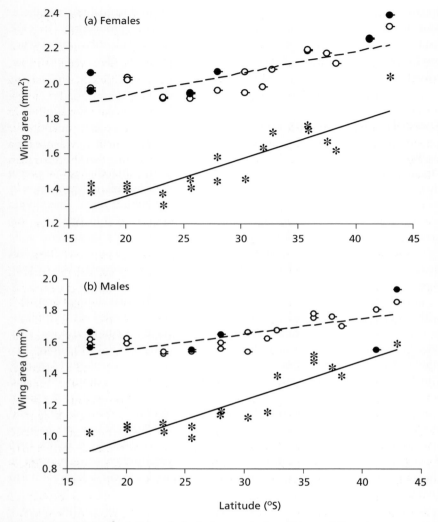

Figure 1.3 Body size in relation to latitude for populations of *Drosophila melanogaster* along the East Coast of Australia. Solid and open circles, flies reared under standard low-density conditions in the laboratory on two different occasions; asterisks, flies collected direct from nature. (Reproduced with permission from James *et al.* 1997.)

nature (Conover & Present 1990; Van't Land *et al.* 1995; Schultz *et al.* 1996; Brown *et al.* 1998; Billerbeck *et al.* 2000). The selective forces at work in the clines may therefore be similar. *D. melanogaster* is a useful model organism for investigating latitudinal clines, because of its short generation time and well-characterized genetics.

In *D. melanogaster*, the genetic and cellular basis of the latitudinal size clines have been compared between different continents, to determine what aspects of the clinal variation are consistent when independently evolved on different occasions. A quantitative genetic analysis of the differences between low and high latitude populations in Australia, South American and South Africa revealed that the genetic basis of the three clines in body size was significantly different, with varying contributions from additive genetic variance, dominance, epistasis and maternal effects (Gilchrist & Partridge 1999). This finding argues that evolution at different gene loci or of different alleles at the same loci occurs when the same quantitative phenotype, in this case size, evolves on different occasions. Further evidence for a variable genetic basis to different clines comes from the finding that the Australian cline in body size is produced almost entirely by latitudinal variation in cell number (James *et al.* 1995), whereas cell size makes a significant contribution to the South American cline (Zwaan *et al.* 2000). This kind of variability suggests that body size itself or some other trait genetically correlated with it, rather than cell size, is the target of selection in relation to latitude.

The repeatability of latitudinal clines in body size in the same species in different geographical regions of the world demonstrates that they are the products of natural selection. In most cases the selective agent(s) producing the cline has not been identified. Many ecological variables, such as day length, population density and temperature, change with latitude, and could therefore be implicated. However, several lines of evidence point to temperature. Firstly, body size has been shown to increase not only with latitude, but also with altitude in *Drosophila* (Stalker & Carson 1948) and a frog *Rana sylvatica* (Berven 1982), and in the cooler periods of the breeding season in *Drosophila* (Stalker & Carson 1949; Tantawy 1964). Secondly, replicated populations of *Drosophila* have been maintained at different temperatures in the laboratory for many generations, and the resulting cold-adapted lines are larger than the warm-adapted ones when reared at a common temperature, irrespective of whether the comparison is conducted in the warm or the cold environment (Anderson 1966, 1973; Cavicchi *et al.* 1985, 1989; Partridge *et al.* 1994a). So far, no other selective agent for the clines has been detected.

It is far from obvious why the body size of ectotherms evolves in response to temperature; the mechanisms altering the relationship between body size and fitness at different temperatures are not fully understood. Especially for the smaller species such as *Drosophila*, regulation of heat exchange with the environment is not plausible, because the organisms are too small to retain or exclude heat from their environment (Stevenson 1985). Nor is resistance to desiccation a plausible explanation. Desiccation resistance in *Drosophila* is in general greater in temperate populations than in tropical ones (Hoffmann & Harshman 1999). However, this could be a consequence rather than a cause of the size variation, humidity shows no consistent pattern of variation in relation to latitude in different continents, and evolution in response to desiccation stress cannot explain thermal evolution of *Drosophila* body size in the laboratory. To solve this problem, the target of selection and the mechanism by which it interacts with temperature to determine fitness requires investigation.

Body size in *Drosophila* is only one of a suite of traits that show latitudinal clines and that evolve in response to laboratory thermal selection. High-latitude and low-temperature populations show increased pre-adult growth efficiency and growth rate (Partridge *et al.* 1994b; Neat *et al.* 1995; Robinson & Partridge 2001, Fig. 1.4), more rapid pre-adult development (James *et al.* 1995; James & Partridge 1995; Van't Land *et al.* 1995, 1999), increased egg size (Azevedo *et al.* 1996) and greater ovariole number (Capy *et al.* 1993; Watada *et al.* 1986; Azevedo *et al.* 1996). All of these characters may be targets of natural selection. Alternatively, some may evolve as correlated responses to selection on others, if there are positive genetic correlations between them. There is abundant evidence for a positive association between body size and adult fitness traits at a single temperature in *Drosophila*. Mating success of males is both phenotypically (Partridge & Farquhar 1981, 1983; Partridge *et al.* 1987a, b) and genetically (Ewing 1961; Wilkinson 1987)

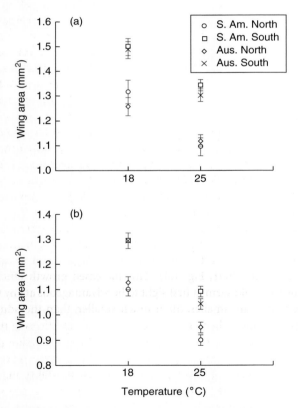

Figure 1.4 Mean wing area and 95% confidence intervals for flies from cline ends in Australia and South America that had been reared on a restricted food supply insufficient for the achievement of full adult body size. (Robinson & Partridge 2001.)

correlated with body size. Strong, positive phenotypic correlations between the size of adult females and their fecundity have been found (e.g. Alpatov 1929b; Sang 1950; Tantawy & Vetukhiv 1960; Partridge *et al.* 1986; Partridge 1988), while genetic correlations have been reported as in general positive, but weak (Robertson 1957; Tantawy 1961; Tantawy & Rakha 1964; Tantawy & El Helw 1966; Hillesheim & Stearns 1992). High positive phenotypic correlations have been found between body size and longevity (e.g. Partridge & Farquhar 1981, 1983; Partridge *et al.* 1986), while genetic correlations have been reported as either positive (Tantawy & Rakha 1964; Tantawy & El Helw 1966; Partridge & Fowler 1992) or absent (Zwaan *et al.* 1995). These findings suggest that large body size is an advantage for adult *Drosophila* and, if selection on size acted only at this stage in the life history, that the trait would be under directional selection for an evolutionary increase. This selection for larger adult body size could be stronger at lower temperatures, and there is some direct evidence for this in *D. melanogaster* females (McCabe & Partridge 1997) and males (Reeve *et al.* 2000).

In contrast to the effect of body size on adult fitness, in the pre-adult period artificial selection for increased adult size has been reported to result in an increased developmental period and lowered survival to adulthood (Robertson 1963; Partridge & Fowler 1992; Santos *et al.* 1992, 1994). There may therefore be overall stabilizing selection on body size through conflicting fitness effects on adults and juveniles. In addition to any effect through adult fitness, larger body size could evolve at lower temperatures because of lowered costs of growth in the pre-adult period. In some support of this idea, the negative correlation between development time and body size induced by thermal selection in nature and in the laboratory is the opposite of the correlation seen with artificial selection for body size, where large size is associated with extended pre-adult development time (Robertson 1960, 1963; Partridge & Fowler 1993). Large selection lines grow at the same rate as controls, but for longer (Partridge *et al.* 1999). In contrast, low-temperature thermal selection lines grow faster with a shorter development time to give a larger adult (Partridge *et al.* 1994b; James & Partridge 1995).

Part of the explanation of the more rapid growth of low temperature populations may be that they also show increased efficiency of growth, in that they convert a fixed amount of food (yeast) consumed into a larger adult (Neat *et al.* 1995; Robinson & Partridge 2001, Fig. 1.4). This increased growth efficiency is surprising, because it would seem at first sight to be advantageous at any temperature. Adult flies in nature are smaller, often much smaller, than the adults produced by the same populations when grown under optimal conditions in the laboratory (Atkinson 1979; James *et al.* 1997), suggesting that food restriction during development in the wild may be common. In addition, where increased growth efficiency increases speed of development, pre-adult mortality may be reduced (Partridge & Fowler 1993), and age of first breeding will be brought forward, an advantage in expanding populations (Lewontin 1965) and where mortality rates are high (Charlesworth 1994). Growth efficiency is plausible as a primary target of selection, since it would be expected to lead to increased growth rate and to reduce development time and/or increase adult body size, all of which are observed. Increased efficiency of food use could also explain increased ovariole number and egg size in the adult.

In addition to its evolutionary effects on body size of ectotherms, temperature during development has a direct effect, with body size increasing as temperature decreases for most species for which the relationship has been investigated (Ray 1960; von Bertalanffy 1960; Precht *et al.* 1973; Atkinson 1994). This direct effect of experimental temperature on body size has been well characterized in several *Drosophila* species (e.g. Alpatov 1930; Imai 1934; Stalker & Carson 1949; Ray 1960; Delcour & Lints 1966; David & Clavel 1967; Powell 1974; Masry & Robertson 1979; Coyne & Beecham 1987) and in *D. melanogaster* results in a peak in body size at about 17°C (Fig. 1.5). *D. melanogaster* grows more efficiently at a lower experimental temperature (Robinson & Partridge 2001, Fig. 1.5). Increased growth efficiency and more rapid growth are also found in high latitude populations of several fish species (Schultz *et al.* 1996; Brown *et al.* 1998; Billerbeck *et al.* 2000).

The norm of reaction of body size to experimental temperature in *D. melanogaster* evolves hardly at all in response either to latitude (e.g. James *et al.* 1997; Morin *et al.* 1999) or to laboratory thermal selection (Partridge *et al.* 1994a). Mean size evolves, with the degree of plasticity changing very little. This finding suggests that the plasticity of body size may not in itself be an adaptive response, but rather a direct, mechanistic response of some aspect of physiology to temperature. This idea is to some extent supported by the finding that temperature influences adult size in *Drosophila* cumulatively with time spent at the experimental temperature during development, rather than acting as a developmental switch (French *et al.* 1998).

The increased growth rate, growth efficiency and body size achieved at lower experimental temperatures may occur because rates of productive and waste metabolic processes scale differently with temperature. A leading source of energy wastage during metabolism is proton leak across the inner membrane of the mitochondrion, which is the single most important contributor to standard metabolic rate (Brand 1990; Rolfe & Brand 1996). There is some evidence that the rate of proton leak decreases sharply with declining temperature (Dufour *et al.* 1996). An unusually steep response of mitochondrial proton leak to temperature could result in greater nutrient availability for growth at lower temperatures, and hence increased growth efficiency. Increased nutrient availability may in turn drive an evolutionary reallocation of resources to growth, leading to genetically increased growth efficiency, growth rate and adult body size. If this idea is correct, then the evolutionary response to low temperature should have an associated disadvantage at higher temperatures, since increased growth efficiency would otherwise be an unconditional advantage at any temperature. So far, no such disadvantage has been identified. One study examined the response of high- and low-latitude populations of *D. melanogaster* to varying levels of larval competition at different experimental temperatures. Body size of high-latitude populations was more sensitive to increas-

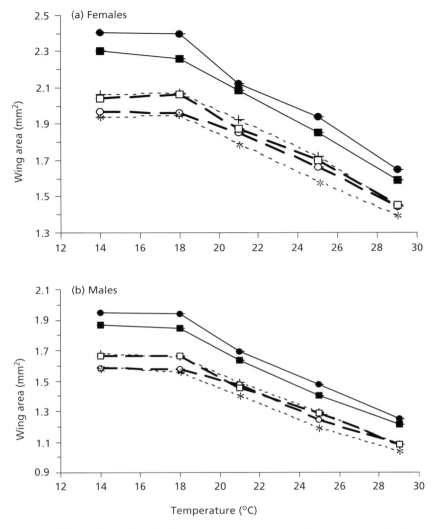

Figure 1.5 Norms of reaction of wing area to experimental rearing temperature for flies from six East Australian populations raised at low larval density with unrestricted food. Solid circles and squares, high-latitude populations; open circles and squares, intermediate-latitude populations; asterisks, low-latitude populations. (Reproduced with permission from James et al. 1997.)

ing larval density, especially at the higher experimental temperature (James & Partridge 1998). However, individuals from high-latitude populations were still larger under all experimental conditions. More severe crowding might have been needed to reverse their advantage.

Future challenges

To understand the plasticity of ectotherm body size to growth temperature, we need to identify the level of organization at which the increase in growth efficiency occurs: is it a feature of mitchondria (perhaps through an altered contribution of proton leak to cellular energy budgets at different temperatures), or is it a feature only of certain tissues or of the whole organism? More generally, identifying the mechanisms underlying the fundamental constraints on what organisms can achieve, and on what combinations of traits are possible, will help us to understand why we see the phenotypes that we see, and why we do not see others. The most direct method for understanding the evolutionary response of ectotherm body size to temperature is likely to be through identification of the genes producing the response to selection. The prospects for success are somewhat increased for model organisms such as *D. melanogaster* for which the complete genome sequence is now available. This should make it more straightforward to go from quantitative trait locus (QTL) mapping to gene. DNA microarray technology can now be used to examine in detail the transcription patterns of all the genes present in a region identified in a QTL analysis. Single nucleotide polymorphisms and other markers can be then used to detect both difference in levels of gene expression and differences in gene sequence.

References

Alpatov, W.W. (1929a) Biometrical studies on variation and races of the honey bee (*Apis melifera*). *Quarterly Review of Biology* **4**, 1–58.

Alpatov, W.W. (1929b) Growth and variation of the larvae of *Drosophila melanogaster*. *Journal of Experimental Zoology* **42**, 407–437.

Alpatov, W.W. (1930) Phenotypical variation in body and cell size of *Drosophila melanogaster*. *Biological Bulletin* **58**, 85–103.

Anderson, W.W. (1966) Genetic divergence in M. Vetukhivís experimental populations of *Drosophila pseudoobscura*. *Genetical Research Cambridge* **7**, 255–266.

Anderson, W.W. (1973) Genetic divergence in body size among experimental populations of *Drosophila pseudoobscura* kept at different temperatures. *Evolution* **27**, 278–284.

Arnett, A.E. & Gotelli, N.J. (1999) Geographic variation in life-history traits of the ant lion, *Myrmeleon immaculatus*: Evolutionary implication of Bergmann's rule. *Evolution* **53**, 1180–1188.

Atkinson, W.D. (1979) A field investigation of larval competition in domestic *Drosophila*. *Journal of Animal Ecology* **48**, 91–102.

Atkinson, D. (1994) Temperature and organism size—a biological law for ectotherms? *Advances in Ecological Research* **25**, 1–58.

Azevedo, R.B.R., French, V. & Partridge, L. (1996) Thermal evolution of egg size in *Drosophila melanogaster*. *Evolution* **50**, 2338–2345.

Berven, K.A. (1982) The genetic basis of altitudinal variation in the wood frog *Rana sylvatica*. I. An experimental analysis of life history traits. *Evolution* **36**, 962–983.

Billerbeck, J.M., Schultz, E.T. & Conover, D.O. (2000) Adaptive variation in energy acquisition and allocation among latitudinal populations of the Atlantic sliverside. *Oecologia* **122**, 210–219.

Brand, M.D. (1990) The contribution of leak of protons across the mitochondrial inner membrane to standard metabolic rate. *Journal of Theoretical Biology* **145**, 267–286.

Brown, J.J., Ehtisham, A. & Conover, D.O. (1998) Variation in larval growth rate among striped bass stocks from different latitudes. *Transactions of the American Fisheries Society* **127**, 598–610.

Bryant, E.H. (1977) Morphometric adaptation of the housefly, *Musca domestica* L., in the United States. *Evolution* **31**, 580–596.

Capy, P., Pla, E. & David, J.R. (1993) Phenotypic and genetic variability of morphometrical traits in natural populations of *Drosophila melanogaster* and *Drosophila simulans*. 1. Geographic variations. *Genetics, Selection and Evolution* **25**, 517–536.

Cavicchi, S., Guerra, D., Giorgi, G. & Pezzoli, C. (1985) Temperature-related divergence in experimental populations of *Drosophila melanogaster*. I. Genetic and developmental basis of wing size and shape variation. *Genetics* **109**, 665–689.

Cavicchi, S., Guerra, D., Natali, V., Pezzoli, C. & G.Giorgi, G. (1989) Temperature-related divergence in experimental populations of *Drosophila melanogaster*. II. Correlation between fitness and body dimensions. *Journal of Evolutionary Biology* **2**, 235–251.

Charlesworth, B. (1990) Optimization models, quantitative genetics and mutation. *Evolution* **44**, 520–538.

Charlesworth, B. (1994). *Evolution in Age-Structured Populations*, 2nd edn. Cambridge University Press, Cambridge.

Charlesworth, B. & Hughes, K.A. (1996) Age-specific inbreeding depression and components of genetic variance in relation to the evolution of senescence. *Proceedings of the National Academy of Sciences USA* **93**, 6140–6145.

Charlesworth, B. & Partridge, L. (1997) Ageing: levelling of the grim reaper *Current Biology* **7**, R440–R442.

Conover, D.O. & Present, T.M.C. (1990) Countergradient variation in growth rate: compensation for length of the growing season among Atlantic silversides from different latitudes. *Oecologia* **83**, 316–324.

Coyne, J.A. & Beecham, E. (1987) Heritability of two morphological characters within and among natural populations of *Drosophila melanogaster*. *Genetics* **117**, 727–737.

David, J.R. & Bocquet, C. (1975) Evolution in a cosmopolitan species: Genetic latitudinal clines in *Drosophila melanogaster* wild populations. *Experientia* **31**, 164–166.

David, J.R. & Clavel, M.-F. (1967) Influence de la temperature subie au cours du developpement sur divers characters biometriques des adultes de *Drosophila melanogaster* Meigen. *Journal of Insect Physiology* **13**, 717–729.

Delcour, J. & Lints, F.A. (1966) Environmental and genetic variations of wing size, cell size, and cell division rate in *Drosophila melanogaster*. *Genetica* **37**, 543–556.

Dufour, S., Rousse, N., Canioni, P. & Diolez, P. (1996) Top-down control analysis of temperature effect on oxidative phosphorylation. *Biochemical Journal* **314**, 743–751.

Engström, G., Liljedahl, L.-E. & Björklund, T. (1992) Expression of genetic and environmental variation during aging. 2. Selection for increased lifespan in. *Drosophila melanogaster*. *Theoretical and Applied Genetics* **85**, 26–32.

Ewing, A.W. (1961) Body size and courtship behaviour in *Drosophila melanogaster*. *Animal Behaviour* **9**, 93–99.

French, V., Feast, M. & Partridge, L. (1998) Body size and cell size in *Drosophila*: the developmental response to temperature. *Journal of Insect Physiology* **44**, 1081–1089.

Gilchrist, A.S. & Partridge, L. (1999) A comparison of the genetic basis of wing size divergence in three parallel body size clines of *Drosophila melanogaster*. *Genetics* **153**, 1775–1787.

Hamilton, W.D. (1966) The molding of senescence by natural selection. *Journal of Theoretical Biology* **12**, 12–45.

Hillesheim, E. & Stearns, S.C. (1992) Correlated responses in life history traits to artificial selection for body weight in *Drosophila melanogaster*. *Evolution* **46**, 745–752.

Hoffmann, A.A. & Harshman, L.G. (1999) Desiccation and starvation resistance in *Drosophila*: patterns of variation at the species, population and intrapopulation levels. *Heredity* **83**, 637–643.

Huey, R.B., Gilchrist, G.W., Carlson, M.L., Berrigan, D. & Serra, L. (2000) Rapid evolution of a geographic cline in size in an introduced fly. *Science* **287**, 308–309.

Hughes, K.A. & Charlesworth, B. (1994) A genetic analysis of senescence in *Drosophila*. *Nature* **367**, 64–66.

Imai, T. (1934) The influence of temperature on variation and inheritance of body dimensions in *Drosophila melanogaster*. *Archives Enwicklungsmechanik* **128**, 634–660.

Imasheva, A.G., Bubli, O.A. & Lazebny, O.E. (1994) Variation in wing length in Eurasian natural populations of *Drosophila melanogaster*. *Heredity* **72**, 508–514.

James, A.C. & Partridge, L. (1995) Thermal evolution of rate of larval development in *Drosophila melanogaster* in laboratory and field populations. *Journal of Evolutionary Biology* **8**, 315–330.

James, A.C. & Partridge, L. (1998) Latitudinal variation in competitive ability in *Drosophila melanogaster*. *American Naturalist* **151**, 530–537.

James, A.C., Azevedo, R.B.R. & Partridge, L. (1995) Genetic and environmental responses to temperature of *Drosophila melanogaster* from a latitudinal cline. *Genetics* **146**, 881–890.

James, A.C., Azevedo, R.B.R. & Partridge, L. (1997) Cellular basis and developmental timing in a size cline of *Drosophila melanogaster*. *Genetics* **140**, 659–666.

Keller, L. & Genoud, M. (1997) Extraordinary lifespans in ants: a test of evolutionary theories of ageing. *Nature* **389**, 958–960.

Kirkwood, T.B.L. & M.R.Rose (1991) Evolution of senscence: late survival sacrificed for reproduction. *Philosophical Transactions of the Royal Society Series B* **332**, 15–24.

Lewontin, R.C. (1965) Selection for colonizing ability. In: *The Genetics of Colonizing Species* (eds H.G. Baker & G.L. Stebbins), pp. 79–94. Academic Press, New York.

Loison, A., FestaBianchet, M., Gaillard, J.M., Jorgenson, J.T. & Jullien, J.M. (1999) Age-specific survival in five populations of ungulates: Evidence of senescence. *Ecology* **8**, 2539–2554.

Long, A.D. & Singh, R.S. (1995) Molecules versus morphology: the detection of selection acting on morphological characters along a cline in *Drosophila melanogaster*. *Heredity* **74**, 569–581.

Lonsdale, D.J. & Levinton, J.S. (1985) Latitudinal differentiation in copepod growth: an adaptation to temperature. *Ecology* **66**, 1397–1407.

Luckinbill, L.S., Arking, R., Clare, M.J., Cirocco, W.C. & Buck, S.A. (1984) Selection for delayed senescence in *Drosophila melanogaster*. *Evolution* **38**, 996–1003.

Masry, A.M. & Robertson, F.W. (1979) Cell size and number in the *Drosophila* wing. III. The influence of temperature differences during development. *Egyptian Journal of Genetics and Cytology* **8**, 71–79.

McCabe, J. & Partridge, L. (1997) An interaction between environmental temperature and genetic variation for body size for the fitness of adult female *Drosophila melanogaster*. *Evolution* **51**, 1164–1174.

Medawar, P.B. (1946) Old age and natural death. *Modern Quarterly* **1**, 30–56.

Medawar, P.B. (1952) *An Unsolved Problem of Biology*. H.K. Lewis, London.

Morin, J.P., Moreteau, B.P., Ètavy, G. & David, J.R. (1999) Divergence of reaction norms of size characters between tropical and temperate populations of *Drosophila melanogaster*. *Journal of Evolutionary Biology* **12**, 329–339.

Mousseau, T.A. (1997) Ecotherms follow the converse to Bergmann's Rule. *Evolution* **51**, 630–632.

Neat, F., Fowler, K., French, V. & Partridge, L.(1995) Thermal evolution of growth efficiency in *Drosophila melanogaster*. *Proceedings of the Royal Society Series B* **260**, 73–78.

Partridge, L. (1988) Lifetime reproductive success in *Drosophila*. In: *Reproductive Success* (ed. T.R. Clutton-Brock), pp. 11–23. Chicago University Press, Chicago.

Partridge, L. (1989) An experimentalist's approach to the role of costs of reproduction in the evolution of life-histories. In: *Towards a More Exact Ecology* (eds P.J. Grubb & J.B. Whittaker), pp. 231–246. Blackwell Scientific Publications, Oxford.

Partridge, L. & Andrews, R. (1985) The effect of reproductive activity on the longevity of male *Drosophila melanogaster* is not caused by an acceleration of ageing. *Journal of Insect Physiology* **31**, 393–395.

Partridge, L. & Barton, N.H. (1993) Optimality, mutation and the evolution of ageing. *Nature* **362**, 305–311.

Partridge, L. & Farquhar, M. (1981) Sexual activity reduces lifespan of male fruitflies. *Nature* **294**, 580–582.

Partridge, L. & Farquhar, M. (1983) Lifetime mating success of male fruitflies (*Drosophila melanogaster*) is related to their size. *Animal Behaviour* **31**, 871–877.

Partridge, L. & Fowler, K. (1992) Direct and correlated responses to selection on age at reproduction in *Drosophila melanogaster*. *Evolution* **46**, 76–91.

Partridge, L. & Fowler, K. (1993) Responses and

correlated responses to artificial selection on thorax length in *Drosophila melanogaster*. *Evolution* **47**, 213–226.

Partridge, L. & French, V. (1996) Why get big in the cold? In: *Animals and Temperature* (eds I.A. Johnston & A.B. Bennett), pp. 265–292. Cambridge University Press, Cambridge.

Partridge, L., Fowler, K., Trevitt, S. & Sharp, W. (1986) An examination of the effects of males on the survival and egg-production rates of female *Drosophila melanogaster*. *Journal of Insect Physiology* **32**, 925–929.

Partridge, L., Hoffmann, A. & Jones, J.S. (1987a) Male size and mating success in *Drosophila melanogaster* and *D. pseudoobscura* under field conditions. *Animal Behaviour* **35**, 468–476.

Partridge, L., Ewing, A. & Chandler, A. (1987b) Male size and mating success in *Drosophila melanogaster*: the roles of male and female behaviour. *Animal Behaviour* **35**, 555–562.

Partridge, L., Barrie, B., Fowler, K. & French, V. (1994a) Evolution and development of body-size and cell-size in *Drosophila melanogaster* in response to temperature. *Evolution* **48**, 1269–1276.

Partridge, L., Barrie, B., Fowler, K. & French, V. (1994b) Thermal evolution of pre-adult life history traits in *Drosophila melanogaster*. *Journal of Evolutionary Biology* **7**, 645–663.

Partridge, L., Prowse, N. & Pignatelli, P. (1999) Another set of responses and correlated responses to selection on age at reproduction in *Drosophila melanogaster*. *Proceedings of the Royal Society of London, Series B* **266**, 255–261.

Pletcher, S.D. & Curtsinger, J.W. (1998) Mortality plateaus and the evolution of senescence: Why are old-age mortality rates so low? *Evolution* **52**, 454–464.

Pletcher, S.D. & Curtsinger, J.W. (2000) The influence of environmentally induced heterogeneity on age-specific genetic variance for mortality rates. *Genetical Research* **75**, 321–329.

Pletcher, S.D., Houle, D. & Curtsinger, J.W. (1998) Age-specific properties of spontaneous mutations affecting mortality in *Drosophila melanogaster*. *Genetics* **148**, 287–303.

Pletcher, S.D., Houle, D. & Curtsinger, J.W. (1999) The evolution of age-specific mortality rates in *Drosophila melanogaster*: Genetic divergence among unselected lines. *Genetics* **153**, 813–823.

Powell, J.R. (1974) Temperature related divergence in *Drosophila* body size. *Journal of Heredity* **65**, 257–258.

Precht, H., Christophersen, J., Hensel, H. & Larcher, W. (1973) *Temperature and Life*. Springer-Verlag, Berlin.

Promislow, D.E.L., Tatar, M., Khazaeli, A.A. & Curtsinger, J.W. (1995) Age-specific patterns of genetic variation in *Drosophila melanogaster*. I. Mortality. *Genetics* **143**, 839–848.

Ray, C. (1960) The application of Bergmann's and Allen's rules to the poikilotherms. *Journal of Morphology* **106**, 289–306.

Reeve, M.W., Fowler, K. & Partridge, L. (2000) Increased body size confers greater fitness at lower experimental temperatures in male *Drosophila melanogaster*. *Journal of Evolutionary Biology* **13**, 836–844.

Ricklefs, R.E. (1998) Evolutionary theories of aging: confirmation of a fundamental prediction, with implications for the genetic basis and evolution of life span. *American Naturalist* **152**, 24–44.

Robertson, F.W. (1957) Studies in quantitative inheritance. XI. Genetic and environmental correlation between body size and egg production in *Drosophila melanogaster*. *Journal of Genetics* **55**, 428–443.

Robertson, F.W. (1960) The ecological genetics of growth in *Drosophila*. 1. Body size and development time on different diets. *Genetical Research Cambridge* **1**, 288–304.

Robertson, F.W. (1963) The ecological genetics of growth in *Drosophila*. VI. The genetic correlation between the duration of the larval period and body size in relation to larval diet. *Genetical Research Cambridge* **4**, 74–92.

Robinson, S.J.W. & Partridge, L. (2001) Temperature and clinal variation in larval growth efficiency in *Drosophila melanogaster*. *Journal of Evolutionary Biology* **14**, 14–21.

Rolfe, D.F.S. & Brand, M.D. (1996) Contribution of mitochondrial proton leak to skeletal muscle respiration and to standard metabolic rate. *American Journal of Physiology* **271**, C1380–C1389.

Roper, C., Pignatelli, P. & Partridge, L. (1993)

Evolutionary effects of selection on age at reproduction in larval and adult *Drosophila melanogaster*. *Evolution* **47**, 445–455.

Rose, M.R. (1984) Laboratory evolution of postponed senescence in *Drosophila melanogaster*. *Evolution* **38**, 1004–1010.

Rose, M.R. & Charlesworth, B. (1981) Genetics of life history in *Drosophila melanogaster*. II Exploratory selection experiments. *Genetics* **97**, 187–196.

Sang, J.H. (1950) Population growth in *Drosophila* cultures. *Biological Reviews* **25**, 188–219.

Santos, M., Fowler, K. & Partridge, L. (1992) On the use of tester stocks to predict the competitive ability of genotypes. *Heredity* **69**, 489–495.

Santos, M., Fowler, K. & Partridge, L. (1994) Gene–environment interaction for body size and larval density in *Drosophila melanogaster*: an investigation of effects on development time, thorax length and adult sex ratio. *Heredity* **72**, 515–521.

Schultz, E.T., Reynolds, K.E. & Conover, D.O. (1996) Countergradient variation in growth among newly hatched *Fundulus heteroclitus*: Geographic differences revealed by common-environment experiments. *Functional Ecology* **10**, 366–374.

Service, P.M., Ochoa, R., Valenzuela, R. & Michieli, C.A. (1998) Cohort size and maximum likelihood estimation of mortality parameters. *Experimental Gerontology* **33**, 331–342.

Sgrò, C.M. & Partridge, L. (1999) A delayed wave of death from reproduction in *Drosophila*. *Science* **286**, 2521–2524.

Shaw, F.H., Promislow, D.E.L., Tatar, M., Hughes, K.A. & Geyer, C.J. (1999) Toward reconciling inferences concerning genetic variation in senescence in *Drosophila melanogaster*. *Genetics* **152**, 553–566.

Stalker, H.D. & Carson, H.L. (1948) An altitudinal transect of *Drosophila robusta* Sturtevant. *Evolution* **2**, 295–305.

Stalker, H.D. & Carson, H.L. (1949) Seasonal variation in the morphology of *Drosophila robusta* Sturtevant. *Evolution* **3**, 330–343.

Stearns, S.C. (1992). *The Evolution of Life Histories*. Oxford University Press. Oxford.

Stevenson, R.D. (1985) Body size and limits to daily range of body temperature in terrestrial ectotherms. *American Naturalist* **125**, 102–117.

Tantawy, A.O. (1961) Effects of temperature on productivity and genetic variance of body size in *Drosophila pseudoobscura*. *Genetics* **46**, 227–238.

Tantawy, A.O. (1964) Studies on natural populations of *Drosophila*. III. Morphological and genetic differences of wing length in *Drosophila melanogaster* and *D. simulans* in relation to season. *Evolution* **18**, 560–570.

Tantawy, A.O. & El Helw, M.R. (1966) Studies on natural populations of *Drosophila*. V. Correlated response to selection in *Drosophila melanogaster*. *Genetics* **53**, 97–110.

Tantawy, A.O. & Rakha, F.A. (1964) Studies on natural populations of *Drosophila*. IV. Genetic variances of and correlations between four characters in *D. melanogaster*. *Genetics* **50**, 1349–1355.

Tantawy, A.O. & Vetukhiv, M.O. (1960) Effects of size on fecundity, longevity and viability in populations of *Drosophila pseudoobscura*. *American Naturalist* **94**, 395–403.

Van't Land, J., Van Putten, P., Villarroel, H., Kamping, A. & Van Delden, W. (1995) Latitudinal variation in wing length and allele frequencies for *Adh* and *Gpdh* in populations of *Drosophila melanogaster* from Ecuador and Chile. Drosophila. *Information Service* **76**, 156.

Van't Land, J., Van Putten, P., Zwaan, B., Kamping, A. & Van Delden, W. (1999) Latitudinal variation in wild populations of *Drosophila melanogaster*: heritabilities and reaction norms. *Journal of Evolutionary Biology* **12**, 222–232.

von Bertalanffy, L. (1960) Principles and theories of growth. In: *Fundamental Aspects of Normal and Malignant Growth* (ed. W.N. Nowinski), pp. 137–259. Elsevier, Amsterdam.

Watada, M., Ohba, S. & Tobari, Y.N. (1986) Genetic differentiation in Japanese populations of *Drosophila simulans* and *Drosophila melanogaster*. II. Morphological variation. *Japan Journal of Genetics* **61**, 469–480.

Wattiaux, J.M. (1968) Cumulative parental age effects in *Drosophila subobscura*. *Evolution* **22**, 406–421.

Wilkinson, G.S. (1987) Equilibrium analysis of sexual selection in *Drosophila melanogaster*. *Evolution* **41**, 11–21.

Williams, G.C. (1957) Pleiotropy, natural selection, and the evolution of senescence. *Evolution* **11**, 398–411.

Williams, G.C. (1966) Natural selection, the cost of reproduction and a refinement of *Lack's Principle*. *American Naturalist* **100**, 687–690.

Zwaan, B., Azevedo, R.B.R., James, A.C., Van't Land, J. & Partridge, L. (2000) Cellular basis of wing size variation in *Drosophila melanogaster*: a comparison of latitudinal clines on two continents. *Heredity* **84**, 338–347.

Zwaan, B., Bijlsma, R. & Hoekstra, R.F. (1995) Direct selection on lifespan in *Drosophila melanogaster*. *Evolution* **49**, 649–659.

Chapter 2
Testing Antonovics' five tenets of ecological genetics: experiments with bacteria at the interface of ecology and genetics

R. E. Lenski

Introduction

When I began graduate school in 1977, I thought that ecology and genetics were completely distinct fields of study. I imagined that I could study ecological patterns and processes in blissful ignorance of genetics and without worrying that evolution would directly impinge on my research. This naïve view was soon dispelled by Janis Antonovics, who taught a wonderful course at Duke University in North Carolina on Ecological Genetics, which I took in 1979. In many treatments of population genetics, natural selection is largely devoid of its ecological context and appears only as an abstract coefficient, S, that operates on gene frequencies, p and q. But Antonovics' course placed selection squarely in its ecological context. Moreover, his course examined the ecological consequences of changing gene frequencies, thus emphasizing the feedback of evolutionary change on ecology.

Antonovics had synthesized his integrated view of ecology and genetics in a provocative paper published a few years earlier (Antonovics 1976). In that paper, he presented the following five tenets:

1 'The ecological amplitude of a species (both within and among communities) has a genetic component. Explaining the abundance and distribution of organisms is basically a genetic problem.' [p. 236]
2 'Forces maintaining species diversity and genetic diversity are similar. An understanding of community structure will come from considering how these kinds of diversity interact.' [p. 238]
3 'Darwinian fitness can be measured in terms of mortality and fecundity of individuals within populations. Adaptation is a dynamic process, operationally definable and not just a "matching" of the individual to the environment.' [p. 239]
4 'Genetic adjustment to environmentally induced changes in fecundity and mortality may be by direct response in the affected age-specific parameters or by compensatory change in other parts of the life history. Adaptation to new environments will result in different genotypes with different life histories.' [p. 240]

Center for Microbial Ecology, Michigan State University, East Lansing, MI 48824, USA.
E-mail: lenski@msu.edu

5 'The distinction between "ecological time" and "evolutionary time" is artificial and misleading. Changes of both kinds may be on any time scale: frequently genetic and ecological changes are simultaneous.' [p. 241]

Antonovics (1976) offered evidence in support of these tenets based largely on studies of plant populations and communities. However, he also stated that the 'generality and usefulness of [the tenets] remain in need of assessment' [p. 236]. Thus, the tenets were intended not as dogma but rather to stimulate new lines of research. They have certainly done so for me. During the past two decades, my colleagues and I have conducted experiments with the bacterium *Escherichia coli* and sometimes its viral predators. The findings from these experiments provide an opportunity to re-examine the generality and utility of Antonovics' five tenets. (These tenets became known among students as the 'ecological geneticist's creed'. Over the years, some of the tenets have changed slightly and several new ones have been added, but in the interest of space I will restrict this paper to the five original tenets.)

The rest of this paper is organized as follows. Each of the next three sections summarizes a different set of experiments with *E. coli*, without explicit reference to Antonovics' tenets. Then I will return to the five tenets and discuss them in light of the experimental findings. I conclude by emphasizing the special importance of integrating ecology and genetics given today's rapid pace of environmental change.

Competition for a limiting resource

In one long-term project, we studied the dynamics of bacterial populations as they mutated and adapted by natural selection to a simple and constant environment (Lenski *et al.* 1991; Lenski & Travisano 1994). Twelve replicate populations were diluted daily into fresh medium in which glucose provided the sole source of energy; the populations were maintained at constant 37°C. Each day, the bacteria grew until they depleted the glucose, and the 1:100 dilution allowed about 6.6 generations of binary fission per day. This experiment has continued now for more than 20 000 bacterial generations (3000 days). At the start of the experiment and at intervals ever since, samples were stored in a freezer at −80°C, where the bacteria exist in a state of suspended animation but can be revived at any time for study. Thus, we have time series for all 12 populations of adaptation to the environment. We can even perform competition experiments between the derived bacteria and the ancestor using a neutral genetic marker that allows us to distinguish strains based on colour. Prior to their competition, both strains are acclimated to the environment in which they compete, in order that we can distinguish genetic adaptation from phenotypic acclimation.

Figure 2.1 shows the evolutionary trajectory for the mean competitive fitness of the derived bacteria relative to their common ancestor. Relative fitness is calculated simply as the ratio of realized growth rates of the derived and ancestral populations as they competed for limiting glucose (Lenski *et al.* 1991). Within a few hundred generations, the bacteria had already begun to improve appreciably, and by genera-

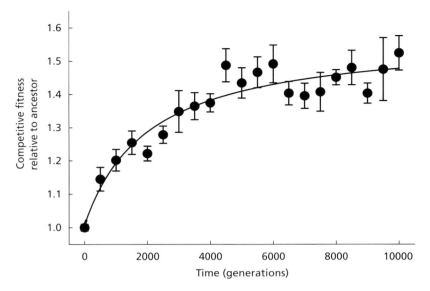

Figure 2.1 Trajectory for competitive fitness relative to the ancestor during 10 000 generations of experimental evolution in *E. coli*. Each point is the mean of 12 replicate populations; the error bars are 95% confidence limits. (Reproduced with permission from Lenski & Travisano (1994) [Copyright 1994 National Academy of Sciences, USA].)

tion 10 000 they were about 50% more fit, on average, then was the ancestor (Lenski & Travisano 1994). The competitiveness of these bacteria is not some fixed attribute of the species, but rather it mutates and evolves.

As the bacteria became better competitors for glucose, they also changed in other respects. One rather surprising change is that the density of bacteria that was reached each day, when the glucose was depleted, declined over time (Fig. 2.2). Evidently, greater fitness and higher population density do not necessarily correspond. Fitness is a dynamic measure that reflects relative rates of change in abundance; it should not be confused with measures of abundance per se. (This point is both obvious and reassuring when one considers that each of our intestines holds more bacteria than there are humans on this planet.) In the evolving populations, the average size of bacterial cells increased, as did the net volume of bacteria produced per unit of glucose (Fig. 2.2). Thus, the efficiency with which the bacteria converted resources into biomass increased (Lenski & Mongold 2000). Average cell size increased in all 12 replicate populations, but the reasons for this change remain unclear. In part, it reflects a positive phenotypic correlation between growth rate and cell size (such that selection for faster growth produces larger cell size, all else equal). However, this correlation cannot fully explain the observed increase in cell size, as the evolved bacteria produce larger cells even when they are forced to grow at the same rate as their ancestor (Mongold & Lenski 1996).

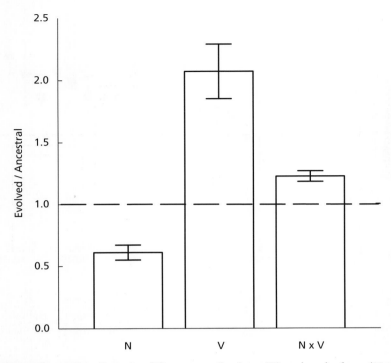

Figure 2.2 Changes in cell number (N), average cell volume (V), and total volume ($N \times V$) after 10 000 generations of evolution in *E. coli*. The height of each bar indicates the mean of 12 independently evolved populations relative to the ancestral value; the error bars are 95% confidence limits. (Data from Lenski & Travisano (1994).)

One does not usually think of bacteria like *E. coli* as having life histories: they seem simply to grow and divide. However, the concept of life history in ecology is not merely a description of morphologically distinct stages but implies an underlying demography that can be projected over time. Bacterial fitness depended on several distinct demographic parameters, even in the simple environment of the long-term experiment. Each day, the populations experienced 'seasons' of feast and famine: first a period of transition from starvation to growth after they were diluted in fresh medium, then a period of exponential growth, followed by another transition phase as the glucose was depleted, and finally a period of resource deprivation. One can ask therefore which life-history traits changed so as to yield the net improvement in competitive fitness (Vasi *et al.* 1994). The bacteria improved primarily in terms of a higher exponential growth rate and a shorter lag upon transfer into fresh medium (Fig. 2.3). However, there was no measurable change in their growth rate at low glucose concentration owing to an increase in K_s, which describes the concentration necessary to grow at half their maximal rate (Fig. 2.3). The bacteria also did not improve in their short-term survival after glucose was

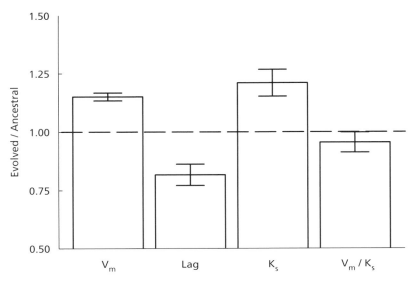

Figure 2.3 Changes in demographic parameters after 2000 generations of evolution in *E. coli*. V_m, maximum growth rate on glucose; Lag, duration of lag phase; K_s, concentration of glucose that supports a growth rate of $V_m/2$; V_m/K_s, proportional growth rate as glucose becomes extremely scarce. The height of each bar indicates the mean of 12 independently evolved populations relative to the ancestral value; the error bars are 95% confidence limits. (Data from Vasi *et al.* (1994).)

exhausted; mortality was not very important in the evolution experiment because the ancestor experienced no measurable mortality during starvation for less than a day. It must be emphasized that this pattern of life-history change depended on the selective environment. We and other groups have studied *E. coli* populations maintained under different conditions, including with low but steady resources and during prolonged starvation (Dykhuizen & Hartl 1981; Zambrano *et al.* 1993; Vasi & Lenski 1999). Such populations also adapted evolutionarily, but they did so by changing other demographic parameters that were more important in those environments.

Another aspect of this experiment that deserves mention is the opportunity for coexistence between ecologically distinct types. Each population was founded by a single genotype, glucose was the sole source of energy provided and limiting to population density, and the physical environment was homogeneous both spatially and temporally (except for daily resource fluctuations), all of which suggest little opportunity for niche diversification. Indeed, bacterial adaptations usually led to competitive exclusion of the previously dominant type—but not always. In one replicate, there evolved a community of two ecotypic 'species' (Rozen & Lenski 2000; see also Rosenzweig *et al.* 1994; Turner *et al.* 1996; Treves *et al.* 1998, for

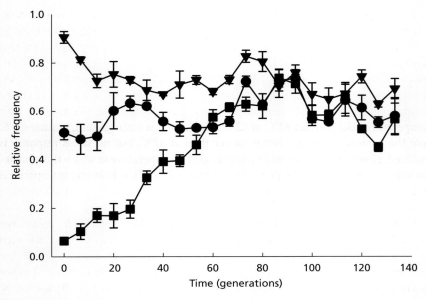

Figure 2.4 Stable equilibrium of two ecotypes that evolved within a population of *E. coli*. The two ecotypes were mixed at three initial ratios; they rapidly converged to an intermediate equilibrium. Each point is the mean of three replicate mixtures; the error bars are standard errors. (Reproduced with permission from Rozen & Lenski (2000).)

similar findings). The stability of their interaction was demonstrated by the fact that each ecotype, when introduced at low frequency, could invade the other type (Fig. 2.4). Based on their demographic parameters in pure culture, one ecotype was clearly superior at exploiting glucose. How did the other type persist in this simple regime? The superior exploiter of glucose secreted some metabolite into the medium that promoted the growth only of the other ecotype. The cross-feeding ecotype also appeared to inhibit the survival of the superior glucose competitor after the glucose was depleted. This two-member community emerged around generation 6000 and has persisted through generation 20 000 and beyond (Rozen & Lenski 2000). During that time, the numerical dominance of the two ecotypes has shifted back and forth several times. The cause of these reversals is not yet known, but a plausible explanation is that the 'ecological' equilibrium changed as each ecotype further adapted genetically to the physical environment or to its competitor. In any case, the organisms themselves have modified the environment such that two ecological strategies can coexist indefinitely.

Changing thermal environments

In the previous study, the populations faced an environment that was both benign and constant over the long-term. The bacteria became better competitors, but their rate of improvement decelerated over time (see Fig. 2.1). In this next study, we took a clone from one of these populations that had adapted for 2000 generations to living on glucose at 37°C, and we used it to found 24 new bacterial populations that were propagated for 2000 more generations in the same medium, but under one of four different thermal regimes (Bennett et al. 1992). One regime was a continuation of the ancestral regime, and it served as a control for the absence of environmental change. Two other regimes were novel but constant environments, with temperature held at either 32°C or 42°C. The former is a benign temperature for *E. coli*; the bacteria grow somewhat slower than at 37°C, but no stress response is induced. However, 42°C is within a degree of the temperature at which these bacteria will cease to grow and begin to die precipitously, and the heat-shock response is induced. The fourth regime was a variable environment with daily alternation between 32°C and 42°C.

Figure 2.5 shows the average rate of genetic adaptation under each regime. For example, the rate of fitness increase for populations that evolved at 32°C was based on competition experiments at 32°C against the common ancestor of all the temperature-selected lines. Genetic adaptation was significantly faster in all of the novel regimes than under the continuation of the ancestral regime (Bennett et al. 1992). Environmental change evidently leads to faster genetic adaptation, provided the change is not so extreme as to cause extinction (see below).

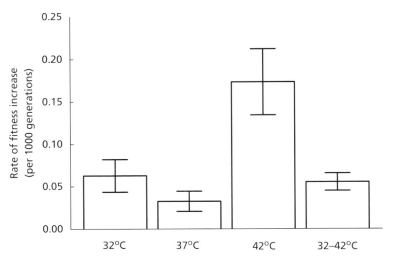

Figure 2.5 Rates of genetic adaptation by *E. coli* to four thermal regimes, measured by improvements in competitive fitness under each regime. For these populations, 37°C is a continuation of the ancestral temperature, 32°C and 42°C are two novel but constant temperatures, and 32–42°C is a novel and variable regime. The height of each bar is the mean of six independent populations; error bars are 95% confidence limits. (Reproduced with permission from Lenski (1995); data from Bennett et al. (1992).)

To examine the thermal specificity of genetic adaptation, we measured the correlated fitness responses of the evolved bacteria across a range of temperatures (Fig. 2.6). There was strong specificity in the responses of the three groups that evolved at constant temperatures (Bennett & Lenski 1993). The populations that evolved at 32°C were better competitors than the 37°C-adapted ancestor at 27°C and 32°C, but they were not significantly better at 22°C or 37°C. The populations that evolved at 37°C improved at 37°C, but this improvement did not carry over to 32°C or 40°C. And the populations that evolved at 42°C were better competitors at 40°C, 41°C, and 42°C; but they were not improved at 37°C, nor had they significantly extended their upper limit for growth. By contrast, the populations that evolved in the variable regime, which alternated between 32°C and 42°C, showed the least specificity of adaptation; this group improved in competition at temperatures ranging from 22°C to 42°C.

Given the specificity of genetic adaptation of the groups that evolved under the novel but constant regimes of 32°C and 42°C, one may ask whether they showed trade-offs—or losses of competitive fitness—at the opposite temperature extreme (Bennett & Lenski 1993). Neither group suffered consistent trade-offs, but within both groups, there was heterogeneity among replicate populations in their correlated responses. In particular, the upper growth limit for some 32°C-adapted

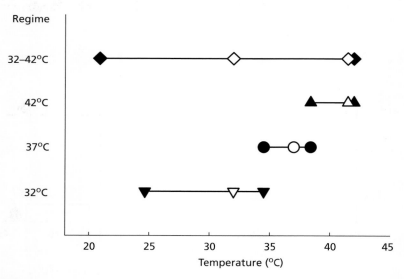

Figure 2.6 The range of temperatures over which *E. coli* populations, adapted to each of four thermal regimes, exhibit improvements in competitive fitness relative to their common ancestor. Open symbols show the temperatures at which populations evolved under each regime. Filled symbols show the approximate range of their fitness improvements. (Reproduced with permission from Lenski (1995); data from Bennett & Lenski (1993).)

populations was shifted down by a degree or so; and one 42°C-adapted population was consistently an inferior competitor at 32°C and below.

None of the changed thermal regimes was so extreme as to cause extinction. That is, all of the regimes were within the ancestor's fundamental thermal niche, defined as the range of temperatures over which the ancestral population could replace itself given the daily dilution into fresh medium (Fig. 2.7). We also studied what happens when the environment changed such that a population found itself above its upper limit for persistence (Mongold *et al.* 1999). These populations declined precipitously by several orders of magnitude, and they usually became extinct. However, a few populations rebounded owing to mutations that extended their fundamental thermal niche. These thermotolerant mutants were consistently inferior competitors within the ancestral niche, even at high temperatures. This trade-off may explain why populations that adapted to 42°C did not show any consistent extension of their upper thermal limit (Bennett & Lenski 1993; Mongold *et al.* 1999). By examining the propensity of populations that previously evolved at different temperatures to give rise to thermotolerant mutants, we sought to test a 'stepping-stone' model, in which populations that evolved at high but non-lethal

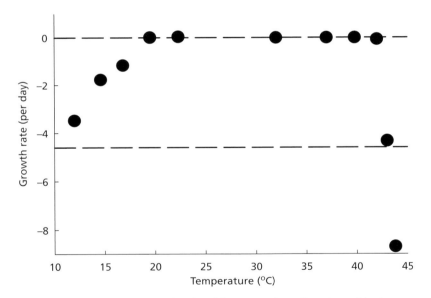

Figure 2.7 The fundamental thermal niche of the ancestral *E. coli* strain used in the experiments on evolutionary adaptation to changing thermal environments. In these experiments, bacteria were diluted 1 : 100 each day into fresh medium. A net growth rate of 0 per day demonstrates population persistence; a net growth rate of –4.6 per day implies dilution without any growth or death. This strain can persist between about 19.5°C and 42°C given these culture conditions. (Reproduced with permission from Bennett & Lenski (1993).)

temperatures may be predisposed to adapt genetically to lethal temperatures (Mongold et al. 1999). We observed a tendency in this direction, but the number of cases in which thermotolerant mutants arose was too small to provide compelling support for this hypothesis.

Interactions with viral predators

Bacteria, including *E. coli*, can be infected by viruses (reviewed by Lenski 1988a). In the case of the lytic viruses discussed below, infection is lethal to the bacterium, and the interaction is effectively that of a predator and its prey (or, more precisely, a parasitoid and its host). We have examined the impact of genetic change on the dynamics of interactions between *E. coli* and several viruses in continuous culture (Lenski & Levin 1985; Lenski 1988b). We also studied how evolution influenced the response of these communities to changes in environmental productivity (Bohannan & Lenski 1997, 1999, 2000).

Figure 2.8 shows the dynamics of *E. coli* and virus T4 in a chemostat during about 500 h. Initially, the viruses held the bacterial population in check at a density far below its resource-limited equilibrium density. Soon, however, T4-resistant

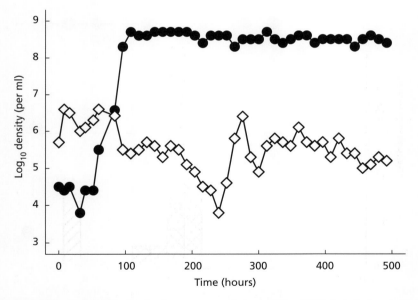

Figure 2.8 Dynamics of interaction between populations of *E. coli* (filled circles) and virus T4 (hollow diamonds) in chemostat culture. The rapid and large increase in the *E. coli* population after about 60 h reflects the emergence of a mutant that is completely resistant to T4 infection. The T4 population persists after the emergence of resistance on a minority population of sensitive cells, which themselves persist owing to their superior competitive ability for glucose. (Reproduced with permission from Lenski & Levin (1985).)

bacterial mutants appeared, and they grew quickly until they reached their resource-limited equilibrium; at that point, the glucose concentration was at a level where the growth of the resistant bacteria matched the rate at which they were washed out of the culture vessel. These mutants were completely resistant, and they did not revert to a sensitive form at any measurable rate (Lenski 1988c). Furthermore, unlike some viruses, T4 did not generate 'host-range' mutants that were able to infect the resistant mutants (Lenski & Levin 1985). Thus, when viruses were introduced to a pure culture of resistant bacteria, the viruses became extinct (Lenski & Levin 1985). However, the virus population did not become extinct following the evolution of resistant mutants (Fig. 2.8).

The T4 population persisted because the resistant bacteria were inferior competitors for glucose (Fig. 2.9), as was shown by mixing resistant and sensitive genotypes in the absence of the virus (Lenski & Levin 1985; Lenski 1988b). Therefore, when resistant mutants increased to their resource-limited equilibrium in the presence of viruses, the amount of glucose was still sufficient to support growth of the sensitive bacteria above the level needed to offset their losses by washout. This excess growth of the sensitive bacteria was exploited by the virus population, which persisted while holding in check the superior competitor. In effect, the community evolved from a two-member food chain to a three-member structure in which a

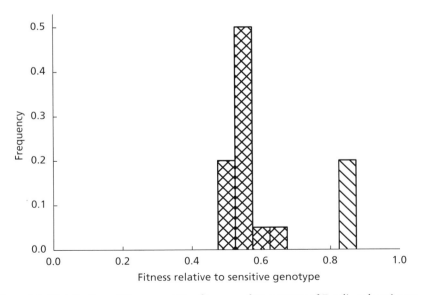

Figure 2.9 Distribution of the competitive fitnesses of 20 mutants of *E. coli*, each resistant to virus T4 infection, relative to their sensitive progenitor. Competitions were performed in glucose without virus. Hatched and cross-hatched fills indicate mutants that were sensitive and cross-resistant to another virus, T7. (Reproduced with permission from Lenski (1988b).)

keystone predator mediated the coexistence of two genotypes at the lower level by preventing competitive exclusion.

Evolution of the interaction increased diversity, and it also fundamentally altered the community's response to environmental change (Bohannan & Lenski 1997, 1999). We manipulated environmental productivity by varying the inflow of glucose to the community. Table 2.1 summarizes the responses to increased productivity of the two-member food chain and the three-member community with virus-mediated coexistence of the sensitive and resistant genotypes. In the simple food chain, the equilibrium density of the T4 predator increased substantially at higher productivity, whereas the bacteria increased only slightly. Moreover, the community was less stable at higher productivity, with increased fluctuations in both populations (Bohannan & Lenski 1997). By contrast, in the three-member community, the total density of the bacteria, but not of the virus, increased at higher productivity. Within the bacteria, only the resistant genotype increased in abundance, essentially drawing down the extra resource such that none of it was available to support faster growth of the sensitive genotype and more viruses. These effects agree well with mathematical models of two- and three-member communities (Leibold 1996), and they show that fundamental changes in both a community's structure and its dynamical properties can arise quickly as a consequence of genetic change.

The trade-off in the bacteria between their competitive ability and resistance to viral infection is an important feature of this interaction. We also studied interactions between *E. coli* and three other lytic viruses (Lenski & Levin 1985). Against virus T5 only, the bacteria suffered no cost of resistance. As a consequence, the sensitive bacteria, and with them the viruses, were driven to extinction following the evolution of resistance. Even in the case of virus T4, the trade-off between bacterial

Table 2.1 Impact of the emergence of an *E. coli* mutant resistant to virus T4 on the community's responses to changes in environmental productivity. The two-member community contains sensitive bacteria and the viral predator; the three-member community includes both sensitive and resistant genotypes and the virus. Productivity was manipulated by varying glucose influx into replicate communities. (Summary of data from Bohannan & Lenski (1997, 1999).)

Effect of increased environmental productivity on:	Two-member community	Three-member community
Density of bacteria	↑	↑↑
Density of sensitive genotype only	NA	↔
Density of resistant genotype only	NA	↑↑
Density of viruses	↑↑	↔

↔, no significant change; ↑, small increase; ↑↑, large increase; NA, distinction between bacterial genotypes is not applicable.

resistance and competitive ability was complex (Lenski 1988b,c). In the absence of T4, one might expect the bacteria would revert to sensitivity. However, evolutionary reversions rarely occurred, probably because mutations to T4 resistance usually knocked out the genes that encode the surface receptor used for viral attachment. Instead, evolving populations of resistant bacteria underwent compensatory changes that significantly reduced the cost of resistance (Lenski 1988c).

Another important feature of the interaction between *E. coli* and virus T4 is that resistance was complete. However, in the case of interactions between *E. coli* and virus T2, the initial mutations conferred only partial resistance (Lenski 1984; Bohannan & Lenski 2000). The ecological consequences of partial resistance are very different from those of complete resistance. With complete resistance, and given the trade-off with competitive ability, the resistant and sensitive genotypes coexisted in the presence of the viral predator. However, partial resistance to virus T2 led to a situation of 'apparent competition', which refers to the indirect interaction between two prey populations that share a predator (Holt 1977; Holt *et al.* 1994). The density of the partially resistant bacteria remained under the control of the virus, and when the environment was sufficiently productive the emergence of the partially resistant genotype caused an increase in the abundance of viruses, more of which were required to hold this prey population in check. The higher density of T2 increased the level of predation on the sensitive bacteria beyond what they could sustain, and they were driven to extinction as a consequence of this feedback involving the shared predator (Fig. 2.10). In less productive environments, by contrast, the sensitive bacteria prevailed owing to their superiority in competition for glucose (Fig. 2.10). Thus, the resulting community structure, and the effect of environmental productivity thereon, depended on whether mutations gave rise to bacteria with partial or complete resistance to the virus.

Interpretation and synthesis

All of our experiments demonstrate 'a genetic component' to 'the abundance and distribution of organisms'—although it was probably a bit *too* strong to call this 'basically a genetic problem' (Tenet 1). The genetic component to organism abundance is dramatically illustrated by the fact that a single mutation, one which conferred resistance to infection by virus T4, changed the density of a bacterial population by several orders of magnitude (see Fig. 2.8). This genetic factor must be placed on an equal footing with the environmental factor that would otherwise control the bacterial population; that is, the density of bacteria following the evolution of resistance was very nearly the same as the sensitive population's density in the absence of any virus. In the case of the interaction between *E. coli* and T5, the resistance mutation was even more important: in addition to the increase in the bacteria to their resource-limited density, the virus population became extinct because the sensitive bacteria had no advantage in competition for glucose. In the case of virus T2, the effect of partial resistance on bacterial abundance was more subtle; partially resistant bacteria sometimes displaced their fully sensitive

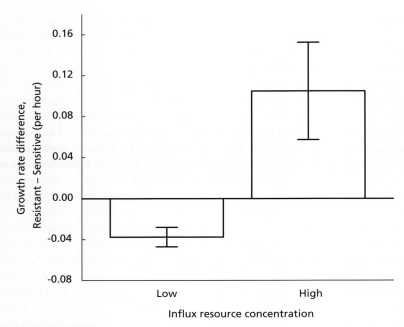

Figure 2.10 Competition between an *E. coli* mutant that is partially resistant to virus T2 and its sensitive progenitor, performed in the presence of T2 at both low and high glucose influxes. The partially resistant genotype prevails at high influx concentration (0.5 μg/mL), whereas the sensitive genotype dominates at low influx concentrations (0.07–0.12 μg/mL). Error bars are 95% confidence intervals. (Data from Bohannan & Lenski 2000.)

progenitors, but the bacterial population remained predator limited (Fig. 2.10). Genetic resistance by plants and animals to consumers (herbivores as well as pathogens and parasites) also may be either partial or complete, depending on the specific mechanism of resistance. As shown by our experiments with bacteria and viruses as well as by mathematical models, this distinction can have profound consequences for the ecological amplitude of the interacting species and the structure of the community in which the interaction is embedded.

Even in our single-species experiments, the abundance of *E. coli* changed as they adapted genetically to growth on glucose. The populations experienced a counterintuitive decline in numbers, but an increase in overall biomass as the average individual bacterium became much larger (see Fig. 2.2). Also, the range of environmental temperatures over which the bacteria persisted—in effect, their distribution—had a genetic component, as occasional mutations arose that allowed bacteria to reproduce at temperatures that were lethal to their progenitors.

The studies of evolving bacteria support the view that 'forces maintaining species diversity and genetic diversity are similar', and they show that 'an understanding of community structure [can] come from considering how these kinds of diversity interact' (Tenet 2). One example is the stable coexistence of two bacterial genotypes on glucose (see Fig. 2.4). As would any stable coexistence, it depended on a trade-off between certain ecological capacities, in this case exploiting glucose and scavenging metabolites. In effect, the genotypes not only partitioned the resources, but through their own activities they produced the resource complexity that allowed them to coexist (Rosenzweig et al. 1994). Other striking examples of the basic similarity between forces maintaining species diversity within communities and genetic diversity within populations come from the interactions between bacteria and viruses (see Fig. 2.8). The coexistence of T4-sensitive and -resistant genotypes mirrors the concept of a keystone predator that maintains species diversity by disproportionately preying on the most competitive species (Paine 1966; Lubchenco 1978). The extinction of virus T5 after resistance evolved emphasizes the importance to coexistence, whether of species or genotypes, of the trade-off between competitive ability and resistance. The dynamical tension between scramble competition for resources and apparent competition mediated by a predator (Holt 1977; Holt et al. 1994) is seen in the dependence on environmental productivity of the fate of mutants partially resistant to virus T2.

All these parallels between the forces that maintain genetic diversity and species diversity reflect interactions that are density dependent. Of course, there are other forces that can maintain genetic diversity within populations, including heterozygote superiority and mutation–selection balance, that have no direct counterparts with respect to maintaining species diversity.

Another interesting aspect of this tenet is understanding how genetic and species diversity interact to influence community structure. The experiments with bacteria and virus T4 illustrate the very strong interaction between these levels of diversity. In the absence of T4, resistant bacteria were rare and inconsequential. In the presence of virus, the resistant genotype not only became numerically dominant but also altered the community's structure and responsiveness to certain kinds of environmental change (see Table 2.1). Prior to the evolution of resistance, the bacterial population was predator limited while the virus population was limited by excess growth of the bacteria, which depended on the productivity of the environment. Thus, in response to increased resources, the growth rate of the bacteria increased, which led to a much higher density of viruses, a slight increase in the bacterial population, and a destabilization of both populations. After the resistant genotype evolved, these mutants drew down the free resource, which reduced the growth rate of sensitive bacteria and thus lowered the equilibrium density of the virus while also stabilizing the interaction. Moreover, the productivity response of the community with genetically diverse prey population was essentially inverted, such that the bacteria but not the viruses increased in density as more resources became available. These experiments show that a community's performance, as well as its structure, can be fundamentally altered by genetic diversity within its members.

All three sets of experiments also show clearly that 'adaptation is a dynamic process' and that 'Darwinian fitness can be measured' as the differential reproductive success of genetically distinct individuals within populations (Tenet 3).

Whether the bacteria adapted to glucose (see Fig. 2.1), to changes in temperature (see Fig. 2.5), or to viral predators (see Fig. 2.8) depended of course on the environment, as did their rate of adaptation. But in every case the bacteria became better adapted to those conditions than were their ancestors, as demonstrated by allowing different genotypes to compete under the relevant regimes. We also tested whether adaptation to one environment led to maladaptation in other environments, which may indicate genetic trade-offs between ecological functions (see Figs 2.3, 2.6, 2.9 and 2.10).

Of course, it is one thing to measure genetic adaptation in simple laboratory systems; it is obviously more difficult to do so in the natural communities that most ecologists study. However, considerable advances have recently been made in studying adaptation even under the most difficult circumstances. These advances are reflected in a long-term study of Darwin's finches (Grant 1999), and more generally by a growing body of work that seeks to develop and apply new methods for studying adaptation (Lande & Arnold 1983; Endler 1986; Harvey & Pagel 1991; Reznick et al. 1997; Thompson 1998, 1999; Clutton-Brock et al. 1999; Huey et al. 2000; Rundle et al. 2000). Regardless of any difficulty, genetic adaptation must be demonstrated before we can move from mere speculation to hypothesis testing about its role in the structure and function of ecological communities. Moreover, by investigating the actual dynamics of genetic adaptation, one obtains information about the time course over which evolution can exert a significant influence on ecological communities. It will certainly be interesting to better understand how the pace of evolutionary change compares with other processes that impact communities, such as climate change, habitat destruction and species introductions.

These experiments also demonstrate that 'adaptation to new environments will result in different genotypes with different life histories', and that 'genetic adjustment . . . may occur by direct response . . . or by compensatory change' (Tenet 4). The bacteria genetically adapted to the limiting resource, temperature regime and to the presence of viral predators. Certain genetic adaptations, but not others, led to trade-offs in different environments. For example, mutants that were resistant to virus T4 were less competitive for glucose, whereas mutants resistant to virus T5 did not incur this cost. Similarly, some populations adapted to high temperature lost fitness at low temperatures, whereas others did not.

Predicting when trade-offs will occur is difficult owing to the inherent genetic and physiological complexity of even simple organisms. Trade-offs may often play a major role in maintaining diversity, but they cannot automatically be assumed to be important and must be experimentally tested (Simms & Rausher 1989; Biere & Antonovics 1996; Kraaijeveld & Godfray 1997). Moreover, even when genetic trade-offs do arise, they are not necessarily immutable constraints on adaptation. Instead, some trade-offs can be ameliorated by compensatory changes that modify the deleterious side-effect of the selected trait without diminishing that trait itself (McKenzie et al. 1982). In our work with bacteria, we have seen compensation for the costs of resistance to both viruses (Lenski 1988c) and antibiotics (not discussed here, but see Bouma & Lenski 1988; Lenski *et al*. 1994), which, in the latter case, has troubling implications for public health (Lenski 1997). Antonovics' attention to compensatory as well as direct change reminds us that the fate of any mutation depends not only on its ecological context but also the genetic 'environment' in which it occurs.

Perhaps Antonovics' most provocative tenet is that 'the distinction between "ecological time" and "evolutionary time" is artificial and misleading', and that 'frequently genetic and ecological changes are simultaneous' (Tenet 5). In the long-term study of bacteria living on glucose, the two time scales seem fairly distinct. Genetic changes in competitiveness, morphology and physiology evolved over hundreds and thousands of generations, but these changes did not prevent us from measuring relative fitness by allowing different strains to compete for several generations (see Figs 2.1–2.3). In fact, evolution was sufficiently slow that we could observe the ecological dynamics leading to the stable coexistence of two ecotypes over a period of a hundred generations or so (see Fig. 2.4). However, when we examined these ecotypes over a much shorter time scale—the seven generations spanning the different 'seasons' (growth phases) within each transfer cycle—we saw that their relative abundance changed depending on the availability of glucose and secondary metabolites in the medium. And over thousands of generations, the identity of the ecotype that was numerically dominant changed repeatedly, a dynamic which may reflect their continuing genetic adaptation to one another or their common environment. Thus, even in this simplest of our experimental systems, one cannot draw any absolutely consistent distinction between ecological and evolutionary scales. More generally, changes in genotype frequencies (at least those that occur by natural selection acting at the level of individuals) necessarily imply differences in demographic parameters. Therefore, it must be true at some level that genetic adaptation has a simultaneous ecological manifestation, even if it is very subtle.

The experiments under different thermal regimes show more obvious convergence in ecological and evolutionary time scales. Genetic adaptation was faster in each of the novel environments than in the continuation of the ancestral environment, indicating that ecological change promotes evolution (see Fig. 2.5). The emergence of mutants that maintained populations at temperatures lethal to their progenitors provides an example in which ecological and genetic changes were essentially simultaneous. Moreover, the ecological fate of the population—extinction or persistence—hinged on a genetic event, illustrating the feedback between ecological and genetic changes.

The most compelling example of the correspondence between ecological and evolutionary time comes from our experiments with bacteria and viruses. Within a matter of a few days, bacterial mutants resistant to viral infection emerged and radically altered the population and community dynamics (see Fig. 2.8 and Table 2.1). By the time the interacting virus and sensitive bacteria went through a few predator–prey cycles, these dynamics were substantially changed by the appearance of resistant mutants (Bohannan & Lenski 1997). In fact, the extreme rapidity of this genetic change made it difficult to study the 'purely ecological' dynamics of this system.

These three sets of experiments with bacteria together illustrate the range of possibilities with respect to distinguishing ecological and evolutionary time scales. On the one hand, in simple and constant environments, genetic change may be sufficiently slow that it does not profoundly undermine one's ability to make ecological inferences. On the other hand, in more complex and novel environments significant genetic change becomes more likely and rapid, and these evolutionary responses will in turn affect the ecological performance of populations and communities. The faster pace of change in more complex and novel environments presumably derives from the fact that organisms living in such environments are generally further from a peak in the fitness landscape than those organisms that live in simpler, more stable environments. The fact that organisms living under more difficult circumstances should often have smaller population sizes may partially offset the tendency for more rapid genetic adaptation in complex and novel environments, especially in extreme cases. For example, only a small proportion of bacterial populations that were challenged with a lethally high temperature produced a thermotolerant mutation before becoming extinct (although those few populations that did survive underwent very rapid change).

A sceptic might say that ecologically significant genetic changes are more likely in bacteria than in plants and animals, owing to the faster generations and larger population sizes of bacteria. However, this view confuses what one can readily measure with what is actually important. Evolutionary changes are, of course, more easily observed in microbes than in plants and animals. But if studies of plants and animals also lasted hundreds and thousands of generations, and given the vast size of many natural populations, who would doubt the potential for widespread genetic change in ecologically important traits? Moreover, there are numerous examples with plants and animals in which ecologically significant genetic changes have occurred within only a few generations, with some of the most conspicuous cases involving interspecific interactions and resistance to chemical agents (McKenzie 1996; Thompson 1998).

Concluding remarks

Life on earth has existed for billions of years, and we can hope that it will continue long into the future. But now and for the foreseeable future, many ecosystems will experience extremely rapid environmental change owing to human activities, including climate change, habitat destruction and the introduction of non-native species (Wilson 1988; Kareiva *et al.* 1993; Simberloff *et al.* 1997). It is precisely such changing circumstances that promote rapid evolution, with its attendant consequences for populations and communities. The on-going extinction of the earth's biota may be similar to major extinctions in the geological past (Raup 1988; Vermeij 1991). Patterns from the fossil record, including time-lagged concordance between rates of extinction and origination of new taxa, indicate that evolutionary change in surviving lineages and communities was often concentrated in and after tumultuous periods (Vermeij 1991; Jablonski & Sepkoski 1996; Erwin 1998; Jackson & Cheetham 1999; Weil & Kirchner 2000). Thus, in our present period of extremely rapid environmental change it becomes especially important that we integrate ecological and genetic studies in an evolutionary framework.

Acknowledgements

I thank Janis Antonovics for being an outstanding and inspirational teacher; and I thank all my collaborators for contributions to this research, especially Al Bennett for leading the project on adaptation to temperature and Brendan Bohannan for his work on interactions of bacteria and viruses. Three reviewers gave many helpful suggestions. The three studies reviewed here have been supported by the US National Science Foundation (DEB-9981397, IBN-9905980, and DEB-9120006).

References

Antonovics, J. (1976) The input from population genetics: 'the new ecological genetics.' *Systematic Botany* **1**, 233–245.

Bennett, A.F. & Lenski, R.E. (1993) Evolutionary adaptation to temperature. II. Thermal niches of experimental lines of *Escherichia coli*. *Evolution* **47**, 1–12.

Bennett, A.F., Lenski, R.E. & Mittler, J.E. (1992) Evolutionary adaptation to temperature. I. Fitness responses of *Escherichia coli* to changes in its thermal environment. *Evolution* **46**, 16–30.

Biere, A. & Antonovics, J. (1996) Sex-specific costs of resistance to the fungal pathogen *Ustilago violacea* (*Microbotryum violaceum*) in *Silene alba*. *Evolution* **50**, 1098–1110.

Bohannan, B.J.M. & Lenski, R.E. (1997) Effect of resource enrichment on a chemostat community of bacteria and bacteriophage. *Ecology* **78**, 2303–2315.

Bohannan, B.J.M. & Lenski, R.E. (1999) Effect of prey heterogeneity on the response of a model food chain to resource enrichment. *American Naturalist* **153**, 73–82.

Bohannan, B.J.M. & Lenski, R.E. (2000) The relative importance of competition and predation varies with productivity in a model community. *American Naturalist* **156**, 329–340.

Bouma, J.E. & Lenski, R.E. (2000) Evolution of a bacteria/plasmid association. *Nature* **335**, 351–352.

Clutton-Brock, T.H., O'Riain, M.J., Brotherton, P.N.M. *et al.* (1999) Selfish sentinels in cooperative mammals. *Science* **284**, 1640–1644.

Dykhuizen, D. & Hartl, D. (1981) Evolution of competitive ability. *Evolution* **35**, 581–594.

Endler, J.A. (1986). *Natural Selection in the Wild*. Princeton University Press, Princeton.

Erwin, D.H. (1998) The end and the beginning: recoveries from mass extinctions. *Trends in Ecology and Evolution* **13**, 344–349.

Grant, P.R. (1999) *Ecology and Evolution of Darwin's Finches*. Princeton University Press, Princeton.

Harvey, P.H. & Pagel, M.D. (1991). *The Comparative Method in Evolutionary Biology*. Oxford University Press, Oxford.

Holt, R.D. (1977) Predation, apparent competition, and the structure of prey communities. *Theoretical Population Biology* **11**, 197–229.

Holt, R.D., Grover, J. & Tilman, D. (1994) Simple rules for interspecific dominance in systems with exploitative and apparent competition. *American Naturalist* **144**, 741–771.

Huey, R.B., Gilchrist, G.W., Carson, M.L., Berrigan, D. & Serra, L. (2000) Rapid evolution of a geographic cline in size in an introduced fly. *Science* **287**, 308–309.

Jablonski, D. & Sepkoski, J.J. (1996) Paleobiology, community ecology, and scales of ecological pattern. *Ecology* **77**, 1367–1378.

Jackson, J.B.C. & Cheetham, A.H. (1999) Tempo

and mode of speciation in the sea. *Trends in Ecology & Evolution* **14**, 72–77.

Kareiva, P.M., Kingsolver, J.G. & Huey, R.M., eds (1993). *Biotic Interactions and Global Change*. Sinauer, Sunderland, MA.

Kraaijeveld, A.R. & Godfray, H.C.J. (1997) Trade-off between parasitoid resistance and larval competitive ability in *Drosophila melanogaster*. *Nature* **389**, 278–280.

Lande, R. & Arnold, S.J. (1983) The measurement of selection on correlated characters. *Evolution* **37**, 1210–1226.

Leibold, M.A. (1996) A graphical model of keystone predators in food webs: trophic regulation of abundance, incidence and diversity patterns in communities. *American Naturalist* **147**, 784–812.

Lenski, R.E. (1984) Two-step resistance by *Escherichia coli* B to bacteriophage T2. *Genetics* **107**, 1–7.

Lenski, R.E. (1988a) Dynamics of interactions between bacteria and virulent phage. *Advances in Microbial Ecology* **10**, 1–44.

Lenski, R.E. (1988b) Experimental studies of pleiotropy and epistasis in *Escherichia coli*. I. Variation in competitive fitness among mutants resistant to virus T4. *Evolution* **42**, 425–432.

Lenski, R.E. (1988c) Experimental studies of pleiotropy and epistasis in *Escherichia coli*. II. Compensation for maladaptive pleiotropic effects associated with resistance to virus T4. *Evolution* **42**, 433–440.

Lenski, R.E. (1995) Evolution in experimental populations of bacteria. In: *Population Genetics of Bacteria* (eds S. Baumberg, J.P.W. Young, E.M.H. Wellington & J.R. Saunders), pp. 193–215. Cambridge University Press, Cambridge.

Lenski, R.E. (1997) The cost of antibiotic resistance—from the perspective of a bacterium. In: *Antibiotic Resistance: Origins, Evolution, Selection and Spread* (eds D.J. Chadwick & J. Goode), pp. 131–151. Wiley, Chichester.

Lenski, R.E. & Levin, B.R. (1985) Constraints on the coevolution of bacteria and virulent phage: a model, some experiments, and predictions for natural communities. *American Naturalist* **125**, 585–602.

Lenski, R.E. & Mongold, J.A. (2000) Cell size, shape, and fitness in evolving populations of bacteria. In: *Scaling in Biology* (eds J.H. Brown & G.B. West), pp. 221–235. Oxford University Press, New York.

Lenski, R.E. & Travisano, M. (1994) Dynamics of adaptation and diversification: a 10,000-generation experiment with bacterial populations. *Proceedings of the National Academy of Sciences of the USA* **91**, 6808–6814.

Lenski, R.E., Rose, M.R., Simpson, S.C. & Tadler, S.C. (1991) Long-term experimental evolution in *Escherichia coli*. I. Adaptation and divergence during 2,000 generations. *American Naturalist* **138**, 1315–1341.

Lenski, R.E., Simpson, S.C. & Nguyen, T.T. (1994) Genetic analysis of a plasmid-encoded, host genotype-specific enhancement of bacterial fitness. *Journal of Bacteriology* **176**, 3140–3147.

Lubchenco, J. (1978) Plant species diversity in a marine intertidal community: importance of herbivore food preference and algal competitive abilities. *American Naturalist* **112**, 23–39.

McKenzie, J.A. (1996). *Ecological and Evolutionary Aspects of Insecticide Resistance*. Academic Press, San Diego.

McKenzie, J.A., Whitten, M.J. & Adena, M.A. (1982) The effect of genetic background on the fitness of diazinon resistance genotypes of the Australian sheep blowfly, *Lucilia cuprina*. *Heredity* **49**, 1–9.

Mongold, J.A. & Lenski, R.E. (1996) Experimental rejection of a nonadaptive explanation for increased cell size in *Escherichia coli*. *Journal of Bacteriology* **178**, 5333–5334.

Mongold, J.A., Bennett, A.F. & Lenski, R.E. (1999) Evolutionary adaptation to temperature. VII. Extension of the upper thermal limit of *Escherichia coli*. *Evolution* **53**, 386–394.

Paine, R.T. (1966) Food web complexity and species diversity. *American Naturalist* **100**, 65–75.

Raup, D.M. (1988) Diversity crises in the geological past. In: *Biodiversity* (ed. E.O. Wilson), pp. 51–57. National Academy Press, Washington, DC.

Reznick, D.N., Shaw, F.H., Rodd, F.H. & Shaw, R.G. (1997) Evaluation of the rate of evolution in natural populations of guppies (*Poecilia reticulata*). *Science* **275**, 1934–1937.

Rosenzweig, R.F., Sharp, R.R., Treves, D.S. &

Adams, J. (1994) Microbial evolution in a simple unstructured environment: genetic differentiation in *Escherichia coli*. *Genetics* **137**, 903–917.

Rozen, D.E. & Lenski, R.E. (2000) Long-term experimental evolution in *Escherichia coli*. VIII. Dynamics of a balanced polymorphism. *American Naturalist* **155**, 24–35.

Rundle, H.D., Nagel, L., Boughman, J.W. & Schluter, D. (2000) Natural selection and parallel speciation in sympatric sticklebacks. *Science* **287**, 306–308.

Simberloff, D., Schmitz, D. & Brown, T., eds (1997). *Strangers in Paradise: Impact and Management of Nonindigenous Species in Florida*. Island Press, Washington, DC.

Simms, E.L. & Rausher, M.D. (1989) The evolution of resistance to herbivory in *Ipomoea purpurea*. II. Natural selection by insects and costs of resistance. *Evolution* **43**, 573–585.

Thompson, J.N. (1998) Rapid evolution as an ecological process. *Trends in Ecology and Evolution* **13**, 329–332.

Thompson, J.N. (1999) The evolution of species interactions. *Science* **284**, 2116–2118.

Treves, D.S., Manning, S. & Adams, J. (1998) Repeated evolution of an acetate-crossfeeding polymorphism in long-term populations of *Escherichia coli*. *Molecular Biology and Evolution* **15**, 789–797.

Turner, P.E., Souza, V. & Lenski, R.E. (1996) Tests of ecological mechanisms promoting the stable coexistence of two bacterial genotypes. *Ecology* **77**, 2119–2129.

Vasi, F.K. & Lenski, R.E. (1999) Ecological strategies and fitness tradeoffs in *Escherichia coli* mutants adapted to prolonged starvation. *Journal of Genetics* **78**, 43–49.

Vasi, F., Travisano, M. & Lenski, R.E. (1994) Long-term experimental evolution in *Escherichia coli*. II. Changes in life-history traits during adaptation to a seasonal environment. *American Naturalist* **144**, 432–456.

Vermeij, G.J. (1991) When biotas meet—understanding biotic interchange. *Science* **253**, 1099–1104.

Weil, A. & Kirchner, J.W. (2000) Delayed biological recovery from extinctions throughout the fossil record. *Nature* **404**, 177–180.

Wilson, E.O., ed. (1988). *Biodiversity*. National Academy Press, Washington, DC.

Zambrano, M.M., Siegele, D.A., Almiron, M., Tormo, A. & Kolter, R. (1993) Microbial competition: *Escherichia coli* mutants that take over stationary phase cultures. *Science* **259**, 1757–1760.

Chapter 3
Sociality and population dynamics

T.H. Clutton-Brock

Introduction

Over the last 30 years, a combination of theoretical models, detailed observational and experimental studies, and quantitative interspecific comparisons has provided a framework for understanding the evolution of reproductive behaviour in animals (Wilson 1975; Stearns 1992; Andersson 1994; Maynard Smith & Szathmary 1995; Krebs & Davies 1997). The strategies used by individuals to locate and defend mates, the degree of polygamy and the development of parental care have been shown to influence the evolution of species differences in many aspects of physiology, anatomy and behaviour (Harvey & Pagel 1991). Since these (and other) aspects of breeding behaviour commonly affect reproductive rates and survival, they are also likely to affect many ecological processes, including rates of population growth, the operation of density-dependent and -independent factors, and the stability of population size.

While the demographic consequences of most aspects of breeding systems have not yet been explored, there is one important exception. Partly because of misconceived claims that territoriality might have evolved to prevent local populations from overexploiting their resources (Wynne-Edwards 1962), a substantial number of vertebrate studies during the 1950s and 1960s focused on the role played by territoriality in limiting population size and regulating density in monogamous birds (Lack 1966; Newton 1998). Four principal conclusions emerged from this body of work. First, that within, as well as across, species, territory size declines and the density of breeding females (or pairs) increases where food is plentiful or competition for territories is intense. For example, across bird species, the size of feeding territories is closely related to body size and diet type (Schoener 1968) while, within species, territory size commonly declines with the availability of resources as well as with rising population density (Lack 1966; Watson & Miller 1971). Second, that even where territory size is adjusted to resource availability, population density is often limited by the availability of breeding territories rather than by resource availability directly (Lack 1966). Third, that differences in territory quality commonly generate large differences in breeding success between individuals, so that a large proportion of recruits are produced by relatively few parents (see Chapter 4 in Newton 1998). Fourth, that populations of territorial species com-

Department of Zoology, University of Cambridge, Downing Street, Cambridge CB2 3EJ, UK

monly consist of a combination of territorial individuals and a larger number of floaters which live in marginal habitat or in the interstices between established territories and replace breeders that die (Krebs 1971). Fluctuations in breeding success or survival may consequently have little effect on the size or growth rate of a territorial population unless they are large enough to remove the entire population of floaters, so that dying residents are no longer replaced (Newton 1998). As a result, territoriality can usually be expected to enhance the stability of breeding populations, especially in core habitat.

Other aspects of reproductive behaviour are likely to have ecological consequences that are at least as far reaching as those of territoriality. This chapter examines the ecological consequences of sociality, concentrating on studies of vertebrates where individuals commonly live in the same, closed group for much of their lives. While the evolution of sociality was one of the original foci of research in behavioural ecology (Bertram 1975, 1978; Pulliam & Caraco 1984), relatively few studies have investigated its ecological consequences. The first section outlines the effects of sociality on competition and describes how group living can focus the effects of resource shortage on individuals that are less capable of competing successfully. As a result, minor changes in population density affecting group size can lead to strong density-dependent changes in reproduction or survival within particular categories of individuals. The next section examines some of the ecological consequences of competition between groups. Where larger groups consistently displace smaller ones, the breeding success and survival of group members may increase with group size. Small groups may show high rates of extinction and the incidence of successful dispersal may be low, since newly founded groups are almost always small and will be at a disadvantage when competing with larger ones. Effects of this kind may be particularly important in cooperative breeders, where breeders rely on assistance from other group members to rear their young.

Sociality also has less direct effects on demographic processes. The existence of long-lasting groups confronts group members with the potential risks of inbreeding, often causing members of one sex to disperse at adolescence (p. 53). Sex differences in dispersal can, in turn, affect many aspects of population demography, including the rate at which vacant territories are occupied and the relationship between population density and the adult sex ratio. Finally, female sociality can permit the development of polygyny, often in association with differences in body size between the sexes. As described below (p. 55), sexual dimorphism may be associated with partial ecological segregation between males and females, as well as with sex differences in mortality. As a consequence of these effects, sociality can have important implications for the management and conservation of vertebrates, some of which are reviewed below (p. 60). The final section speculates on the demographic effects of other aspects of the breeding systems in other groups of organisms and predicts the development of comparative studies of population dynamics.

Sociality and competition within groups

Where offspring of one or both sexes settle in the range used by their parents, forming closed groups of stable membership, increasing group size commonly generates competition for limited resources between group members. As groups grow in size, their members need to travel further each day to find adequate food, and the costs of feeding increase (Waser 1977). Unless range or territory size rises with group size (Macdonald 1983), resources per head are likely to decline and aggressive disputes may increase in frequency (Hoogland 1979). Interference between group members may also extend to reproduction, increasing the extent of reproductive suppression (see below) and the incidence of infanticide (Packer *et al.* 1988). Sociality may also accelerate the transmission of disease or parasites between group members (Dobson & Poole 1998) while rising group size may increase the frequency and prevalence of parasites and endemic diseases (Hoogland 1979). In conjunction, these effects commonly lead to density-dependent processes at the group level, generating reductions in breeding success or survival in larger groups (Clutton-Brock *et al.* 1982a; Van Schaik *et al.* 1983).

Where increasing group size depresses reproductive success or survival, younger, weaker or more subordinate individuals are often more strongly affected than older or more dominant ones (Dittus 1977, 1979; Clutton-Brock *et al.* 1984). For example, in red deer, juvenile males show lower survival than juvenile females (Clutton-Brock *et al.* 1985a) and rising group size depresses the survival of males more than that of females (Clutton-Brock *et al.* 1985a). However, relationships between group size and breeding success or survival are often complex. In red deer, increasing group size depresses juvenile survival when population density is low but, as population density increases and the ranges of neighbouring groups overlap to a greater extent, this effect weakens and disappears (Fig. 3.1) (Clutton-Brock *et al.* 1987).

In general, the tendency for sociality to focus intragroup competition on the weakest individuals can be expected to advance and intensify density-dependent changes in survival and breeding success and thus to enhance population stability in social species. However, there may also be cases where sociality has the opposite effect. For example, as group size increases and populations become more aggregated in their distribution, the number of susceptible hosts that any infected individual will contact is likely to increase, leading to a reduction in the threshold for establishment of the pathogen in the host population (Dobson & Poole 1998). Under these circumstances, sociality may tend to reduce population stability. This is especially likely if competition with older and more dominant individuals impairs the development of immune responses in younger animals, increasing the chance that some group members may be susceptible to infection.

In a minority of vertebrates, reproductive competition between females has led to the suppression of reproductive activity in all females except for one dominant individual. In some social mammals, subordinate females are capable of reproduction but fail to breed while, in others, their reproductive system is inactive

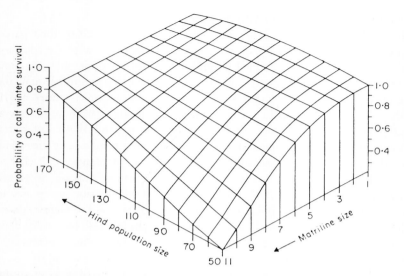

Figure 3.1 Logistic curves fitted to demonstrate how the overwinter survival of red deer calves changes with increasing population density and rising matriline size. At high population density, the effects of matriline size disappear. (Reproduced with permission from Clutton-Brock et al. 1987.)

and they are (temporarily) incapable of breeding (Creel & Waser 1997; Faulkes & Abbott 1997; Moehlman & Hofer 1997). Where only a single female breeds, the population growth rate will be determined principally by the number of separate breeding groups, and any factors that reduce group number are likely to reduce both the growth rate and the size of the population. Under these circumstances, the fecundity and reproductive rate of breeding females may be unlikely to vary with population density, while the risk of local population extinction is likely to depend on the number of groups rather than the density of individuals.

Sociality and competition between groups

In many social vertebrates, larger groups can displace smaller ones and may, in some cases, drive them from their territories (Wrangham 1980). Where these benefits of large group size exceed the costs of increased competition for resources between group members, breeding success and survival may initially increase with group size, leading to inverse density dependence or 'Allee' effects at the group level (Dobson & Poole 1998; Clutton-Brock et al. 1999) (Fig. 3.2).

Inverse density dependence at the group level is likely to be particularly strong in cooperative breeders. In these species, most group members do not breed themselves but rather help the dominant female to feed and protect her young (Stacey & Koenig 1990; Solomon & French 1997). The development of cooperation varies

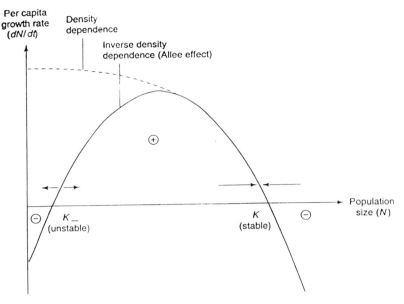

Figure 3.2 Illustration of the Allee effect, from a simple mathematical model of population dynamics. The per capita growth rate (dN/dt) is negative above the carrying capacity (K) and positive below. However, in the presence of an Allee effect, it also decreases below a given population size, and can even become negative below a critical population threshold ($K\_$). When a population displaying this type of population dynamics is driven below the critical threshold, the low, sometimes negative, per capita growth rate may lead it to extinction. (Reproduced with permission from Courchamp et al. 1999b.)

widely, from cases where some parents are assisted by young of the previous year, but are responsible for most of the care of their own broods or litters, to obligate cooperative breeders where adults cannot breed without helpers, and parents are responsible for a relatively small proportion of rearing costs. Where cooperation is not highly developed, the contribution of helpers to the parents' fitness is often small, the presence of helpers may depress the survival of the parents' offspring territories and breeding success or survival may decline in larger groups (Komdeur 1994a,b; Woodroffe 2000). In contrast, where cooperation is highly developed, strong positive correlations between group size and the survival of juveniles and/or adults are common, and Allee effects are pronounced (Macdonald 1979; Moehlman 1979; Rood 1990) (Fig. 3.3).

In some obligate cooperate breeders, reproductive strategies appear to be adapted to the need for assistants. For example, in Seychelles' warblers, females are more likely than males to remain in their parents' territory while, in African wild dogs, males are more likely to remain and assist their parents. In both species, breeders apparently adjust the sex ratio of their offspring to their need for helpers. In Seychelles' warblers, pairs with few helpers that occupy good territories produce

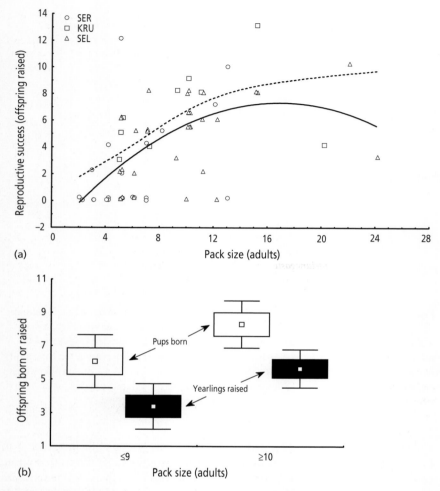

Figure 3.3 (a) Offspring raised to one year of age as a function of pack size for African wild dogs in three ecosystems. SEL = Selous Game Reserve, KRU = Kruger National Park, SER = Serengeti National Park. (b) Number of offspring born (open bars) and raised to one year of age by African wild dogs in packs greater or smaller than the median pack size. Data from the Selous Game Reserve. Bars show mean ± one standard error, whiskers ± two standard errors (reproduced with permission from Creel & Creel 2001).

female-biased broods (Komdeur et al. 1997) while, in African wild dogs, primiparous females (which normally lack helpers) produce male-biased litters and multiparous females produce female-biased litters (Creel et al. 1998).

Inverse density dependence at the group level can have important consequences for population dynamics. If group size is reduced by density-independent factors or stochastic events, the low breeding success and survival associated with small group size may delay the recovery of smaller groups or cause them to continue to decline to extinction. Rates of group extinction are consequently likely to be relatively high in cooperative breeders (Courchamp *et al.* 1999a,b) and populations may spend much of their time below the carrying capacity of their environments. Empirical studies of several cooperative mammals show that smaller groups have an increased risk of extinction and indicate that group extinction rates may be unusually high (Clutton-Brock *et al.* 1998). Long-term studies of wild populations are lacking but populations might be expected to show greater instability than in species where breeding success and survival decline with increasing group size.

Sociality and dispersal

Sociality can also influence population dynamics through its effects on dispersal. Species characterized for breeding success or survival in small groups commonly show low rates of successful dispersal because colonizers are at a disadvantage when competing with larger groups and are unlikely to establish new breeding units successfully. Females are often reluctant to leave their natal group and may only disperse when evicted by dominant breeders (e.g. Clutton-Brock *et al.* 1998) or when they can disperse in small groups. As a result, populations are commonly slow to recolonize territories where groups have become extinct, leading to spatial variation in population density (Clutton-Brock *et al.* 1999). For example, after 2 years of drought, seven territorial groups of meerkats in our Kalahari study population became extinct (Clutton-Brock *et al.* 1999). Food availability, condition and breeding success all increased during the next 3 years but, although the size of the remaining groups increased, the number of groups did not rise, and the seven territories remained unoccupied (Fig. 3.4). Since most meerkat groups contain a single breeding female, the reduction in numbers of groups depressed the population growth rate for several years after the initial reduction in group numbers.

Inverse density dependence may also affect the ecological factors constraining group size. In many cooperative species where subordinates are suppressed, they can be viewed as queuing for opportunities to breed or disperse (Kokko & Johnstone 1999). Since the benefits of queuing will usually decrease with queue length, the net benefits of dispersal are likely to increase with group size. As a result, the upper limits of group size may be limited by group size-dependent dispersal rates rather than by resource abundance *in situ*, and feeding competition may be rare (Barnard 2000).

The formation of closed breeding groups also confronts adolescents with the potential risks of inbreeding (Greenwood 1980; Jimenez 1994). Studies of a wide range of vertebrates now provide evidence of behavioural adaptations that reduce these risks. In particular, where daughters grow up in territories defended by their fathers, daughters commonly avoid mating with their father, defer sexual maturity until his death, or disperse (Wolff 1992; McNutt 1996; Cooney 1999). Group members that lack unrelated breeding members of both sexes may cease to breed until immigration occurs (e.g. O'Riain 2001).

In many social vertebrates, sex differences in the frequency of dispersal are

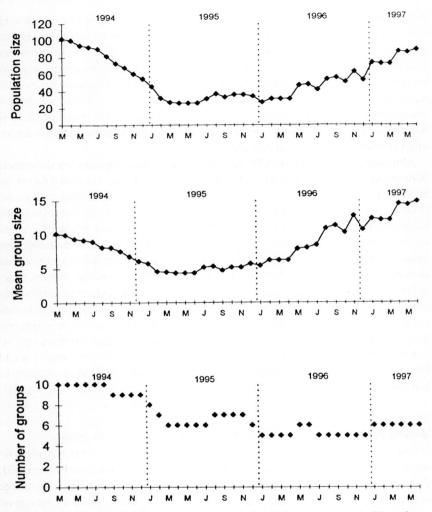

Figure 3.4 Demographic changes in meerkat populations following a period of drought (1994/5): (a) population size; (b) mean group size; and (c) number of groups in study area. (Reproduced with permission from Clutton-Brock et al. 1999a.)

related to the relative probability of close inbreeding. In most social mammals, breeding females are typically recruited from animals born in the same group, while males disperse to join other groups or establish new ones with dispersing females. In virtually all of these species, the breeding tenure of established males is sufficiently short and the development of females sufficiently slow that fathers rarely occupy breeding roles by the time their daughters reach maturity (Clutton-Brock 1991). Females that commonly remain in their natal group typically avoid mating with males reared in the group (McNutt 1996; Cooney 1999), forcing males born in the group to disperse to breed. In contrast, in a few species (including equids, some primates and at least one carnivore), the breeding tenure of individual males or of male kin groups exceeds the average age of females at first breeding. As a result, females are forced to disperse in order to avoid inbreeding, and males may remain in their natal group (Clutton-Brock 1989, 1991). Similar sex differences in dispersal exist in social birds, although, here, female-biased dispersal is more frequent than male-biased dispersal (Greenwood 1980), possibly because females are commonly able to breed in the second year of life while male tenure often extends over more than one season.

Although they have received little attention from demographers, sex differences in dispersal are likely to have important effects on population dynamics. As yet, we know little about relationships between population density and rates of dispersal in either sex, but recent studies suggest that their consequences may be substantial. For example, in red deer (where females typically adopt ranges overlapping those of their mothers and older sisters, and males disperse to join bachelor groups between the ages of 2 and 4 years), emigration rates increase with population density in males but not in females (Fig. 3.5a) (Clutton-Brock et al. 1997b). Since males also avoid immigrating into local populations where female density is high (Fig. 3.5f), changes in emigration rates of males with increasing population density contribute to the development of strong female biases in the adult sex ratio on populations close to carrying capacity which can have important implications for management (see p. 58). It is not known whether equivalent patterns occur in species where females disperse at adolescence but, if so, they might be expected to generate local biases in the adult sex ratio that favour males.

Mating systems, sexual dimorphism and population dynamics

Where females live in groups, individual males can commonly monopolize breeding access to several females, and polygyny often develops (Clutton-Brock 1989). The demographic consequences of polygyny are diverse and include delays in age at first breeding in males; intense competition for female groups, often associated with high rates of injury; a shortening of the reproductive lifespan of males relative to that of females; and a reduction in effective population size (Clutton-Brock 1988; Clutton-Brock 1989; Clutton-Brock & Lonergan 1994).

Polygyny also has important effects on the evolution of sex differences in growth and size, which can have far-reaching consequences. In many polygynous vertebrates, male mating success increases with body size, while size has weaker effects on female breeding success (Trivers 1972). As a result, selection has favoured the evolution of larger body size in males than females (Trivers 1972). Since larger adult size is usually associated with increases in early growth rates (before and after birth), selection for male mating success has commonly led to sex differences in early growth rates (Clutton-Brock et al. 1982b). For example, in red deer, the breeding success of males is related to their body size, and adult size is, in turn, related to early growth, with the result that birth weight exerts a strong effect on breeding

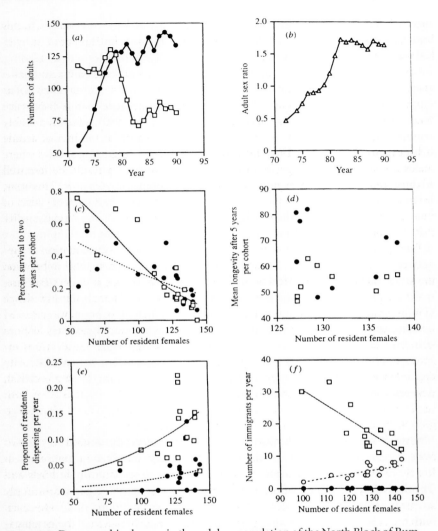

Figure 3.5 Demographic changes in the red deer population of the North Block of Rum, following the cessation of culling in 1972. (a) Numbers of adults (≥2 years) of each sex resident in the study area 1972–95: □ males, ● females. (b) The sex ratio (females : males) of resident adults (≥2 years). (c) Percentage survival to 24 months for members of each cohort of calves: □males, ● females; and (d) mean longevity of males and females reaching two years calculated across surviving members of each cohort. (e) Proportion of natal males (□) and females (●) of 2–6 years emigrating (permanently) from the study area each year. (f) Number of animals of all ages immigrating into the study area per year as permanent residents (□ males, ● females) or rut immigrants (males only, ○), immigrating temporarily during the rut. (Reproduced with permission from Clutton-Brock et al. 1998.)

success in males (Kruuk *et al.* 1999b). As in many other polygynous mammals, this has led to the evolution of sex differences in growth rates both before and after birth, as well as in adult body size.

Sex differences in size and growth can have important energetic and demographic consequences. In grazing mammals, metabolic requirements may scale as (approximately) Body Weight$^{0.72}$ but feeding rates (calculated as rates of forage intake per minute grazing) can show a considerably lower scaling factor, possibly around Body Weight$^{0.33}$, the scaling factor for the breadth of the incisor arcade (Clutton-Brock & Harvey 1983; Illius & Gordon 1987). In grazing species where males are substantially larger than females, males may, as a result, be less well adapted than females to exploit swards where biomass is very low. This may, perhaps, explain why male ungulates commonly spend an increased proportion of their grazing time on swards of higher biomasses but lower nutritional quality (Clutton-Brock *et al.* 1982b).

Ecological segregation between the sexes can have a variety of ecological consequences that, as yet, have been little explored. In some cases, competition between the sexes may be asymmetrical with the result that, as female numbers increase, males are progressively excluded from preferred resources. There is some evidence that this situation occurs in red deer: after part of the Rum population was released from culling and population density increased, males made progressively less use than females of areas that were previously strongly selected by both sexes (Clutton-Brock *et al.* 1987). Sexual segregation may also affect the operation of density dependence. For example, where competition between the sexes is asymmetrical, changes in the density of females (rather than in the combined density of both sexes) may be primarily responsible for causing density-dependent changes in growth, breeding success and survival (Clutton-Brock & Albon 1989).

Sex differences in growth also have important effects for survival in the two sexes. Higher pre-dispersal mortality rates have been recorded in male than in female juveniles in a considerable number of polygynous, dimorphic birds and mammals. Sex differences of this kind in mammals were, at one time, commonly attributed to the fact that males are the heterogametic sex (Trivers 1972). However, both the existence of similar differences in birds (where females are the heterogametic sex) and the existence of interspecific correlations between the extent of sex differences in mortality and sexual size dimorphism (Clutton-Brock *et al.* 1985a) suggests that they are a consequence of the greater energetic requirements of growing males in size dimorphic species.

The magnitude of sex differences in survival commonly increases with the severity of environmental conditions. In red deer, for example, mortality of juvenile males relative to juvenile females increases with rising population density (Fig. 3.6) and tends to be high when winter weather conditions are unfavourable (Clutton-Brock *et al.* 1985b; Clutton-Brock *et al.* 1991). In some mammals, these differences even extend back into the gestation period; for example, in red deer, the sex ratio of calves at birth (% males) declines with increasing population density (Fig. 3.7a) and (when density is controlled) is low after unfavourable winters following con-

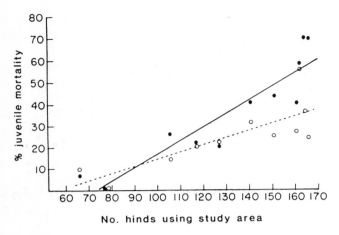

Figure 3.6 Mortality in juvenile red deer during the first 2 years of life among males (●) and females (○) born between 1971 and 1982 in the red deer population of the North Block of Rum. (Reproduced with permission from Clutton-Brock & Albon 1989.)

ception (Kruuk et al. 1999a,b) (Fig. 3.7b). Since these reductions in sex ratios at birth are associated with reduced fecundity, they are probably caused by differential mortality *in utero*, and a substantial number of studies of domestic animals and humans suggest rates of mortality are commonly higher in male fetuses compared to females (McMillen 1979).

While there is clear evidence of an association between sex differences in juvenile mortality and size dimorphism, it is less obvious that sex differences in mortality increase with sexual dimorphism in adults. Although survival rates are commonly lower in adult males than adult females in many polygynous vertebrates (Glucksmann 1974; Clutton-Brock et al. 1982b), there is little evidence that the extent of sex differences in adult mortality are consistently related to the degree of sexual dimorphism in body size (Promislow 1992; Loison et al. 1999). There are at least two explanations of this apparent contrast in sex differences in mortality between juveniles and adults. First, the costs of sexually selected increases in body size may be greater during the period of growth than in adulthood. Alternatively, the effects of increasing adult size on the mortality of adult males may be obscured by those of other aspects of the breeding system that influence male survival, such as the frequency of adult dispersal, the intensity of fight for females, or the timing and extent of condition loss during the mating season.

Sex differences in survival and emigration rates can lead to biases in the adult sex ratio. For example, after the red deer population of the North Block of Rum was released from culling, female numbers increased and male numbers declined with the result that the adult sex ratio changed from a strong male bias to a 2:1 bias in favour of females (see Fig. 3.5a). Similar biases in the adult sex ratio have been

SOCIALITY AND POPULATION DYNAMICS

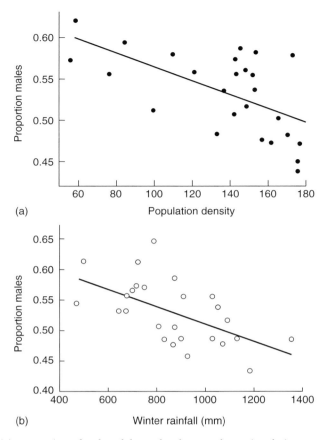

Figure 3.7 (a) Proportion of male red deer calves born each year in relation to population density in the North Block of Rum after correcting for effect of winter rainfall. Regression coefficient -0.080 (± 0.026 SE), $t = -3.12$, d.f. = 23, $P = 0.005$; $r = -0.552$. (b) Proportion of male red deer calves born each year in relation to winter rainfall in the North Block of Rum (November to January, in millimetres) after correcting for the effect of density. Regression coefficient -0.0128 (± 0.046 SE), $t = -3.03$, d.f. = 23, $P = 0.006$; $r = -0.543$. (Reproduced with permission from Kruuk et al. 1999a, b.)

recorded in a number of other ungulate populations (Clutton-Brock et al. 1982b, 1991, 1997a).

Biases in the adult sex ratio can have important consequences for the dynamics of populations. Where sex ratio biases are extreme, they may lead to reductions in female fecundity if receptive females are unable to locate mature males or if stimulation by more than one male is necessary to induce oestrus and receptivity in females. For example, following heavy exploitation of males, populations of African elephants showed both strong biases in the adult sex ratio and a reduction in female fecundity (Poole 1989; Dobson & Poole 1998). Similarly, in Soay sheep, where male stimulation plays a role in inducing female oestrus, the timing of conception is delayed after mating seasons when the ratio of males to females is low as a result of heavy mortality of males in the previous year (Clutton-Brock et al. 1992).

Biased adult sex ratios can also affect the age distribution of breeding success among males. In red deer, for example, biased adult sex ratios are associated with reductions in the period for which mature males hold harems, increases in the proportion of males holding harems and reductions in the average age of harem-holding males (Clutton-Brock et al. 1997b). Similarly, in Soay sheep, biased adult sex ratios are associated with an increase in the proportion of females mated by young males and with changes in the distribution of male success (Stevenson & Bancroft 1995; Pemberton et al. 1996). These effects are likely to broaden genetic representation in the next generation and to reduce the tendency for biased adult sex ratios to diminish the effective population size (Pemberton et al. 1996).

Sociality and conservation

Sociality also has important implication for conservation and management. Among social vertebrates, the area of a group's range or territory increases with the size and energetic requirements of groups (Milton & May 1976; Harvey & Clutton-Brock 1981). Where endangered species live in relict habitat patches, they are commonly at risk when they cross patch boundaries. As a result, group size and home-range area, rather than population density, may be the principal determinants of extinction risk (Fig. 3.8) (Woodroffe & Ginsberg 1998) and reserve sizes need to be adjusted so as to minimize the frequency with which breeding groups leave the reserve. Similarly, where breeding success and survival increase with group size, successful conservation may require measures to raise or maintain the size of groups. If small groups are likely to be at a disadvantage in competition with larger ones, reintroduction programmes may need to concentrate on the release of established breeding groups, rather than individual animals.

Sociality also has implications for the management of abundant species. Where groups defend territories, and population growth rates vary between groups, the imposition of average culling rates across all groups (the normal practice) will cause some groups to be overculled and some underculled (Milner-Gulland et al. 2000). Where immigration rates are low, this too, is likely to lead to the reduction in size of less productive groups, to spatial heterogeneity in relative population density, and to a reduction in maximum sustainable yields (Clutton-Brock & Albon 1989).

Sex differences in mortality and emigration have their own implications for management. Where managers are principally concerned to maximize their annual offtake of mature males, rather than their maximum yield of both sexes (as in many ungulate species), female-biased adult sex ratios can substantially reduce economic yield. Where male mortality or dispersal are density dependent (and especially

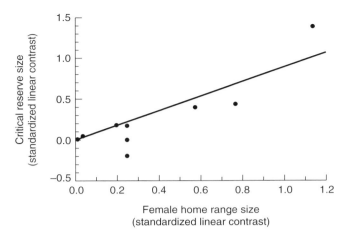

Figure 3.8 Relation between the (log) critical size of reserves necessary to maintain populations and (log) female home range size, calculated for 10 species of large carnivore. $r^2 = 0.84$, $F_{1,28} = 42.1$, $P < 0.005$. The effect remains strong after controlling for the (non-significant) effect of population density ($t = 4.00$, $P = 0.005$). (Reproduced with permission from Woodroffe & Ginsberg 1998.)

where they are affected principally by female density), managers may increase their potential offtake of males by maintaining female numbers well below ecological carrying capacity. For example, calculations for red deer populations incorporating density-dependent sex differences in survival but not in emigration show that the number of mature stags that can be culled per year rises by nearly 30% as annual culls imposed on females are raised from 2% to 16% (Clutton-Brock & Lonergan 1994) (Fig. 3.9).

Discussion

Like territoriality, sociality has far-reaching effects on demographic processes and population dynamics. Where resources are limited, competition between group members may focus the effects of shortages on weaker group members, strengthening strong density-dependent processes. In contrast, where intergroup competition is intense and favours large groups, reproductive success and survival may decline in small groups. Effects of this kind are particularly likely to be important in groups that breed, feed or defend themselves cooperatively, where members of small groups commonly suffer severe disadvantages. Inverse density dependence at the group level is likely to influence population dynamics in at least two ways. First, it may raise the probability that smaller groups will become extinct and, by doing so, may increase the risk of local population extinction. Second, it may reduce the frequency of successful dispersal, leading to long delays in the occupation of habitat left vacant by the extinction of occupying groups.

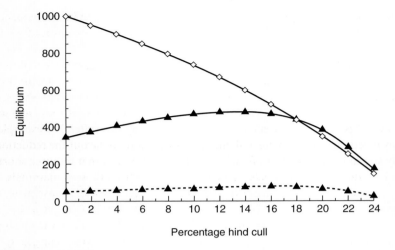

Figure 3.9 Numbers of female red deer (hinds) (◇) and males (stags) (▲) of different age categories (hinds; all animals ≥1 year old; stags ≥1 year old; --▲-- = 5 years old) expected at equilibrium before the onset of the annual cull. The size of unculled population is scaled to 1000 hinds. The model assumes that reproduction and survival change with the density of hinds and allows hind culls to vary. All stags are culled at age 5, giving the maximum sustainable yield for each hind density. (Reproduced with permission from Clutton-Brock & Lonergan 1994.)

Sociality can also have less direct effects on population dynamics through its influence on sex differences in dispersal, mating systems and sexual dimorphism. Species living in long-lasting, stable groups typically show pronounced sex biases in dispersal, combined with behavioural mechanisms that reduce the chance of close inbreeding. Female sociality facilitates the evolution of polygyny, which is commonly associated with sex differences in growth and survival that can generate strong biases in the adult sex ratio. In extreme cases, these can affect the fecundity of females as well as the potential benefits of different management strategies.

The examples used in this chapter have been drawn principally from social mammals—partly because these are the studies with which I have been most closely associated and partly because social mammals commonly form closed groups, so that the demographic consequences of sociality are likely to be strong. However, most of the processes described here have their counterparts in many other social organisms. For example, evidence that rising group size may focus the effects of resource shortage on weaker individuals is available from birds, fish and insects as well as mammals (Clutton-Brock 1991; Godfray 1994). Inverse density dependence at the group level may occur in many eusocial insects, as well as in cooperative birds (Wilson 1971). Female sociality leads to the development of polygyny and the evolution of sexual dimorphism in many invertebrates as well as in other groups of vertebrates (Andersson 1994). Indeed, it is worth considering whether some of the processes described in this chapter could have parallels in plant populations where related conspecifics or clones form monospecific stands that compete for space or resources with their neighbours.

This is not to say that the consequences of sociality will be the same in different social organisms. It may well be that demographic effects of sociality are more pronounced in birds and mammals where many social species live in closed, stable groups. Patterns of dispersal vary widely (Greenwood 1980) while the characteristics that affect competitive success within groups may also differ. For example, in many invertebrates that mate in flight, sexual selection can favour the reduction of body size in males relative to females (Andersson 1994) so that resource shortage may have quite different effects on males and females from those in mammals.

The conclusion to be drawn from studies of the demographic consequences of territoriality and sociality is that a knowledge of individual breeding strategies can offer insight into the demographic processes underlying population dynamics. Eventually, an understanding of these processes may provide a basis for generalizations and predictions about the population dynamics and the effects of ecological change in different groups of organisms. We have not yet reached this stage. The demographic consequences of many important aspects of breeding systems have yet to be explored, and empirical studies that can relate contrasting breeding strategies to population stability are still scarce (Shea *et al.* 1994; Saether *et al.* 1996). Nevertheless, the links between breeding systems and demographic processes are sufficiently obvious to encourage behavioural ecologists to investigate the demographic consequences of different behavioural strategies and population biologists to parameterize models of population dynamics for specific breeding systems.

References

Andersson, M. (1994). *Sexual Selection*. Princeton. Princeton University Press.

Barnard, J. (2000) *Costs and benefits of group foraging in cooperatively breeding meerkats*. PhD Thesis, Cambridge.

Bertram, B.C.R. (1975) Social factors influencing reproduction in wild lions. *Journal of Zoology* **177**, 463–482.

Bertram, B.C.R. (1978) Living in groups: predators and prey. In: *Behavioural Ecology* (eds J.R. Krebs & N.B. Davies), pp. 64–96. Blackwell Scientific Publications, Oxford.

Clutton-Brock, T.H. (1988) Reproductive success. *Studies of Individual Variation in Contrasting Breeding Systems*. Chicago University Press, Chicago.

Clutton-Brock, T.H. (1989) Mammalian mating systems. *Proceedings of the Royal Society, London, Series B* **236**, 339–372.

Clutton-Brock, T.H. (1991) *The Evolution of Parental Care*. Princeton University Press, Princeton, NJ.

Clutton-Brock, T.H. (1989) Female transfer and inbreeding avoidance in social mammals. *Nature* **337**, 70–72.

Clutton-Brock, T.H. & Albon, S.D. (1989). *Red Deer in the Highlands*. Blackwell Scientific Publications, Oxford.

Clutton-Brock, T.H. & Harvey, P.H. (1983) The functional significance of variation in body size among mammals. In: *Advances in the Study of Mammalian Behavior* (eds J.F. Eisenberg & D.G. Kleiman), pp. 632–663. The American Society of Mammalogists, Washington D.C.

Clutton-Brock, T.H. & Lonergan, M.E. (1994) Culling regimes and sex ratio biases in Highland red deer. *Journal of Applied Ecology* **31**, 521–527.

Clutton-Brock, T.H., Albon, S.D. & Guinness, F.E.

(1982a) Competition between female relatives in a matrilocal mammal. *Nature* **300**, 178–180.

Clutton-Brock, T.H., Guinness, F.E. & Albon, S.D. (1982b) *Red Deer: Behavior and Ecology of Two Sexes*. Edinburgh University Press, Edinburgh.

Clutton-Brock, T.H., Guinness, F.E. & Albon, S.D. (1984) Individuals and populations: the effects of social behaviour on population dynamics in deer. *Proceedings of the Royal Society, Edinburgh* **83B**, 275–290.

Clutton-Brock, T.H., Albon, S.D. & Guinness, F.E. (1985a) Parental investment and sex differences in juvenile mortality in birds and mammals. *Nature* **313**, 131–133.

Clutton-Brock, T.H., Major, M. & Guinness, F.E. (1985b) Population regulation in male and female red deer. *Journal of Animal Ecology* **54**, 831–846.

Clutton-Brock, T.H., Albon, S.D. & Guinness, F.E. (1987) Interactions between population density and maternal characteristics affecting fecundity and survival in red deer. *Journal of Animal Ecology* **56**, 857–871.

Clutton-Brock, T.H., Price, O.F., Albon, S.D. & Jewell, P.A. (1991) Persistent instability and population regulation in Soay sheep. *Journal of Animal Ecology* **60**, 593–608.

Clutton-Brock, T.H., Price, O.F., Albon, S.D. & Jewell, P.A. (1992) Early development and population fluctuations in Soay sheep. *Journal of Animal Ecology* **61**, 381–396.

Clutton-Brock, T.H., Illius, A., Wilson, K., Grenfell, B.T., MacColl, A.D. & Albon, S.D. (1997a) Stability and instability in ungulate populations: an empirical analysis. *American Naturalist* **149**, 195–219.

Clutton-Brock, T.H., Rose, K.E. & Guinness, F.E. (1997b) Density-related changes in sexual selection in red deer. *Proceedings of the Royal Society, London, Series B* **264**, 1509–1516.

Clutton-Brock, T.H., Brotherton, P.N.M., Smith, R. et al. (1998) Infanticide and expulsion of females in a cooperative mammal. *Proceedings of the Royal Society, London, Series B* **265**, 2291–2295.

Clutton-Brock, T.H., MacColl, A.D.C., Chadwick, P., Gaynor, D., Kansky, R. & Skinner, J.D. (1999a) Reproduction and survival of suricates (*Suricata Suricatta*) in the southern Kalahari. *African Journal of Ecology* **37**, 69–80.

Clutton-Brock, T.H., Gaynor, D., McIlrath, G.M. et al. (1999b) Predation, group size and mortality in a cooperative mongoose, *Suricata suricatta*. *Journal of Animal Ecology* **68**, 672–683.

Clutton-Brock, T.H., Iason, G.R. & Guiness, F.E. (1987) Sexual segregation and density-related changes in habitat use in male and female red deer (*Ceruus elaphus*). *Journal of Zoology, London* **211**, 275–289.

Cooney, R. (1999) Cooperative breeding and reproductive skew in the Damaraland mole-rat. Cambridge.

Courchamp, F., Grenfell, B. & Clutton-Brock, T.H. (1999a) Population dynamics of obligate cooperators. *Proceedings of the Royal Society, London, Series B* **266**, 557–664.

Courchamp, F., Clutton-Brock, T.H. & Grenfell, B. (1999b) Inverse density dependence and the Allee effect. *Trends in Ecology and Evolution* **14**, 405–410.

Creel, S.R. & Waser, P.M. (1997) Variation in reproductive suppression among dwarf mongooses: interplay between mechanisms and evolution. In: *Cooperative Breeding in Mammals* (eds N.G. Solomon, & J.A. French), pp. 150–170. Cambridge University Press, Cambridge.

Creel, S., Creel, N.M. & Monfort, S.L. (1998) Birth order, estrogens and sex-ratio adaptation in African wild dogs (*Lycaon pictus*). *Animal Reproduction Science* **53**, 315–320.

Creel, S. & Creel, N.M. (2001) *The African Wild Dog: Behavior, Ecology and Conservation*. Princeton University Press, Princeton.

Dittus, W.P.J. (1977) The social regulation of population density and age–sex distribution in the toque monkey. *Behaviour* **63**, 281–322.

Dittus, W.J.P. (1979) The evolution of behaviours regulating density and age specific sex ratios in a primate population. *Behaviour* **69**, 255–301.

Dobson, A. & Poole, J. (1998) Conspecific aggregation and conservation biology. In: *Behavioural Ecology and Conservation Biology* (ed. T. Caro,), pp. 193–208. Oxford University Press, Oxford.

Faulkes, C.G. & Abbott, D.H. (1997) The physiology of a reproductive dictatorship: regulation of male and female reproduction by a single-breeding female in colonies of naked mole-rats. In: *Cooperative Breeding in Mammals* (eds N.G. Solomon & J.A. French), pp. 268–301. Cambridge University Press, Cambridge.

Glucksmann, A. (1974) Sexual dimorphism in mammals. *Biological Reviews of the Cambridge Philosophical Society* **49**, 423–475.

Godfray, C. (1994). *Parasitoids: Behavioural and Evolutionary Ecology*. Princeton University Press, Princeton, NJ.

Greenwood, P.J. (1980) Mating systems, philopatry and dispersal in birds and mammals. *Animal Behaviour* **28**, 1140–1162.

Harvey, P.H. & Clutton-Brock, T.H. (1981) Primate home-range size, metabolite needs and ecology. *Behavioural Ecology and Sociobiology* **8**, 151–155.

Harvey, P.H. & Pagel, M. (1991). *The Comparative Method in Evolutionary Biology*. Oxford University Press, Oxford.

Hoogland, J.L. (1979) Aggression, ectoparasitism and other possible costs of prairie dogs (Sciuridae: *Cynomys* spp.) coloniality. *Behaviour* **69**, 1–35.

Illius, A.W. & Gordon, I.J. (1987) The allometry of food intake in grazing ruminants. *Journal of Animal Ecology* **56**, 989–1000.

Jimenez, J.A., Hughes, K.A., Alaks, G., Graham, L. & Lacy, R.C. (1994) An experimental study of inbreeding depression in a natural habitat. *Science* **266**, 271–273.

Kokko, H. & Johnstone, R.A. (1999) Social queuing in animal societies: a dynamic model of reproductive skew. *Proceedings of the Royal Society, London, Series B* **266**, 571–578.

Komdeur, J. (1994a) The effect of kinship on helping in the cooperative breeding Seychelles warbler, *Acrocephalus seychellensis*. *Proceedings of the Royal Society, London, Series B* **256**, 47–52.

Komdeur, J. (1994b) Experimental evidence for helping and hindering by previous offspring in the cooperative breeding Seychelles warbler, *Acrocephalus sechellensis*. *Behavioural Ecology and Sociobiology* **34**, 175–186.

Komdeur, J., Daan, S., Tinbergen, J. & Mateman, C. (1997) Extreme adaptive modification in the sex ratio of the Seychelles warblers' eggs. *Nature* **385**, 522–525.

Krebs, J.R. (1971) Territory and breeding density in the great tit, *Parus major*. *Ecology* **52**, 2–22.

Krebs, J.R. & Davies, N.B. (1997) *Behavioral Ecology: an Evolutionary Approach*. Blackwell Science Inc, Boston, MA.

Kruuk, L.E.B., Clutton-Brock, T.H., Albon, S.D., Pemberton, J.M. & Guinness, F.E. (1999a) Population density affects sex ratio variation in red deer. *Nature* **399**, 459–461.

Kruuk, L.E.B., Clutton-Brock, T.H., Rose, K.E. & Guinness, F.E. (1999b) Early determinants of lifetime reproductive success differ between the sexes in red deer. *Proceedings of the Royal Society, London, Series B* **266**, 1655–1661.

Lack, D. (1966). *Populations Studies of Birds*. Oxford University Press, Oxford.

Loison, A., Festa-Bianchet, M., Gaillard, J.M., Jorgenson, J.T. & Jullien, J.M. (1999) Age-specific survival in five populations of ungulates: evidence of senescence. *Ecology* **80**, 2539–2554.

Macdonald, D.W. (1979) 'Helpers' in fox society. *Nature* **282**, 69–71.

Macdonald, D.W. (1983) The ecology of carnivore social behaviour. *Nature* **301**, 379–384.

Maynard Smith, J. & Szathmary, E. (1995). *The Major Transitions in Evolution*. Oxford University Press, Oxford.

McMillen, M.M. (1979) Differential mortality by sex in fetal and neonatal deaths. *Science* **204**, 89–91.

McNutt, J.W. (1996) Sex biased dispersal in African wild dogs, *Lycaon pictus*. *Animal Behaviour* **52**, 1067–1077.

Milner-Gulland, E.J., Coulson, T.N. & Clutton-Brock, T.H. (2000) On harvesting a structured ungulate population. *Oikos* **88**, 592–602.

Milton, K. & May, M.C. (1976) Body weight, diet and home range area in primates. *Nature* **259**, 459–462.

Moehlman, P.D. (1979) Jackal helpers and pup survival. *Nature* **277**, 382–383.

Moehlman, P.D. & Hofer, H. (1997) Cooperative breeding, reproductive suppression and body mass in canids. In: *Cooperative Breeding in Mammals* (eds N.G. Solomon & J.A. French,), pp. 34–75. Cambridge University Press, Cambridge.

Newton, I. (1998). *Population Limitation in Birds*. Academic Press, London.

O'Riain, M.J., Bennett, M.C., Brotherton, P.N.M., McIlrath, G.M. & Clutton-Brock, T.M. (2001) Reproductive suppression and inbreeding avoidance in wild populations of cooperatively breeding meerkats (*Suricata suricatta*). *Behavioural Ecology and Sociobiology* **48**, 471–477.

Packer, C., Herbst, L., Pusey, A.E. *et al.* (1988)

Reproductive success of lions. In: *Reproductive Success: Studies of Individual Variations in Contrasting Breeding Systems* (ed. T.H. Clutton-Brock,), pp. 367–383. University of Chicago Press, Chicago.

Pemberton, J.M., Smith, J.A., Coulson, T.N. *et al.* (1996) The maintenance of genetic polymorphism in small island populations: large mammals in the Hebrides. *Philosophical Transactions of the Royal Society of London, Series B* **351**, 745–752.

Poole, J.H. (1989) The effects of poaching on the age structure and social and reproductive patterns of east African elephant populations. In: the Ivory Trade and the Future of the African Elephant. Ivory Trade Review Group. IUCN, Gland, Switzerland.

Promislow, D.E.L. (1992) Costs of sexual selection in natural populations of mammals. *Proceedings of the Royal Society, Series B* **247**, 203–210.

Pulliam, H.R. & Caraco, T. (1984) Living in groups: is there an optimal group size?. In: *Behavioural Ecology* (eds J.R. Krebs & N.B. Davies) pp. 122–147. Blackwell Scientific Publications, Oxford.

Rood, J.P. (1990) Group size, survival, reproduction and routes to breeding in dwarf mongooses. *Animal Behaviour* **39**, 566–572.

Saether, B.E., Ringsby, T.H. & Roskaft, E. (1996) Life-history variation, population processes and priorities in species conservation: towards a reunion of research paradigms. *Oikos* **77**, 217–226.

Schoener, T.W. (1968) Sizes of feeding territories among birds. *Ecology* **49**, 704–726.

Shea, K., Rees, M. & Wood, S.N. (1994) Trade-offs, elasticities and the comparative method. *Journal of Animal Ecology* **82**, 951–957.

Solomon, N.G. & French, J.A. (1997). *Cooperative Breeding in Mammals*. Cambridge. Cambridge University Press.

Stacey, P.B. & Koenig, W.D. (1990). *Cooperative Breeding in Birds*. Cambridge University Press, Cambridge.

Stearns, S.C. (1992). *The Evolution of Life Histories*. Oxford University Press, Oxford.

Stevenson, I.R. & Bancroft, D.R. (1995) Fluctuating trade-offs favour precocial maturity in male Soay sheep. *Proceedings of the Royal Society, London, Series B* **262**, 267–275.

Trivers, R.L. (1972) Parental investment and sexual selection. In: *Sexual Selection and the Descent of Man* (ed. B. Campbell), pp. 136–179. Aldine, Chicago, IL.

Van Schaik, C.P., Van Noordwijk, M.A., Boer, R.J. & Tunkelaar, I. (1983) The effect of group size on time budgets and social behaviour in wild longtailed macaques (*Macaca fascicularis*). *Behavioural Ecology and Sociobiology* **13**, 173–181.

Waser, P. (1977) Feeding, ranging and group size in the mangahey *Cercocebus albigena*. In: *Primate Ecology* (ed. T.H. Clutton-Brock). Academic Press, London.

Watson, A. & Miller, R.G. (1971) Territory size and aggression in a fluctuating red grouse population. *Journal of Animal Ecology* **40**, 367–383.

Wilson, E.O. (1971). *The Insect Societies*. Belknap Press, Cambridge, MA.

Wilson, E.O. (1975) *Sociobiology, The Modern Synthesis.* Belknap Press, Cambridge, Mass.

Wolff, J.O. (1992) Parents suppress reproduction and stimulate dispersal in opposite sex juvenile white-footed mice. *Nature* **359**, 409–410.

Woodroffe, R. & Macdonald, D.W. (2000) Helpers provide no detectable benefits in the European badger (*Meles meles*). *Journal of Zoology* **250**, 113–118.

Woodroffe, R. & Ginsberg, J.R. (1998) Edge effects and the extinction of populations inside protected areas. *Science* **280**, 2126–2128.

Wrangham, R.W. (1980) An ecological mode of female-bonded primate groups. *Behaviour* **75**, 262–300.

Wynne-Edwards, V.C. (1962). *Animal Dispersion in Relation to Social Behaviour*. Oliver and Boyd, Edinburgh.

Chapter 4
Studies of the reproduction, longevity and movements of individual animals

I. Newton

Introduction

The development of methods for marking animals, so that they could be individually recognized throughout their lives, began about a century ago. It opened the door to studies on the reproduction, longevity and movements of individual animals on a scale that would not otherwise have been possible. In time, it provided estimates of the demographic parameters of many species, together with detailed studies of individual life histories. Although marks or tags have now been applied to a wide range of animals, from large insects to large vertebrates, it is birds that have been most widely studied, mainly from the application of leg bands. It is therefore from research on birds that I shall draw my examples for this chapter.

Studies based on the marking of individuals depend critically on markers that: (i) remain attached and readable for as long as necessary; (ii) do not affect the lives of the wearers in any significant way. In the case of bird bands, some earlier materials and designs resulted in the thinning and loss of bands, especially in large and long-lived species, thereby reducing survival estimates. These problems of ring construction were rectified several decades ago, and have not been significant in more recent studies. Whether metal leg bands adversely affect the behaviour of birds is harder to assess because of the impossibility of gaining similar (control) data from unmarked birds, and because other kinds of marks (e.g. wing tags or radio tags) are unlikely to be less disruptive. Earlier findings that red colour bands enhanced the mating success and survival of captive zebra finches *Poephila guttata* (Burley 1985) were not confirmed by later studies using a different experimental design (Ratcliffe & Boag 1987). For the time being, therefore, in the absence of evidence to the contrary, researchers can only assume that single light-weight metal leg bands have negligible or no effects on the lives of the birds they study.

In this chapter, I shall be concerned with three aspects, namely: (i) age-related trends in breeding and survival; (ii) lifetime reproductive success; and (iii) dispersal patterns. On each aspect, I shall start by describing the findings from my own

Centre for Ecology and Hydrology, Monks Wood, Abbots Ripton, Huntingdon, Cambridgeshire PE28 2LS

research on the Eurasian sparrowhawk *Accipiter nisus* studied over a 25-year period in a 200-km² area in south Scotland (Newton 1986, 1988, 1989b), and then draw comparisons with other species in order to attempt some generalizations. The sparrowhawk is a small bird of prey, which breeds in forest and woodland over much of Eurasia, and preys upon other birds. Individuals can live up to 10 years (although hardly any do so), can start breeding in their first, second or third year of life, and each year thereafter can raise up to six young. In south Scotland, the birds were resident year round (i.e. non-migratory), and breeding density remained relatively stable over the study period.

Age-related performance

The ease with which banded birds of some species can be followed year after year in the same area has enabled an examination of the changes in annual survival and breeding success that occur during the course of the average lifespan (Clutton-Brock 1988; Newton 1989a; Martin 1995). Such patterns are important in understanding the evolution of life histories, particularly senescence (Holmes & Austad 1995; Ricklefs 1998), and also in population demography (Fisher 1958).

Some patterns for the sparrowhawk are shown in Fig. 4.1, in which both annual survival and aspects of breeding performance improve during the first few years of life and then deteriorate in later life. These patterns, like those produced for other birds, are based on the pooled data for all individuals represented in each age group. However, the same trend in breeding success was followed by individuals examined in successive age groups, 1–2, 2–3, 3–4, etc. (Newton & Rothery 1998). In fact the overall trend could be attributed entirely to individuals following this trend, and no evidence was found for an alternative explanation, that birds of different performance levels were differentially represented at different ages. An improvement in performance in the early years of life has been documented now in many different wild bird species (Newton 1989a; Saether 1990; Fig. 4.2), but a deterioration in later life in relatively few because it requires a longer study. However, the extent of changes in both reproduction and survival through the lifespan varies greatly between species (compare the shapes of curves in Fig. 4.2). Such differences depend partly on the maximum potential lifespans and reproductive rates of the species concerned.

Initial improvements with age in breeding and survival have been variously attributed to physical maturation, to increased experience of foraging or breeding, to age and experience of mate, or to improvement of breeding site and social status (review Newton 1989a). In particular, many birds have been found to feed or breed more efficiently with increasing experience (the latter can affect breeding independently of age: for Arctic skua *Stercorarius parasiticus*, see Davis 1976; for western gull *Larus occidentalis*, see Pyle *et al.* 1991; for great skua *Stercorarius skua*, see Ratcliffe *et al.* 1998), and also to improve in dominance status which gives greater access to resources (for Eurasian oystercatcher *Haematopus ostralegus*, see Goss-

LIFE HISTORY STUDIES OF INDIVIDUAL ANIMALS

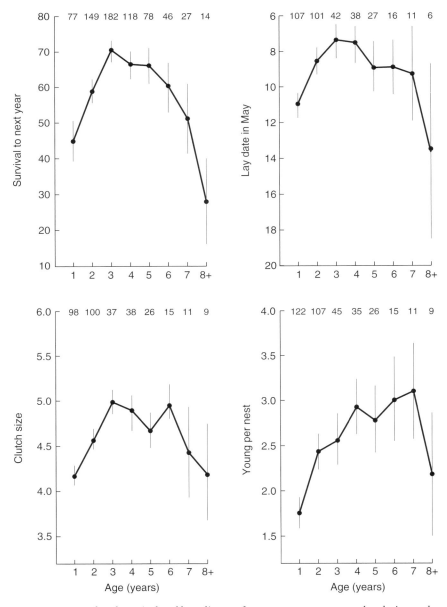

Figure 4.1 Age-related survival and breeding performance among sparrowhawks in south Scotland. Recorded data corrected for variation associated with year and (in the case of breeding performance) with territory. Numbers show sample sizes at each age group. (Reproduced with permission from Newton 1988.)

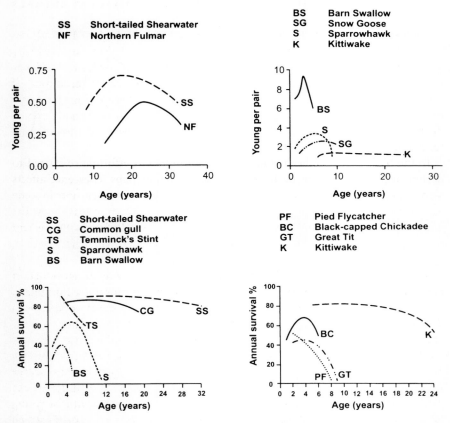

Figure 4.2 Mean annual survival and breeding success (young per female) in relation to age for different bird species, based on the annual identification of individuals in intensively studied local populations. The curves were calculated from quadratic regressions fitted to the annual data points, weighted according to sample sizes. In some species estimates were based only on birds ringed as nestlings, and hence of precisely known ages. But in short-tailed Shearwater *Puffinus tenuirostris*, reproductive success was estimated from year of first breeding, and in Temminck's stint *Calidris temminckii*, common gull *Larus canus*, short-tailed Shearwater *Puffinus tenuirostris* and black-legged kittiwake *Rissa tridactyla*, survival was estimated from year of first breeding, assuming that all individuals started at the mean age for their populations. Survival curves based on data in Perrins (1979) and McCleery & Perrins (1989) for great tit *Parus major*; Loery et al. (1987) for black-capped chickadee *P. atricapillus*; Sternberg (1989) for European pied flycatcher *Ficedula hypoleuca*; Møller & Lope (1999) for barn swallow *Hirundo rustica*; Hildén (1978) for Temminck's stint *Calidris temminckii*; Newton (1988) for Eurasian sparrowhawk *Accipiter nisus*; Rattiste & Lilleleht (1987) for common gull *Larus canus*; Aebischer & Coulson (1990) for black-legged kittiwake *Rissa tridactyla*; and Bradley et al. (1989) and Wooller et al. (1989) for short-tailed Shearwater *Puffinus tenuirostris*. Reproductive curves based on Møller & Lope (1999) for barn swallow *Hirundo rustica*; Rockwell et al. (1993) for snow goose *Anser caerulescens*; Newton (1988) for sparrowhawk *Accipiter nisus*; Aebischer & Coulson (1990) for black-legged kittiwake *Rissa tridactyla*; Wooller et al. (1990) for short-tailed Shearwater *Puffinus tenuirostris*; Ollason & Dunnet (1988) for northern fulmer *Fulmarus glacialis*.

Custard *et al.* 1984; for herring gull *Larus argentatus*, see Monaghan 1980). It has also been suggested that young breeders may show reproductive restraint in order to enhance their subsequent performance (Curio 1983), but there is little evidence for this view, which is in any case hard to test unequivocally. Deterioration of breeding and survival rates in later life, in the face of greater experience, can only be ascribed to senescence, defined as a decrease in age-dependent residual reproductive value, and usually associated with declining efficiency of physiological function.

In the birds that have been studied, not all aspects of performance started to decline at the same age, or declined at the same rate. Perhaps senescence affects some aspects of performance before others, or is offset by experience more in some aspects than in others. There is a need for studies that attempt to tease apart the effects of experience from declining body function, and also to find to what extent senescence, as reflected in declining survival rate, reflects past reproductive effort, and to what extent it occurs regardless of past reproduction.

The shape of the curves depicting age-related changes in annual reproduction and survival are likely to vary within species, according to prevailing conditions. Thus, in some bird species the effects of food shortages on breeding have been shown to be more marked in young birds than in middle-aged ones, thus accentuating the age-related trend over that found in good feeding conditions (for Brandt's cormorant *Phalacrocorax penicillatus*, see Boekelheide & Ainley 1989; for western gull *Larus occidentalis*, see Sydeman *et al.* 1991; for great skua *Stercorarius skua*, see Ratcliffe *et al.* 1998). The implication is that the performance of different age groups is differentially depressed by food shortages or competition, with the young age groups most affected. Hence, the patterns in Figs 4.1 and 4.2, most of which refer to numerically stable populations, are likely to alter to some extent with changes in competition and other conditions.

Knowledge of age-related changes in reproduction and survival in turn enable the calculation of residual reproductive value, the number of young that individuals of different ages can expect to produce during the rest of their lives (Fisher 1958). Because in the sparrowhawk both survival and breeding rates peak in mid-life, so does residual reproductive value. The average 1-year-old female could expect to produce 3.1 young during the rest of her life (or 4.8 if she bred in her first year), but a female that lived to 4 years could expect to produce 8.2 young during the rest of her life, while one that reached 9 years could expect to produce only 2.1 further young (Newton & Rothery 1998). The lower mean residual reproductive value of young birds, compared to middle-aged ones, is perhaps counter-intuitive, but it results from the low mean survival prospects of young birds. Although the merits of residual reproductive value for use in population and life-history modelling were well appreciated by Fisher (1958), and have been re-emphasized by later authors (Caswell 1982; Goodman 1982; Stearns 1992; Charlesworth 1994; Rousset 1999), it is only in recent years that it has become possible to calculate such values for a range of different organisms, including at least one other bird species (Møller & Lope 1999).

Lifetime reproductive success

For several bird species, subject to long-term study, it has now been possible to follow a representative sample of individuals throughout their natural lifespans, and hence to measure lifetime reproductive success (LRS), i.e. the total numbers of young raised by individuals during their entire lives (Clutton-Brock 1988; Newton 1989a). Interest in such longitudinal studies has grown rapidly with the realization that lifetime reproductive rates could provide good approximations of Darwinian fitness, i.e. of the contributions that particular types of individuals make to future gene pools. Despite the great theoretical importance of fitness, which is discussed in almost every recent textbook on evolution, it has remained one of the most elusive measures in biology. To illustrate the issues involved, I shall again start with my own work on the sparrowhawk *Accipiter nisus* L.

Over the 25-year period, I recorded the lifetime fledgling productions for 194 different breeding females (Fig. 4.3). These birds were representative, in their respective lifespans, of the female population as a whole (Newton 1986). They spent 1–8 years (mean 2.3 years) as breeders, and raised from 0 to 24 young (mean 5.0 young) during their lives. They thus showed enormous variation in their lifetime productions. The peak at 3–5 young was because most birds raised only one brood per lifetime, and most broods contained 3–5 young. About 17% of females that laid eggs failed to produce any young, even though they laid in up to four different years. About 43% variation in lifetime fledgling production could be attributed to individual variation in lifespan (1–10 years) and about 49% to variation in length of breeding life (lifespan minus age of first breeding, 1–8 years) (Fig. 4.3).

By incorporating information on prebreeding mortality at different ages, it was possible to calculate the lifetime productions of a generation of female fledglings, including those which died before breeding (Newton 1988). About 72% of female young died in years 1–3 of life without having bred, another 5% produced eggs but no young, and only 23% produced young, but in greatly varying numbers (Fig. 4.4). Overall, then, less than one-quarter of individuals in one generation of fledglings contributed to the next generation, and the most productive 5% of fledglings (or 20% of breeders) produced more than half the next generation of young. Similar measures of lifetime reproductive success are now available for the breeders of more than 20 bird species, and for several mammal species (Clutton-Brock 1988; Newton 1989a).

That only a tiny fraction of eggs that are laid survive to produce mature adults is a familiar idea in fish or other animals that produce large numbers of eggs. But this recent bird research reveals an additional point; namely that only a proportion of individuals in any one generation produce any mature young. In a stable population of blue tits *Parus caeruleus* L., a species with high annual reproductive and mortality rates, only 3% of fledglings in one generation produced half of the young in the next generation. In species with lower reproductive rates, some 6–9% of individuals produced half the next generation (Table 4.1). At the other end of the spectrum, the proportion of fledglings that themselves produced no young varied

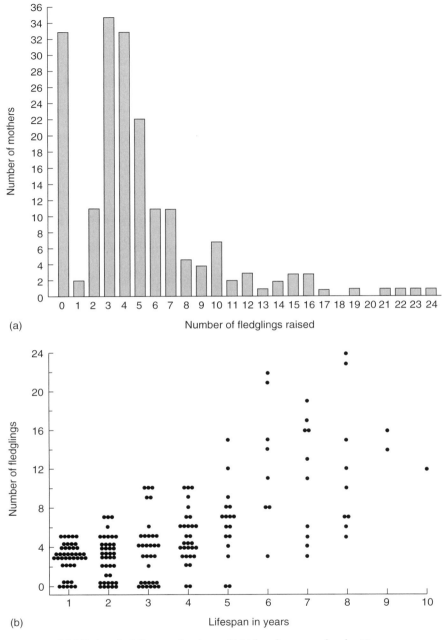

Figure 4.3 (a) Lifetime fledgling productions of 194 female sparrowhawks. Mean per female = 5.03 young. From Newton (1989a). (b) Lifetime fledgling productions of the same 194 females in relation to their lifespan. Each spot represents the number of fledglings raised by an individual female. Relationship between lifetime production (y) and lifespan (x_1): $y = 0.127 + 1.426x_1$, $r^2 = 0.425$, $P = 0.0001$. Relationship between lifetime production (y) and duration of breeding life (= lifespan minus age at first breeding, x_3): $y = 0.854 + 1.850x_3$, $r^2 = 0.498$, $P = 0.0001$. (Reproduced with permission from Newton 1989b.)

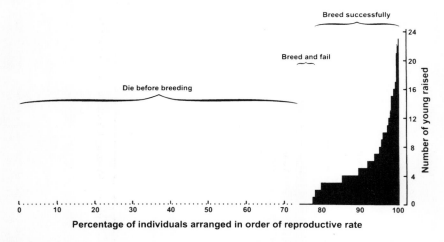

Figure 4.4 Total lifetime fledgling productions for a generation of female sparrowhawks including birds that died before they could breed. Birds arranged on a percentage scale in order of their lifetime productions. (Updated from Newton 1988.)

in different species between 66 and 87% (Table 4.1). This unproductive component comprised mainly individuals that died before having a chance to breed, but also some that attempted to breed but failed. In any population that remains demographically fairly stable (as do some bird species in stable environments), the proportion of productive individuals is likely to remain more or less constant through time. But this proportion may of course alter during a period of population change (for higher LRS values in Soay sheep *Ovis aries* living during a period of population growth than in those living during a period of stability or decline, see Coltman et al. 1999).

The importance of LRS as a measure of individual performance is twofold. First, it combines two key measures of performance, namely survival and success at individual breeding attempts, into a single overall measure of performance; and secondly, it reveals more than any other measure the full extent of individual variation in reproductive success. Barely perceptible differences in the annual success of individuals can become substantial when repeated over whole life spans of variable duration. Lifetime studies also provide a comparison on equal terms between the sexes of species which have polygamous mating systems, and in which reproduction may be confined to a smaller part of the lifespan in one sex than in the other. LRS could thus provide a better basis for estimating Darwinian fitness than any other measure yet available, facilitating more firmly based studies of selection and life-history strategies.

One of the findings to emerge following the advent of DNA analyses was that the social partner of a female bird is not always the father of all her offspring (Birkhead & Møller 1992). Such extra-pair fertilization can distort observational measures of

LIFE HISTORY STUDIES OF INDIVIDUAL ANIMALS

Table 4.1 Proportion of any one generation of bird fledglings that (a) die before they can breed, (b) breed but fail to produce young, (c) breed successfully, and (d) produce half the young in the next generation.

	Proportion of fledglings that:			
Species	(a) die before breeding	(b) breed but fail to produce young	(c) produce one or more young	(d) produce half the next generation
Blue tit *Parus caeruleus*	86	1	13	3
Common kingfisher *Alcedo atthis*	81	1	18	6
Splendid fairywren *Malurus splendens*	78	7	15	–
European pied flycatcher *Ficedula hypoleuca*	77	1	22	–
Meadow pipit *Anthus pratensis*	72	4	22	7
Eurasian sparrowhawk *Accipiter nisus*	72	5	23	5
Ural owl *Strix uralensis*	72	1	27	6
Osprey *Pandion haliaetus*	71	2	24	6
Black-legged kittiwake *Rissa tridactyla*	60	4	36	–
Barnacle goose *Branta leucopsis*	42	24	34	9

Details from studies by A.A. Dhondt, H. Hötker, M. Bunzel & J. Druke; I. Newton, S. Postpupalsky, P. Saurola, M. Owen & J. Black; H. Sternberg, I. Rowley & E. Russell; and J.C. Coulson, all reported in Newton (1989a).

the LRS of male birds (note the records in Figs 4.3 and 4.4 are for females only). However, the fact that many birds can be easily trapped enables blood samples to be taken for the checking of parentage by DNA fingerprinting, thus providing a means of eliminating this potential error in measures of individual reproductive success. The same is also true for some other kinds of animals.

Dispersal

Although important in population processes, the movements of individual animals have generally proved difficult to study. This is partly because, without marking, it is usually impossible to distinguish residents from non-residents, or to know the source of immigrants or the destination of emigrants. Birds banded in particular localities can be reported elsewhere by other people, which means that information can be gained on their movements in a way not possible for most other animals, except by extensive radio-tracking.

I shall concentrate here on dispersal, because this type of movement is performed by almost all animals and has profound effects on the distribution, density and genetic structure of populations (e.g. Johnson & Gaines 1990; Holt & McPeek 1996). For present purposes, it is useful to distinguish: (i) natal dispersal, measured by the linear distances between natal and first breeding sites; (ii) breeding dispersal, measured by the distances between the breeding sites of successive years; and (iii) non-breeding dispersal, measured by the distances between the wintering sites of successive years. These three categories are appropriate, not only for resident populations, but also for migratory ones which spend their breeding and non-breeding periods in different regions. In general, individual birds move much greater distances between birth site and breeding site than between the breeding or wintering sites of successive years (e.g. Newton 1986; Paradis *et al.* 1998).

In studies of dispersal, the data from national banding schemes have proved especially useful, because the records from birds banded and recovered in different localities can be pooled. This provides for each species a database which can be assumed to give a fairly representative overall picture of dispersal distances. Although the natal sites are known precisely for birds ringed as chicks, the assumption usually has to be made that individuals of breeding age recovered in the breeding season were in fact nesting, or had the potential to nest, at the localities where they were reported. Collectively, such records reflect the settling patterns of individuals with respect to natal site, regardless of their movements in the period between ringing and recovery which remain unknown.

Natal dispersal

The recovery pattern for sparrowhawks in Britain implies that most individuals settled to breed within 20 km of their natal site, but that some settled at much greater distances (maximum 120 km). In fact, the density of recoveries declined in approximately exponential manner in concentric bands out from the central natal

LIFE HISTORY STUDIES OF INDIVIDUAL ANIMALS

site (Fig. 4.5). No directional preferences were apparent, and recoveries came from all sectors of the compass.

This type of skewed settling pattern has been found for all species of birds that have been studied (e.g. Paradis *et al.* 1998), the main difference between species being one of scale, with some species dispersing over much longer distances than others. Thus, in the same area of Germany, 90% of young blue tits *Parus caeruleus*

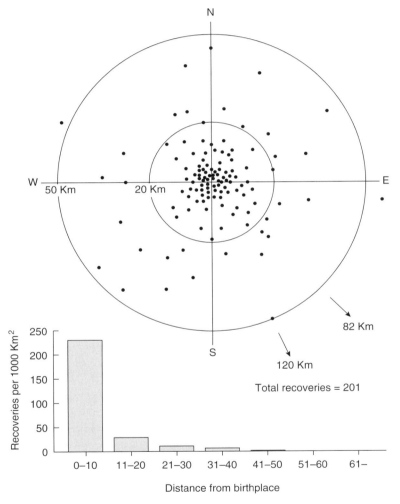

Figure 4.5 Locations of sparrowhawks recovered in the breeding season, shown in relation to the natal site (centre), and density of recoveries in concentric circles out from the natal site. Based on recoveries obtained in the British Banding Scheme. (Reproduced with permission from Newton 1979.)

settled to breed within 4 km of where they hatched, 90% of wood nuthatches *Sitta europaea* settled within 9 km, and 90% of European pied flycatchers *Ficedula hypoleuca* within 17 km. These figures meant that 90% of the nestlings from any one source locality subsequently bred within surrounding circular areas of about 50 km^2, 250 km^2 and 1000 km^2, respectively, in the three species (Berndt & Sternberg 1968). Even in such small birds, however, occasional individuals settled at far greater distances. Moreover, within species the settling pattern may differ somewhat from year to year or from region to region (see Sokolov 1997 for pied flycatcher), depending on circumstances including the density of the population and the patchiness of habitat in the region concerned. In many bird species, one sex moves further from the natal site than the other, although the overlap between the sexes is great (Greenwood 1980).

Research has revealed some of the factors that influence the movements of individuals. In some studies, young fledged late in the season dispersed further, on average, than young fledged earlier (for Eurasian tree sparrow *Passer montanus*, see Pinowski 1965; for blue tit *Parus caeruleus* see Dhondt & Hublé 1968; for European pied flycatcher *Ficedula hypoleuca*, see Sokolov 1997; for sparrowhawk *Accipiter nisus*, see Newton & Rothery 2000). In other studies, natal dispersal distances varied from year to year inversely with the numbers of young produced (Nilsson 1989), or increased through time as population densities increased (for Eurasian sparrowhawk *Accipiter nisus*, see Wyllie & Newton 1991; for lesser kestrel *Falco naumanni*, see Negro et al. 1997). These various findings imply that dispersal distances may be density dependent, influenced by competition for resources, such as territories, roosts or nest sites, food or mates (for experimental evidence, see Arcese 1989).

Breeding dispersal

One of the earliest findings from bird banding was that individuals tended to stay in, or return to, areas where they had previously bred. Many individuals were found to occupy the same territories in successive years, while others moved to different territories nearby. The site fidelity of adult birds has been shown mostly by counting the proportion of marked individuals present in a particular area in one year that returned there the next. On this basis, any birds that moved outside the area would be missed, and (because of movements) the smaller the area the lower the likely observed return rate, all else being equal. None the less, in some studies the proportion of individuals returning to particular areas was so high that, allowing for known mortality rates, few if any individuals could have moved elsewhere (Table 4.2). In fact, as shown by general band recoveries, the settling pattern in breeding dispersal is skewed like that of natal dispersal, but over shorter distances (e.g. Paradis et al. 1998). In some of the species in Table 4.2, notably the yellow-breasted chat *Icteria virens*, return rates were low, indicating that many surviving individuals left the study area to nest elsewhere.

For some species, more detailed studies have revealed the factors that underlie year-to-year territory changes, as well as the consequences of such moves for the

Table 4.2 Annual return rates of migrant birds to specific study areas in successive breeding seasons. The data are drawn from studies in which attempts were made to identify all individuals present. Records for different years are pooled: a bird ringed in one year and recovered in the next is counted as 1 'bird year', and a bird ringed in one year and recovered in the next 2 years is counted as 2 'bird-years', and so on. Size of study area: * <1 km^2 or <1 km of coast, ** 1–10 km^2 or 1–10 km of coast, *** 10–100 km^2 or 10–100 km of coast.

		Males		Females		
Species	Location	Number of 'bird-years'	% returned	Number of 'bird-years'	% returned	Source
Passerines						
Great reed warbler *Acrocephalus arundinaceus*	Sweden**	126	52	140	59	Bensch & Hasselquist 1991
Willow warbler *Phylloscopus trochilus*	Finland*	122	41	97	17	Tiainen 1983
Willow warbler *Phylloscopus trochilus*	England**	307	41	–	–	Lawn 1994
Willow warbler *Phylloscopus trochilus*	England**	176	30	133	17	Pratt & Peach 1991
Willow warbler *Phylloscopus trochilus*	Scotland**	48	46	50	34	da Prato & da Prato 1983
Marsh warbler *Acrocephalus palustris*	England*	41	44	–	–	Kelsey 1989
Wood warbler *Phylloscopus sibilatrix*	England*	29	28	–	–	Norman 1994
Greater whitethroat *Sylvia communis*	Scotland**	28	28	29	21	da Prato & da Prato 1983
Yellow-breasted chat *Icteria virens*	Indiana*	18	11	29	0	Thompson & Nolan 1973
American redstart *Setophaga ruticilla*	New Hampshire**	134	30	48	19	Holmes & Sherry 1992
Black-throated blue warbler *Dendroica caerulescens*	New Hampshire**	49	39	50	36	Holmes & Sherry 1992
House wren *Troglodytes aedon*	Illinois**	643	38	1468	23	Drilling & Thompson 1988
Bobolink *Dolichonyx oryzivorus*	New York**	85	44	86	25	Gavin & Bollinger 1988
Others						
Semi-palmated plover *Charadrius semipalmatus*	Manitoba	127	59	126	41	Flynn *et al.* 1999
Black kite *Milvus migrans*	Spain***	142	83	143	90	Forero *et al.* 1999

These figures should be viewed in relation to annual survival rates which tend to lie in the region of 40–50% for small passerines and 60–90% for various shorebirds, large waterfowls and large raptors.

individuals concerned. In south Scotland, about one-third of all breeding sparrowhawks surviving from the previous year changed territories. This was true for both sexes, but whereas most males that moved were found in neighbouring territories, many females moved over longer distances. In this respect, therefore, females showed less site fidelity than males.

The territory-changing behaviour of sparrowhawks from one year to the next varied according to their age and previous nest success (Fig. 4.6). In general, birds showed progressively greater site fidelity, and less tendency to change territories, with increasing age. This trend was evident in females even up to the oldest age groups of 7–10 years but, because few birds survived that long, samples of such old birds were small. In addition, in all age groups, birds that failed in their breeding were more likely to change territories for the next year than were birds that succeeded (Fig. 4.6).

As a group, the 289 female sparrowhawks that stayed on the same territory from one year to the next showed high nest success, but no improvement between years (86% of nests successful in the first year as against 79% in the following year). In contrast, the 112 females that changed territories bred better after the move than before (45% of nests successful in the first year against 64% in the following year, $\chi_1^2 = 8.71, P = 0.003$). However, after such a change, nest success was still significantly lower (64% of 112 nests) than in birds that had stayed on the same territory (79% of 289 nests, $\chi_1^2 = 9.64, P = 0.002$). The tendency to improvement in nest success after a change of territory was also apparent in males, although the samples were much smaller.

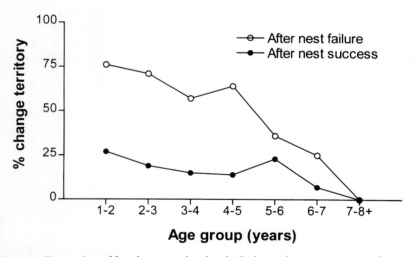

Figure 4.6 Proportion of female sparrowhawks which changed nesting territories between years, according to age and previous nest success. (Reproduced with permission from Newton 2000.)

This tendency to improvement was largely because territories varied in quality, and those birds that had poor territories (with a history of poor occupancy and nest success) mostly failed in their breeding and moved to a better territory in the next year. These trends were apparent in both first-year and older birds, even though in most years a greater proportion of first-year than of older birds was found in the poorer territories (Newton 1991). Overall, therefore, birds tended to stay on high-grade territories, but to move away from low-grade territories where failures were more frequent. These findings, which were apparent in both sexes, helped to explain why, over a period of years, territories were occupied non-randomly: some more often than expected by chance at the population levels found and others less (Newton 1991). In addition, it may be inferred that territory changes resulted in improved breeding success, not just in the females involved, but in the population as a whole. Such changes also ensured that in each year the population was concentrated in the better parts of the habitat available.

Sparrowhawks that stayed on the same territory from one year to the next also retained the same mate more often than did birds that changed territories. The difference between groups was significant, with overall figures of mate retention of 55% per year in stayers and 17% per year in movers, respectively (Table 4.3). Moreover, all moves in which the pair stayed together were to an adjacent territory in the same wood (about 0.5 km). In some instances where a bird had a different mate, the original mate was known to be alive and breeding elsewhere, whereas in other such instances the original mate was known to be dead, but most were of unknown status.

These findings on sparrowhawks apply to many other bird species that have been studied using bands, the commonest findings being:
1 sex differences in dispersal distances within species (e.g. Greenwood 1980; Gavin & Bollinger 1988; Payne & Payne 1993), possibly related to sex differences in territory acquisition and defence;
2 a tendency for greater site fidelity with increasing age (Newton & Marquiss 1982; Harvey et al. 1984; Newton 1993; Aebischer 1995; Morton 1997; Bried & Jou-

Table 4.3 Year-to-year fidelity to mate in relation to fidelity to territory in sparrowhawks.

	Number of birds that in the second year had				
	Same territory		Different territory		Significance of variation between categories
	Same mate	Different mate	Same mate	Different mate	
Males	20	20	2	10	$\chi^2_1 = 4.20, P = 0.040$
Females	20	13	2	10	$\chi^2_1 = 6.80, P = 0.009$

Differences between sexes were not significant. Birds tended to reunite with the previous mate when they remained on the same territory.

ventin 1998; Forero *et al.* 1999), possibly related to increasing benefits through life of site familiarity;

3 a greater tendency to change territories after a breeding failure than after a success (Newton & Marquiss 1982; Beletsky & Orians 1991; Payne & Payne 1993), possibly related to site quality;

4 a tendency to move to better territories with increasing age (e.g. Hildén 1979; Matthysen 1990; Montalvo & Potti 1992), possibly related to increase in status and competitive ability;

5 a strong tendency for a change of territory to be associated with a change of mate (e.g. Johnston & Ryder 1987; Bradley *et al.* 1990).

The advantage of changing both territory and mate simultaneously is that it allows for the possibility that breeding failure could be either site related or mate related. Conversely, the advantage of retaining territory and mate after successful breeding is presumably that the chance of breeding well next time is greater than if either is changed. In some long-lived species, this is more than simply sticking with a winning combination, because mean breeding success increases with pair duration. The longer partners have been together, the more likely they are to raise young, and if they divorce, they may lose one or more years until they find and breed successfully for the first time with a new mate (for examples see Black 1996). In such species, therefore, experience of mate affects performance independently of age. Some long-lived swans and albatrosses have been known to keep the same partners for more than 20 years.

Non-breeding dispersal

In non-migratory populations, such as the sparrowhawk in Britain, this is not normally an issue, because individuals usually stay in the same general area year round. But some migratory species have shown a remarkable propensity to return, not only to the same breeding localities in successive years, but also to the same wintering localities (Table 4.4). Again, the proportions of individuals that returned to some study areas were so high that they must have included most of the survivors.

Nomadic species

The species in Tables 4.2 and 4.4 are not typical of all birds. Such precise homing behaviour is likely to be favoured only where habitats remain fairly stable from year to year, and where returning birds could expect to survive and reproduce. It would not be expected in species that depend on unpredictable habitats or food supplies, which are available in different areas in different years (e.g. Cohen & Levin 1991). This is the case in some waterbirds affected by fluctuating water levels (Johnson & Grier 1988), in desert species affected by irregular rainfall (Serventy & Whittell 1976), in boreal finches that exploit sporadic tree-seed crops (Svärdson 1957; Newton 1972) and in some predatory birds that exploit locally abundant rodents (Newton 1979). The local population densities of such species often fluctuate greatly from year to year, in line with fluctuating habitat or food supplies. The

Table 4.4 Annual return rates of migrant birds to specific study areas in successive winters. The data are drawn from detailed studies in which attempts were made to identify all individuals present. Records for different years are pooled: a bird ringed in one year and recovered in the next is counted as 1 'bird-year', and a bird ringed in one year and recovered in the next 2 years is counted as 2 'bird-years', and so on. Size of study area: *<1 km² or <1 km of coast, **1–10 km² or 1–10 km of coast, ***10–1000 km² or 10–1000 km of coast.

Species	Location	Number of 'bird-years'[1]	% returned	Source
Passerines				
Marsh warbler *Acrocephalus palustris*	Zambia*	17	47	Kelsey 1989
Greenish warbler *Phylloscopus trochiloides*	India**	25	52	Price 1981
Great reed warbler *Acrocephalus arundinaceus*	Malaysia, area 1*	62	50	Nisbet & Medway 1972
	Malaysia, area 2*	83	37	Nisbet & Medway 1972
American redstart *Setophaga ruticilla*	Jamaica	111	51	Holmes & Sherry 1992
Black-throated blue warbler *Dendroica caerulescens*	Jamaica	57	46	Holmes & Sherry 1992
Prairie warbler *Dendroica discolor*	Puerto Rico*	25	40	Staicer 1992
Cape May warbler *Dendroica tigrina*	Puerto Rico*	8	38	Staicer 1992
Northern Parula warbler *Parula americana*	Puerto Rico*	65	49	Staicer 1992
Shorebirds				
Ruddy turnstone *Arenaria interpres*	Scotland**	42	95	Metcalfe & Furness 1985
Green sandpiper *Tringa ochropus*	England***	115	84	Smith *et al.* 1992
Eurasian oystercatcher *Haematopus ostralegus*				
adults	England**	734	89	Goss-Custard *et al.* 1982
2–4-year-olds	England**	475	83	Goss-Custard *et al.* 1982
Purple sandpiper *Calidris maritima*				
adults	Helgoland**	117	85	Dierschke 1998
first-year birds	Helgoland***	30	63	Dierschke 1998
Waterfowl				
Bewick's swan *Cygnus columbianus bewickii*	England**	690	67	Rees 1987
Canada goose *Branta canadensis*	Minnesota*	271	78	Raveling 1979
Barnacle goose *Branta leucopsis*	Scotland**	540	76	Percival 1991
Greater white-fronted goose *Anser albifrons*	Scotland***	531	85	Wilson *et al.* 1991
Bean goose *Anser fabalis*	Sweden**	25	60	Nilsson & Persson 1991
Snow goose *Anser caerulescens*	Texas—Louisiana*	77	86	Prevett & MacInnes 1980
Harlequin duck *Histrionicus histrionicus*				
males	British Columbia**	82	77	Robertson & Cooke 1999
females	British Columbia**	66	62	Robertson & Cooke 1999
Bufflehead *Bucephala albeola*				
males	Maryland*	91	26	Limpert 1980
females	Maryland*	37	11	Limpert 1980

[1] These figures should be viewed in relation to annual survival rates which tend to be in the region of 40–50% for small passerines, and 60–90% for various shorebirds, large waterfowl and large raptors.

speed with which local numbers increase in response to improving conditions has led to the view that such species are nomadic, with individuals concentrating in different areas in different years, wherever conditions are good at the time. Because individuals tend not to return to the same areas year after year, they cannot be re-caught by resident banders operating in the same areas every year (and hence did not figure in Tables 4.2 and 4.4). Knowledge of their movements is instead mostly dependent on occasional band recoveries reported by members of the public. Such band recoveries have now confirmed that some individual birds of species that exploit 'boom-and-bust' food supplies do indeed breed in widely separated areas in different years. The data mostly concern rodent-eaters and seed-eaters, some species of which show exceptionally long natal and breeding dispersal.

Among rodent-eaters, most information concerns Tengmalm's owl *Aegolius funereus*, which nests readily in boxes and has been studied at many localities in northern Europe. Both sexes tend to stay in the same localities if vole densities remain high, with adults moving no more than about 5 km between nest boxes used in successive years. But if vole densities crash, females move much longer distances, many having shifted 100–580 km between nesting sites in different years (Fig. 4.7). Females that moved such long distances between breeding seasons also changed mates. In contrast, few long movements were observed in males: most moved no more than 5 km (as did females during vole peaks), but one moved 21 km between breeding sites, and two moved more than 100 km in years of low vole numbers (Fig. 4.7). The greater residency of males has been attributed to their need to guard cavity nest sites which are scarce in their conifer forest habitat, while their smaller size makes them better able than females to catch small birds, and hence to survive (without breeding) through low vole conditions (Korpimäki et al. 1987).

In other rodent-eating species, both sexes may be inferred to move long distances, because areas can become almost deserted (or reoccupied) from one year to the next. Although the chances of recording marked individuals at places far apart are low, movements of up to several hundred kilometres between one breeding site and another have been recorded in the young and adults of some other species of rodent-eating owls and diurnal raptors (for short-eared owl *Asio flammeus*, see Village 1987; for long-eared owl *A. otus*, see Marks et al. 1994; for great grey owl *Strix nebulosa*, see Duncan 1992; for diurnal raptors, see Newton 1979).

Among tree-seed eaters, extraordinarily long natal and breeding dispersal distances have been recorded from common crossbills *Loxia curvirostra* in western Eurasia, where the main food plant is Norway spruce *Picea abies*. The seed crops of this tree species fluctuate greatly from year to year, but poor crops in one region may coincide with good crops in another region (Svärdson 1957). Crossbills therefore make one major movement each year, in June to August, leaving areas where previous crops are coming to an end, and concentrating in mainly different areas where developing crops are good (Newton 1972). Most recorded movements were more than 2000 km, and one adult was found in consecutive breeding seasons at

LIFE HISTORY STUDIES OF INDIVIDUAL ANIMALS

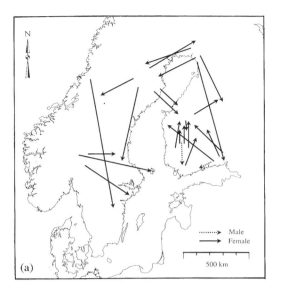

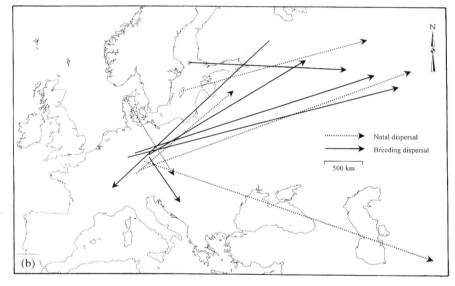

Figure 4.7 (a) Movements of individual Tengmalm's owls *Aegolius funereus* between nesting sites in Fennoscandia in different years. (Compiled from information in Löfgren *et al.* 1986, Korpimäki *et al.* 1987, and Sonerud *et al.* 1988.) (b) Movements of individual common crossbills *Loxia curvirostra* between natal and breeding sites (dashed lines) and between the breeding sites of different years (continuous lines) in Eurasia, based on individuals ringed and recovered in different breeding seasons (taken as January to April, Newton 1972). (Compiled from information in Schloss 1984, and from records in the Swiss and Finnish Bird Banding Schemes.)

localities 3170 km apart (Fig. 4.7). Again, however, because the records are based mainly on band recoveries supplied by members of the public, they carry the assumption that birds found in the breeding season were in fact nesting at the locality concerned. As spruce virtually never crops well in the same area 2 years running (e.g. Svärdson 1957), it is perhaps not surprising that, in spruce areas, no records have emerged of ringed crossbills staying more than 1 year at a time. This behaviour contrasts with that of crossbills in pine areas, where the crops are more consistent and where many ringed individuals were proved to remain from year to year (Senar et al. 1993). Other finches that depend on tree seeds have also shown displacements of more than 500 km between their breeding sites in successive years, or between their wintering sites in successive years (Newton 1972). They contrast with closely related finch species which feed on the more consistent seed crops of herbaceous plants, and show relatively short natal and breeding dispersal.

In both the raptors and the seed-eaters, these ring recoveries give some idea of the huge areas over which individuals roam in order to find food supplies sufficient for reproduction. With their ability to move rapidly over large areas, these species can exploit a specialist niche in a way that other, more sedentary animals could not. But because of the sporadic nature of their habitats or food supplies, huge or widely separated habitat areas are required to accommodate their populations long term.

Achievements and challenges

The marking of individual animals, and in particular of birds, has contributed much to our knowledge of the behaviour and life histories of individuals, and through this to a wider ecological understanding. We now have estimates of annual survival rates for a wide range of bird species and, in some species, also on how these annual rates vary with age and environmental conditions. When linked with measures of reproductive rates, survival studies have provided a level of demographic understanding for birds unmatched by that for any other vertebrate group. A striking finding is the degree of inequality in individual lifespan and reproductive success within a population, which in turn ensures enormous variation in the contributions that individuals make to future generations.

Long-term studies of marked individuals have also highlighted the extent of age-related variation in reproduction and survival. In contrast to earlier findings, senescence has emerged as a significant factor in avian life histories, affecting both reproduction and survival. Such studies have reinforced the need for further work on life histories, of the relationship between reproduction and survival at different ages, and of the ultimate and proximate factors involved. This should in turn promote the development of life-history theory and the understanding of senescence (for recent discussion, see Ricklefs 1998).

Senility has proved difficult to demonstrate in wild birds, as in most animals. This is because most species can be aged only when young, so a prolonged study is needed to give individuals of known old age, but also because in most wild populations so few individuals reach old age that a very large sample of young individuals is needed to provide enough old ones for study.

So far, almost all studies of senescence in free-living organisms have done little more than demonstrate age-specific mortality, reproduction and reproductive value. Such studies provide no clue to the underlying physiological mechanisms of senescence. Suggestions include: (i) an inability in later life to repair damaged strands of DNA; (ii) a deterioration in the immune system, with a consequent increase in disease; and (iii) the negative effects of mutations at late stages of the life cycle (Rose 1991; Partridge & Barton 1993). Studies of these aspects may seem beyond the fieldworker, but important clues on the mechanisms of ageing could be gained from studies of parasitism and other adverse factors in relation to age (see Møller & de Lope 1999). Again, birds could prove useful in such studies because, compared to mammals, they have higher survival rates relative to body size (Williams 1957; Promislow 1991; Ricklefs 1998).

Without studies of marked individuals, we would know much less than we do about migration patterns, and particularly about dispersal, despite its importance in population ecology and genetics. Existing studies have revealed some of the factors that influence the distances moved by individuals between natal site and breeding sites, or between successive breeding or wintering sites. Such movements vary with population density and other factors promoting competition, as well as with gender, age and other features of the individual. Migrating bird species that depend on predictable habitats and food supplies show remarkable homing behaviour, returning year after year to the same breeding and wintering localities, often to the same territories. This applies to both short-distance and long-distance migrants which may breed and winter on different continents. In contrast, species that depend on unpredictable habitats and food supplies can range over huge areas, and band recoveries have revealed that some individuals have moved hundreds or thousands of kilometres between natal site and breeding site or between the breeding sites of different years.

The accumulation of individual data from marked birds and other animals has also led to the development of new statistical methodology and software for analysing survival and other aspects of demography (notably Lebreton *et al*. 1992). Significant practical innovations for estimating survival rates include the programs SURGE (Pradel *et al*. 1990) and MARK (White & Burnham 1999). Many other recent developments in the analysis of bird banding data have been reported in the proceedings of Euring Conferences (e.g. Lebreton & North 1993; Baillie *et al*. 1999), including analysis of dispersal and other movement patterns (Nichols & Kaiser 1999). Moreover, models incorporating individual behaviour and performance are now being used to predict how populations might respond to specific environmental changes (e.g. Goss-Custard *et al*. 1995). Hence, for the foreseeable future, the marking of individual birds and other animals is likely to provide continued growth in our understanding of individual performance and movements, as well as of the wider fields of population ecology and life-history evolution.

Acknowledgements

I am grateful to Nancy Huntly and two anonymous referees for helpful comments on the manuscript.

References

Aebischer, N.J. (1995) Philopatry and colony fidelity of Shags *Phalacrocorax aristotelis* on the east coast of Britain. *Ibis* **137**, 11–18.

Aebischer, N.J. & Coulson, J.C. (1990) Survival of the Kittiwake in relation to sex, year, breeding experience and position in the colony. *Journal of Animal Ecology* **59**, 1063–1071.

Arcese, P. (1989) Intrasexual competition, mating system and natal dispersal in song sparrows. *Animal Behaviour* **38**, 958–979.

Baillie, S.R., North, P.M. & Gosler, A.G. (1999) Large-scale studies of marked birds. Proceedings of the EURING 97 Conference. *Bird Study* **46** (Suppl.), S1–S308.

Beletsky, L.D. & Orians, G.H. (1991) Effects of breeding experience and familiarity on site fidelity in female red-winged blackbirds. *Ecology* **72**, 787–796.

Bensch, S. & Hasselquist, D. (1991) Territory infidelity in the polygynous great reed warbler *Acrocephalus arundinaceus*: the effect of variation in territory attractiveness. *Journal of Animal Ecology* **60**, 857–871.

Berndt, R. & Sternberg, H. (1968) Terms, studies and experiments on the problems of bird dispersion. *Ibis* **110**, 256–269.

Birkhead, T. & Møller, A.P. (1992). *Sperm Competition in Birds* Academic Press, London.

Black, J.M., ed. (1996). *Partnerships in Birds. The Study of Monogamy*. University Press, Oxford.

Boekelheide, R.J. & Ainley, D.G. (1989) Age, resource availability and breeding effort in the Brandt's Cormorant. *Auk* **106**, 389–401.

Bradley, J.S., Wooller, R.D., Shira, I.J. & Serventy, D.L. (1989) Age-dependent survival of breeding short-tailed shearwaters *Puffinus tenuirostris*. *Journal of Animal Ecology* **58**, 175–188.

Bradley, J.S., Wooller, R.D., Skira, R.J. & Serventy, D.L. (1990) The influence of mate retention and divorce upon reproductive success in short-tailed shearwaters *Puffinus tenuirostris*. *Journal of Animal Ecology* **59**, 487–496.

Bried, J. & Jouventin, P. (1998) Why do lesser sheathbills *Chionis minor* switch territory? *Journal of Avian Biology* **29**, 257–265.

Burley, N. (1985) Leg-band colour and mortality patterns in captive breeding populations of Zebra Finches. *Auk* **102**, 647–651.

Caswell, H. (1982) Optimal life histories and the maximisation of reproductive value: a general theorem for complex life cycles. *Ecology* **63**, 1218–1222.

Charlesworth, B. (1994). *Evolution in Age-Structured Populations*, 2nd edn. University Press, Cambridge.

Clutton-Brock, T., ed. (1988). *Reproductive Success*. Chicago University Press, Chicago, IL.

Cohen, D. & Levin, S.A. (1991) Dispersal in patchy environments: the effects of temporal and spatial structure. *Theoretical Population Biology* **39**, 63–99.

Coltman, D.W., Smith, J.A., Bancroft, D.P. *et al.* (1999) Density dependent variation in lifetime breeding success and natural and sexual selection in Soay Rams. *American Naturalist* **154**, 730–746.

Curio, E. (1983) Why do young birds reproduce less well? *Ibis* **121**, 400–404.

Davis, J.W. (1976) Breeding success and experience in the arctic skua *Stercorarius parasiticus*. *Journal of Animal Ecology* **45**, 531–536.

Dhondt, A. & Hublé, J. (1968) Fledging date and sex in relation to dispersal in young tits. *Bird Study* **15**, 127–134.

Dierschke, V. (1998) Site fidelity and survival of purple sandpipers *Calidris maritima* at Helgoland (SE North Sea). *Ringing and Migration* **19**, 41–48.

Drilling, N.E. & Thompson, L.F. (1988) Natal and breeding dispersal in house wrens (*Troglodytes aedon*). *Auk* **105**, 480–491.

Duncan, J.R. (1992) *Influence of prey abundance and snow cover on Great Grey Owl breeding dispersal.* PhD Thesis, University of Manitoba, Winnipeg.

Fisher, R.A. (1958) *The Genetical Theory of Natural Selection*, 2nd edn. Oxford University Press, Oxford.

Flynn, L., Nol, E. & Zharikov, Y. (1999) Philopatry, nest-site tenacity, and mate fidelity of Semipalmated Plovers. *Journal of Avian Biology* **30**, 47–55.

Forero, M.G., Donázar, J.A., Blas, J. & Hiraldo, F. (1999) Causes and consequences of territory change and breeding dispersal distance in the black kite. *Ecology* **80**, 1298–1310.

Gavin, T.A. & Bollinger, E.K. (1988) Reproductive correlates of breeding site fidelity in bobolinks (*Dolichonyx oryzivorus*). *Ecology* **69**, 96–103.

Goodman, D. (1982) Optimal life histories, optimal rotation, and the value of reproductive value. *American Naturalist* **119**, 803–823.

Goss-Custard, J.D., Le, V., dit. Durell, S.E.A., Sitters, H.P. & Swinfen, R. (1982) Age-structure and survival of a wintering population of oystercatchers. *Bird Study* **29**, 83–98.

Goss-Custard, J.D., Clarke, R.T. & Durell, S.E.A. (1984) Rates of food intake and aggression of Oystercatchers *Haematopus ostralegus* on the most and least preferred mussel *Mytilus edulis* beds of the Exe Estuary. *Journal of Animal Ecology* **53**, 233–245.

Goss-Custard, J.D., Caldow, R.W.G., Clarke, R.T. & West, A.D. (1995) Deriving population parameters from individual variations in foraging behaviour II. Model tests and population parameters. *Journal of Animal Ecology* **64**, 277–289.

Greenwood, P.J. (1980) Mating systems, philopatry and dispersal in birds and mammals. *Animal. Behaviour* **28**, 1140–1162.

Harvey, P.H., Greenwood, P.J., Campbell, B. & Stenning, M.J. (1984) Breeding dispersal of the pied flycatcher (*Ficedula hypoleuca*). *Journal of Animal Ecology* **53**, 727–736.

Hildén, O. (1978) Population dynamics in Temminck's stint *Calidris temminckii*. *Oikos* **30**, 17–28.

Hildén, O. (1979) Territoriality and site tenacity of Temminck's Stint *Calidris temminckii*. *Ornis Fennica*. **56**, 56–74.

Holmes, D.H. & Austad, S.N. (1995) The evolution of avian senescence patterns: implications for understanding primary aging processes. *American Zoologist* **35**, 307–317.

Holmes, R.T. & Sherry, T.W. (1992) Site fidelity of migratory warblers in temperate breeding and neotropical wintering areas: applications for population dynamics, habitat selection, and conservation. In: *Ecology and Conservation of Neotropical Migrant Landbirds* (eds J.M.Hagan, III & D.W.Johnston), pp. 563–575. Smithsonian Institution Press, Washington.

Holt, R.D. & McPeek, M.A. (1996) Chaotic population dynamics favors the evolution of dispersal. *American Naturalist* **148**, 709–718.

Johnson, M.L. & Gaines, M.S. (1990) Evolution of dispersal: theoretical models and empirical tests using birds and mammals. *Annual Review of Ecological Systematics* **21**, 449–480.

Johnson, V.H. & Ryder, J.P. (1987) Divorce in larids: a review. *Colonial Waterbirds* **10**, 16–26.

Johnston, D.H. & Grier, J.W. (1988) Determinants of breeding distributions of ducks. *Wildlife Monographs* **100**, 1–37.

Kelsey, M.G. (1989) A comparison of the song and territorial behaviour of a long-distance migrant, the Marsh Warbler *Acrocephalus palustris*, in summer and winter. *Ibis* **131**, 403–414.

Korpimäki, E., Lagerström, M. & Saurola, P. (1987) Field evidence for nomadism in Tengmalm's Owl *Aegolius funereus*. *Ornis Scandinavica* **18**, 1–4.

Lawn, M.R. (1994) Site fidelity and annual survival of territorial male Willow Warblers *Phylloscopus trochilus* at four adjacent sites in Surrey. *Ringing and Migration* **15**, 1–7.

Lebreton, J.-D. & North, P.M., eds (1993) *Marked Individuals in the Study of Bird Population*. Birkhäuser Verlag, Basel.

Lebreton, J.D., Burnham, K.P., Clobert, J. & Anderson, D.R. (1992) Modelling survival and testing biological hypotheses using marked animals: a unified approach with case studies. *Ecological Monographs* **62**, 67–118.

Limpert, R.J. (1980) Homing success of adult buffleheads to a Maryland wintering site. *Journal of Wildlife Management* **44**, 905–908.

Loery, G., Pollock, K.H., Nichols, J.D. & Hines, J.E. (1987) Age-specificity of black-capped chickadee survival rates: analysis of capture–recapture data. *Ecology* **68**, 1038–1044.

Löfgren, O., Hörnfeldt, B. & Carlsson, B.-G. (1986) Site tenacity and nomadism in Tengmalm's Owl (*Aegolius funereus* L) in relation to cyclic food production. *Oecologia* **69**, 321–326.

Marks, J.S., Evans, D.L. & Holt, D.W. (1994) *Long-eared Owl*. The American Ornithologists Union, Washington, DC, Number 133, pp. 1–24.

Martin, K. (1995) Pattern and mechanisms for age-dependent reproduction and survival in birds. *American Zoologist* **35**, 340–348.

Matthysen, E. (1990) Behavioural and ecological correlates of territory quality in the Eurasian Nuthatch (*Sitta europaea*). *Auk* **107**, 86–95.

McCleery, R.H. & Perrins, C.M. (1989) Great Tit. In: *Lifetime Reproduction in Birds* (ed. I. Newton), pp. 35–53. Academic Press, London.

Metcalfe, N.B. & Furness, R.W. (1985) Survival, winter population stability and site-fidelity in the Turnstone *Arenarea interpres*. *Bird Study* **32**, 207–214.

Møller, A.P. & de Lope, F. (1999) Senescence in a short-lived migratory bird: age-dependent morphology, migration, reproduction and parasitism. *Journal of Animal Ecology* **68**, 163–171.

Monaghan, P. (1980) Dominance and dispersal between feeding sites in the herring gull (*Larus argentatus*). *Animal Behaviour* **28**, 521–527.

Montalvo, S. & Potti, J. (1992) Breeding dispersal in Spanish Pied Flycatchers *Ficedula hypoleuca*. *Ornis Scandinavica* **23**, 491–498.

Morton, M.L. (1997) Natal and breeding dispersal in the mountain white-crowned sparrow *Zonotrichia leucophrys oriantha*. *Ardea* **85**, 145–154.

Negro, J.J., Hiraldo, F. & Donázar, J.A. (1997) Causes of natal dispersal in the lesser kestrel: inbreeding avoidance or resource competition? *Journal of Animal Ecology* **66**, 640–648.

Newton, I. (1972). *Finches*. Collins, London.

Newton, I. (1979). *Population Ecology of Raptors*. Poyser, Berkhamsted.

Newton, I. (1986). *The Sparrowhawk*, T. & A.D. Poyser, Calton.

Newton, I. (1988) Age and reproduction in the Sparrowhawk. In: *Reproductive Success* (ed. T.H. Clutton Brock), pp. 201–219. Chicago University Press, Chicago, IL.

Newton, I., ed. (1989a). *Lifetime Reproduction in Birds*. Academic Press, London.

Newton, I. (1989b) Sparrowhawk. In: *Lifetime Reproduction in Birds* (ed. I. Newton), pp. 279–296. Academic Press, London.

Newton, I. (1991) Habitat variation and population regulation in sparrowhawks. *Ibis* **133** (Suppl. 1), 76–88.

Newton, I. (1993) Age and site fidelity in female sparrowhawks *Accipiter nisus*. *Animal Behaviour* **46**, 161–168.

Newton, I. (2001) Causes and consequences of breeding dispersal in the Sparrowhawk. *Ardea* **89**, in press.

Newton, I. & Marquiss, M. (1982) Fidelity to breeding area and mate in sparrowhawks *Accipiter nisus*. *Journal of Animal Ecology* **51**, 327–341.

Newton, I. & Rothery, P. (1998) Age-related trends in the breeding success of individual female Sparrowhawks *Accipiter nisus*. *Ardea* **86**, 21–31.

Newton, I. & Rothery, P. (2000) Post-fledging recovery and dispersal of ringed Eurasian Sparrowhawks *Accipiter nisus*. *Journal of Avian Biology* **31**, 226–236.

Nichols, J.D. & Kaiser, A. (1999) Quantitative studies of bird movement: a methodological review. *Bird Study* **46** (Suppl.), S289–S298.

Nilsson, J.-A. (1989) Causes and consequences of natal dispersal in the marsh tit, *Parus palustris*. *Journal of Animal Ecology* **58**, 619–636.

Nilsson, L. & Persson, H. (1991) Site tenacity and turnover rate of staging and wintering bean geese *Anser fabalis* in southern Sweden. *Wildfowl* **42**, 53–59.

Nisbet, I.C.T. & Medway, L. (1972) Dispersion, population ecology and migration of Eastern great reed warblers *Acrocephalus orientalis* wintering in Malaysia. *Ibis* **114**, 451–494.

Norman, S.C. (1994) Dispersal and return rates of willow warbler *Phylloscopus trochilus* in relation to age, sex and season. *Ringing and Migration* **15**, 8–16.

Ollason, J.C. & Dunnet, G.M. (1988) Variation in breeding success in fulmars. In: *Reproductive Success* (ed. T.H. Clutton-Brock), pp. 263–278. Chicago University Press, Chicago, IL.

Paradis, E., Baillie, S.R., Sutherland, W.J. & Gregory, R.D. (1998) Patterns of natal and breeding dispersal in birds. *Journal of Animal Ecology* **67**, 518–536.

Partridge, L. & Barton, N.H. (1993) Optimality, mutation and the evolution of ageing. *Nature* **362**, 305–311.

Payne, R.B. & Payne, L.L. (1993) Breeding dispersal in Indigo Buntings: circumstances and consequences for breeding success and population structure. *Condor* **95**, 1–24.

Percival, S.M. (1991) The population structure of

Greenland Barnacle Geese *Branta leucopsis* on the winter grounds on Islay. *Ibis* **133**, 357–364.

Perrins, C.M. (1979). *British Tits*. Collins, London.

Pinowski, J. (1965) Overcrowding as one of the causes of dispersal of young tree sparrows. *Bird Study* **12**, 27–33.

Pradel, R.J., Clobert, J. & Lebreton, J.-D. (1990) Recent developments for the analysis of multiple capture–recapture data sets: an example concerning two blue tit populations. *Ring* **13**, 193–204.

da Prato, S.R.D. & da Prato, E.S. (1983) Movements of whitethroats *Sylvia communis* ringed in the British Isles. *Ringing and Migration* **4**, 193–210.

Pratt, A. & Peach, W. (1991) Site tenacity and annual survival of a willow warbler *Phylloscopus trochilus* population in southern England. *Ringing and Migration* **12**, 128–134.

Prevett, J.P. & MacInnes, C.D. (1980) Family and other social groups in snow geese. *Wildlife Monographs* **71**, 1–46.

Price, T. (1981) The ecology of the greenish warbler *Phylloscopus trochiloides* in its winter quarters. *Ibis* **123**, 131–144.

Promislow, D.E.L. (1991) Senescence in natural populations of mammals. *Evolution* **45**, 1869–1887.

Pyle, P., Spear, I.B., Sydeman, W.J. & Ainley, D.G. (1991) The effects of experience and age on the breeding performance of western gulls. *Auk* **108**, 25–33.

Ratcliffe, L.M. & Boag, P.T. (1987) Effects of colour bands on male competition and sexual attractiveness in zebra finches (*Poephila guttata*). *Canadian Journal of Zoology* **65**, 333–338.

Ratcliffe, N., Furness, R.W. & Hamer, K.C. (1998) The interactive effects of age and food supply on the breeding ecology of great skuas. *Journal of Animal Ecology* **67**, 853–862.

Rattiste, K. & Lilleleht, V. (1987) Population ecology of the common gull *Larus canus* in Estonia. *Ornis Fennica* **64**, 25–26.

Raveling, D.G. (1979) Traditional use of migration and winter roost sites by canada geese. *Journal of Wildlife Management* **43**, 229–235.

Rees, E.C. (1987) Conflict of choice within pairs of Bewick's swans regarding their migratory movement to and from the wintering grounds. *Animal Behaviour* **35**, 1685–1693.

Ricklefs, R.E. (1998) Evolutionary theories of aging: confirmation of a fundamental prediction with implications for the genetic basis and evolution of lifespan. *American Naturalist* **152**, 24–44.

Robertson, G. & Cooke, F. (1999) Winter philopatry in migratory waterfowl. *Auk* **116**, 20–34.

Rockwell, R.F., Cooch, E.G., Thompson, C.B. & Cooke, F. (1993) Age and reproductive success in female lesser snow geese, experience, senescence and the cost of philopatry. *Journal of Animal Ecology* **62**, 323–333.

Rose, M.R. (1991). *Evolutionary Biology of Ageing*. Oxford University Press, Oxford.

Rousset, F. (1999) Reproductive value vs sources and sinks. *Oikos* **86**, 591–596.

Saether, B.-E. (1990) Age-specific variation in reproductive performance of birds. *Current Ornithology* **7**, 251–283.

Schloss, W. (1984) Ringfunde des Fichtenkreuzschnabels (*Loxia curvirostra*). *Auspicium* **7**, 257–284.

Senar, J.C., Borras, A., Cabrera, T. & Cabrera, J. (1993) Testing for the relationship between coniferous crop stability and common crossbill residence. *Journal of Field Ornithology* **64**, 464–469.

Serventy, D.L. & Whittell, H.M. (1976) *Birds of Western Australia*, 5th edn. University of Western Australia Press, Perth.

Smith, K.W., Reed, J.M. & Trevis, B.E. (1992) Habitat use and site fidelity of green sandpipers *Tringa ochropus* wintering in southern England. *Bird Study* **39**, 155–164.

Sokolov, L.V. (1997) Philopatry of migratory birds. *Physiological and General Biology Reviews* **11**, 1–58.

Sonerud, G.A., Solheim, R. & Prestrud, K. (1988) Dispersal of Tengmalm's Owl *Aegolius funereus* in relation to prey availability and nesting success. *Ornis Scandinavica* **19**, 175–181.

Staicer, C.A. (1992) Social behaviour of the northern parula, Cape May warbler, and Prairie warbler wintering in second-growth forest in southwestern Puerto Rico. In: *Ecology and Conservation of Neotropical Migrant Landbirds* (eds J.M. Hagen & D.W. Johnston), pp. 308–320. Smithsonian Institution Press, Washington.

Stearns, S.C. (1992). *The Evolution of Life Histories*. Oxford University Press, Oxford.

Sternberg, H. (1989) Pied flycatcher. In: *Lifetime Reproduction in Birds* (ed. I. Newton), pp. 55–74. Academic Press, London.

Svärdson, G. (1957) The 'invasion' type of bird migration. *British Birds* **50**, 314–343.

Sydeman, W.J., Penniman, J.F., Penniman, T.F., Pyle, P. & Ainley, D.G. (1991) Breeding performance in the western gull, effect of parental age, timing of breeding and year in relation to food supply. *Journal of Animal Ecology* **60**, 135–149.

Thompson, C.F. & Nolan, V. (1973) Population biology of the yellow-breasted chat (*Icteria virens* L.) in southern Indiana. *Ecological Monographs* **43**, 145–171.

Tiainen, J. (1983) Dynamics of a local population of the willow warbler *Phylloscopus trochilus* in southern Finland. *Ornis Scandinavica* **14**, 1–15.

Village, A. (1987) Numbers, territory size and turnover of short-eared owls *Asio flammeus* in relation to vole abundance. *Ornis Scandinavica* **18**, 198–204.

White, G.C. & Burnham, K.P. (1999) Program MARK: survival estimates from populations of marked animals. *Bird Study* **46** (Suppl.), S120–S139.

Williams, G.C. (1957) Pleiotrophy, natural selection, and the evolution of senescence. *Evolution* **11**, 398–411.

Wilson, H.J., Norriss, D.W., Walsh, A., Fox, A.D. & Stroud, D.A. (1991) Winter site fidelity in Greenland white-fronted geese *Anser albifrons flavirostris*, implications for conservation and management. *Ardea* **79**, 287–294.

Wooller, R.D., Bradley, J.S., Skira, I.J. & Serventy, D.L. (1989) Short-tailed shearwater. In: *Lifetime Reproduction in Birds* (ed. I. Newton), pp. 405–417. Academic Press, London.

Wooller, R.D., Bradley, J.S., Skira, I.J. & Serventy, D.L. (1990) Reproductive success of short-tailed shearwaters *Puffinus tenuirostris* in relation to their age and breeding experience. *Journal of Animal Ecology* **59**, 161–170.

Wyllie, I. & Newton, I. (1991) Demography of an increasing population of sparrowhawks. *Journal of Animal Ecology* **60**, 749–766.

Part 2
Functional and community ecology

Chapter 5
Specificity, links and networks in the control of diversity in plant and microbial communities

A.H. Fitter

Introduction

A long-running debate in community ecology is that which addresses the control of diversity in plant communities. Since there are so few ways in which plants can differentiate their use of resources, attention has focused on two main mechanisms: spatial and temporal differentiation (Chesson 1985; Fitter 1987) and the impact of grazing animals (Pacala & Crawley 1992). Sometimes the latter has been expanded to take into account antagonistic interactions generally, including the role of pathogens (Burdon 1987). The general message is that such impacts reduce the ability of potentially dominant species to monopolize resources and so make it possible for subordinate species to survive. There are numerous demonstrations of the consequences of reductions in grazing pressure, leading to loss of diversity in plant communities. One that was well documented was the introduction of myxomatosis into Britain in 1953, which resulted in large areas of chalk grassland, one of the most diverse north European plant communities, reverting to species-poor, coarse grassland or scrub, and eventually woodland (Thomas 1960).

Most of the debate has ignored another group of interactions that have at least as large an effect, but one that is less easily studied, namely those between plants and soil micro-organisms. These include pathogenic and other parasitic interactions, but also a number of mutualistic relationships of great significance, as well as indirect effects such as those that alter nutrient cycling rates and so change competitive balance among plant species. In this paper, I explore the importance of some of these microbe–plant interactions as determinants of the structure, dynamics and diversity of plant communities. I shall concentrate on mycorrhizal symbioses, as the most widespread of all plant–microbe symbioses, and because mutualistic symbioses highlight the phenomena most strongly. Many of the examples refer to the most widespread of these symbioses, which forms between fungi in the order Glomales (Zygomycotina) and probably around two-thirds of all plant species, and is variously known as the arbuscular, vesicular–arbuscular or Glomalean mycor-

Department of Biology, University of York, York YO10 5YW, UK

rhiza; here I refer to it as mycorrhiza and to other types by their full name (e.g. ectomycorrhiza, ericoid mycorrhiza).

Impacts of soil micro-organisms on plant community structure

Changing the soil microflora can have a powerful impact on the plant community. A dramatic example was the introduction of the oomycete *Phytophthora cinnamomi* to Australia, a generalist root pathogen to which native *Eucalyptus* spp. proved to have little resistance. The density of *Eucalyptus* in many areas was reduced to very low levels, and these reductions have persisted for 30 years (Weste & Ashton 1994). Such natural experiments are highly indicative; controlled experiments to reveal the details of the impact of soil micro-organisms are more difficult. Adding microbes to soil is rarely undertaken, if only because of the potential danger of replicating a disaster such as that of *P. cinnamomi*. Eliminating taxa is the usual method, but there are no specific agents that can be used to remove target taxa from soil communities without affecting others. Fungicides have been widely used. For example, Newsham *et al.* (1995c) added benomyl to a lichen-rich grassland for 3 years. One impact was unrelated to changes in the soil microflora: the abundant lichen *Cladonia rangiformis* was eliminated by the fungicide and the moss *Ceratodon purpureus* became very common in consequence. However, there were numerous other changes in plant species composition in the benomyl-treated plots: *Erodium cicutarium* and *Crepis capillaris* (both mycorrhizal species) declined in abundance, while *Rumex acetosella* and *Arenaria serpyllifolia*, two non-mycorrhizal plants, increased, in the latter case by 200% (Fig. 5.1a). Among the mycorrhizal species at the site, there was a direct negative relationship between the change in frequency and the reduction in mycorrhizal colonization brought about by benomyl (Fig. 5.1b).

Another long-term benomyl application study was reported by Hartnett & Wilson (1999). They added the fungicide to a tallgrass prairie over a period of 5 years, and found that mycorrhizal colonization of roots was reduced from an average of 10–20% of root length to <5%. The dominant grasses were *Andropogon scoparius*, *A. gerardi* and *Sorghastrum nutans*. All had previously been shown to be highly dependent on mycorrhizal colonization; all were reduced in abundance by benomyl application; and the consequence was a marked increase in a range of previously suppressed species that were less dependent on mycorrhizal fungi. Species richness, evenness and diversity of the community all increased, although canopy cover was unaffected. In this case, the fungi permitted a small group of species to dominate the community and so reduced diversity.

Several of these biocide studies have indicated that mycorrhizal fungi may be especially important in the control of plant community structure. However, there is also evidence that other fungi, and probably also bacteria, may be key players too. For example, another tallgrass prairie study showed that potentially pathogenic fungi might determine the performance of two grass species (Holah & Alexander 1999). Mills and Bever (1998) explained negative feedback of two grass species,

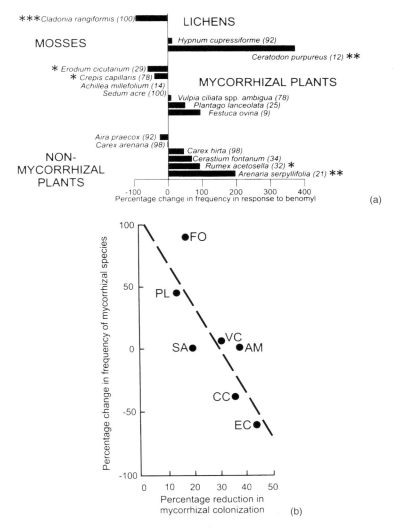

Figure 5.1 (a) The effect of benomyl application on the frequency of principal lichen, moss, mycorrhizal and non-mycorrhizal higher plant species in a lichen-rich plant community at Mildenhall, UK. Significant responses in frequency to benomyl application are indicated by *, $P<0.05$; **, $P<0.01$; ***, $P<0.001$. Percentage frequencies of each species in control plots are shown in brackets. (b) The relationship between percentage changes in the frequencies of mycorrhizal plant species and percentage reductions in mycorrhizal colonization in response to benomyl application to a lichen-rich plant community at Mildenhall, UK. EC, *Erodium cicutarium*; CC, *Crepis capillaris*; AM, *Achillea millefolium*; VC, *Vulpia ciliata* ssp. *ambigua*; FO, *Festuca ovina*; PL, *Plantago lanceolata*; SA, *Sedum acre*. (Reproduced with permission from Newsham et al. 1995c.)

Danthonia spicata and *Panicum sphaerocarpon*, on their own performance in terms of the build-up of specific soil pathogens, notably in the oomycete genus *Pythium*. *Pythium* spp. were also invoked by Packer and Clay (2000) to explain the poor survival of seedlings of *Prunus serotina* close to parent trees, an analogous pattern to that postulated by Janzen (1970) to be caused by seed predators in tropical moist forest. Similarly, there is strong evidence that the survival of *Ammophila arenaria* on sand dunes depends on sand build-up because this allows the plants to escape from pathogens (van der Putten *et al.* 1993).

Additive experiments can only realistically be undertaken by reconstructing communities. Grime *et al.* (1987) planted mixtures of plant species with and without a mycorrhizal fungal inoculum. Diversity was greatest in the mycorrhizal communities, largely because the competitive ability of a range of subordinate species was enhanced relative to that of the dominant grass species, *Festuca ovina*, which was less dependent on mycorrhizal colonization. That experiment simply addressed the question as to whether the presence of mycorrhizal fungi affected plant diversity. Van der Heijden *et al.* (1998) set up communities of old-field plants with varying numbers (1, 2, 4, 8, 14) of species of mycorrhizal fungi, drawn randomly from a pool of 23 species cultured from an old field ecosystem at Guelph, Canada. They therefore were testing the importance of the diversity of the fungal population. They found that a number of performance measures of both the plant and fungal community were strongly enhanced by increasing the number of mycorrhizal fungal species (Fig. 5.2). This is *prima facie* evidence that fungal diversity directly affects plant performance.

However, this type of experimental design has proved controversial, with numerous authors arguing that it merely reflects the fact that species differ in their ecological impact, and that larger pools of species are more likely to contain the most effective species. This phenomenon has been termed sampling effect. On this basis, Wardle (1999) argued that the results of van der Heijden *et al.* (1998) were artefactual; the authors responded vigorously to this charge, pointing out that Wardle assumed that it was possible for a single fungus to be equally effective at promoting the growth of all plants in a community (van der Heijden *et al.* 1999), whereas it is increasingly being demonstrated that mycorrhizal fungi are host selective. Therefore, they suggested, there can be no 'most effective' fungus, as required by the sampling effect argument. A further piece of evidence that supports van der Heijden *et al.*'s view is that the variance associated with plant performance is not a function of number of fungal species. If sampling effect were responsible for the response in Fig. 5.2, the variance should be greater at low fungal diversity (where communities will either contain the 'super-fungus' or not) than at high diversity.

The control of microbial diversity

If microbial, and especially mycorrhizal, fungal diversity can have such a large effect on the composition of plant communities, we need to understand what determines that diversity. This is problematic, as we cannot yet quantify it. It is

PLANT–MICROBE INTERACTIONS AND DIVERSITY

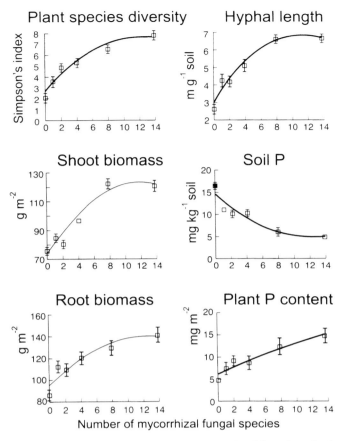

Figure 5.2 The relationship between the number of mycorrhizal fungal species in an inoculum and a range of plant, fungal and soil variables in reconstructed old-field communities. The bars represent 1 SE. (Reproduced with permission from van der Heijden et al. 1998.)

widely held that soils harbour extremely high levels of undescribed diversity (Kennedy 1999; Tiedje et al. 1999). Molecular techniques suggest that a very high proportion of soil bacteria cannot be cultured by the standard techniques of microbiology (Torsvik et al. 1996). Standard descriptions of the diversity of soil communities are therefore certainly serious underestimates of their diversity. The same stricture applies to mycorrhizal fungi. About 150 species have been reliably described, the process depending on the characterization of the spores, since the vegetative characters show little variation. However, almost every survey of mycorrhizal diversity that has been published lists a number of spore types that remained unidentified, suggesting that the true morphological diversity is significantly greater. Increasingly, workers are also using DNA sequence data to characterize diversity, and that too reveals taxa that are not morphologically identifiable.

A key question in the study of microbial diversity is to determine whether it is only local (α) diversity that is high, or whether larger-scale, β and γ diversity is also underestimated. Finlay has strongly argued that the global diversity of free-living microbial species is low, and that a significant fraction of that global diversity can usually be recovered in a single habitat (Finlay & Clarke 1999). For example, Finlay *et al.* (1999) recorded all the ciliates in a lake in Australia. They listed 85 species, all of which were previously described, and all but one of which were known to occur in northern Europe. They conclude that because such organisms are easily dispersed, local and global diversity are essentially the same. Could the same be true for soil micro-organisms?

The soil is a complex medium, offering vastly greater opportunity for microhabitat differentiation, than freshwater. It is also apparent that many soil microbes are powerfully affected by the identity of the other organisms in the community, notably plant species; this is true of free-living as well as symbiotic microbes. The rhizosphere of different, coexisting plant species may support quite distinct microbial populations. Westover *et al.* (1997) showed that when various grassland species occurred as monocultures, their rhizospheres were microbially distinct; when the same species grew juxtaposed at the same site, the rhizosphere population was either intermediate between the two monocultures or resembled one or the other. This implies that diversity will be large. Certainly, some groups of soil organisms do have very high levels of undescribed diversity at all scales. When Bloemers *et al.* (1997) undertook an inventory of nematodes in a rainforest soil in Cameroon, they described 431 species in 194 genera; 90% of these species were previously undescribed.

Nevertheless, mycorrhizal fungi appear to provide a paradox. With around 150 described species, surveys consistently show that 20–25 species can be found in a single habitat. The mean number of species revealed by such surveys, which include single-habitat studies, ecosystem studies and regional studies is 16 (Table 5.1) and there appears to be no evidence that species number increases as the area sampled increases, which, if true, would be a distinctive feature of mycorrhizal fungal communities. Techniques used include both morphological spore surveys and molecular characterization of fungi in plant roots. In almost all of these studies, there was clear evidence that the list of species was not complete. Morton *et al.* (1995) showed that successive rounds of trap culturing would reveal additional species, but very few investigators have used such an approach. It seems reasonable, therefore, to suggest that it is possible to find 10–20% of all known arbuscular mycorrhizal fungi in most habitats, and certainly some species turn up consistently in samples from very different ecosystems separated by large distances. This could imply that mycorrhizal fungi, like free-living protists, are ubiquitous and show little ecological differentiation, which may well be true of a small number of species such as *Glomus mosseae*, which shows almost no genetic variation in samples collected from as far apart as England and Indonesia (Lloyd-MacGilp *et al.* 1996; Fig. 5.3). However, if there are really only a few hundred species of mycorrhizal fungi that colonize perhaps 200 000 species of plants (Trappe 1987), then they must

Table 5.1 Examples of counts of number of species of arbuscular mycorrhizal fungi (Glomales) based on either spore collections or trap cultures, found in single habitats, ecosystems or entire regions. The study marked * used molecular techniques.

Single habitat			Ecosystem			Region		
Habitat	No.	Source	System	No.	Source	Region	No.	Source
Grassland, USA	27	Bever et al. 1996	Cacao, Venezuela	8	Cuenca and Meneses, 1996	Poland	21	Blaszkowski, 1989
Old field, USA	24	Bever, unpublished, in Morton et al., 1995	Disturbed rainforest, Mexico	16	Guadarrama and Alvarez Sanchez, 1999	Nutrient-poor, soils, Venezuela	24	Cuenca et al. 1998
Old Meadow, Canada	13	Hamel et al. 1994	Wheat, USA	13	Hettrick and Bloom 1983	Atlantic coastal dunes, USA	23	Koske 1987
Tallgrass prairie, USA	20	Hettrick and Bloom 1983	Old field succession to forest, USA	25	Johnson et al. 1991	Apple orchards, USA	43	Miller et al. 1985
Sand dune, USA	17	Koske and Morton, unpublished, in Morton et al. 1995	Sand dunes, USA	17	Koske and Gemma 1997	Ando soils, Japan	16	Saito and Vargas 1991
Desert, USA	11	Morton et al. 1995	Sand dunes, USA	6	Koske and Halvorson 1981			
Desert, USA	10	Morton et al. 1995	Lake dunes, USA	14	Koske and Tews 1987			
Old field, USA	23	van der Heijden et al. 1998	Turf grass, USA	19	Koske et al. 1997			
Poplar plantation, USA	10,12	Walker et al. 1982	Wetlands, USA	9	Miller and Bever 1999			
Woodland, UK	13	*Helgason et al. 1998 (and unpublished)	Dunes, Brazil	12	Stürmer and Bellei 1994			
			Mesquite scrub, USA	11	Stutz and Morton 1996			
			Sand dunes, USA	9	Tews and Koske 1986			
Mean	17			14			25	

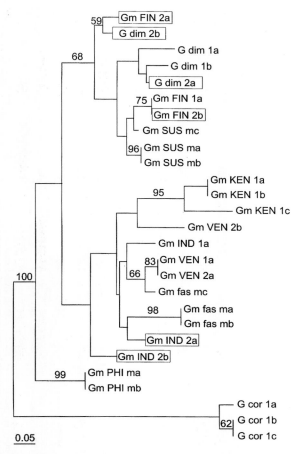

Figure 5.3 Phylogenetic tree of the internal transcribed spacer (ITS) and 5.8S regions of a range of spores of *Glomus* spp. The analysis is based on insertions and deletions only. Bootstrap values greater than 50% are shown. Gm, *G. mosseae*; G dim, *G. dimorphicum*; G fas, *G. fasciculatum*; G cor, *G. coronatum*; FIN, Finland; SUS, Sussex, UK; KEN, Kenya; IND, Indonesia; VEN, Venezuela; PHI, Philippines.

display little or no specificity, which would be unusual in a symbiotic group. Rhizobium bacteria, in many ways an ecologically similar group, once thought to be non-specific, are now known to have high levels of specificity (Albrecht *et al.* 1999).

There is now increasing evidence that mycorrhizal fungi are also strongly host dependent. This is well established for ectomycorrhizal fungi. For example, Massicotte *et al.* (1999) planted five tree species in mixtures on three forest soils. They recovered 55 distinct host–fungus combinations from these soils including 18 different ectomycorrhizal morphotypes, but only four of these were specific to a particular plant and fungus. Both Sanders and Fitter (1992) and Bever *et al.* (1997)

PLANT–MICROBE INTERACTIONS AND DIVERSITY

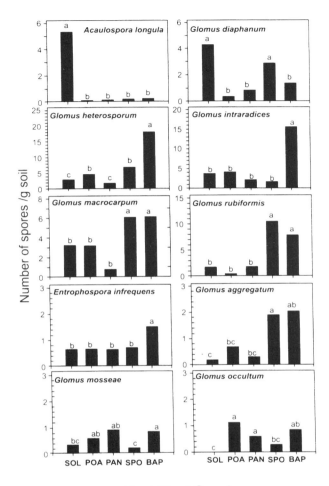

Figure 5.4 Mean spore densities of 10 arbuscular mycorrhizal (AM) fungal species isolated from field soil samples that varied significantly in abundance among five host plant species. Bars with different letters indicate means that are significantly different. SOL, *Solidago missouriensis* Nutt.; POA, *Poa pratensis* L.; PAN, *Panicum virgatum* L.; SPO, *Sporobolus heterolepis* (A. Gray); BAP, *Baptisia bracteata* Muhl. ex Ell. (Reproduced with permission from Eom *et al.* 2000.)

used trap cultures to recover spores of various arbuscular mycorrhizal fungi from field sites. They both used plants from the site as the bait plant on which the fungi could grow. In each case, the species of fungi recovered varied with the identity of the bait plant. Similarly, Eom *et al.* (2000) found very different numbers of spores of various fungal species when they sampled soil from within monospecific stands of five dominant species on the Konza Prairie, Kansas (Fig. 5.4). For example, one fungal species, *Acaulospora longula*, was almost exclusively found in the soil around *Solidago missouriensis*.

The idea that mycorrhizal symbioses exhibit some degree of specificity is increasingly well founded, although it is certain that this is not an absolute phenomenon in many cases. In culture, many mycorrhizal fungi are highly promiscuous, readily colonizing almost any host species which they encounter, but this may be a tautology: since only some taxa in any habitat are easily cultured, those that are cultured are likely to have the least demanding colonization requirements. The ones that are not culturable by present techniques may well be those that colonize few plant species and only under certain conditions. In our own studies in a deciduous woodland (Pretty Wood) in Yorkshire, we have been able to identify at least a dozen species of mycorrhizal fungi, using both morphological and molecular tools. Yet, despite several years of intensive effort, we have succeeded in bringing fewer than half of these into culture, and the most easily cultured species is one that is rare in the roots of plants and also whose spores we have never observed in the field. This is not an uncommon experience, and strongly suggests that the widely used laboratory mycorrhizal cultures represent a group of species that may be wholly unrepresentative of the group as a whole. Even where cultures of a mycorrhizal fungus exist, it is often the case that isolation has only been achieved on few occasions, suggesting that the genotypes in culture may be distinct from those that are encountered in field samples and yet are hard to culture (C. Walker, personal communication).

Multiple colonization of the roots of a single plant by several different mycorrhizal fungi is common. Clapp *et al.* (1995), using molecular techniques, showed that fungi belonging to three different genera (*Glomus*, *Acaulospora* and *Scutellospora*) were present in the roots of single plants of bluebell *Hyacinthoides non-scripta*, in Pretty Wood. In the same wood, Merryweather and Fitter (1998a) demonstrated that there were distinct morphotypes, identifiable again to the same three genera, in single bluebell roots, and that two distinct morphotypes could physically coexist, for example in the inner and outer cortex of the same piece of root. There was also a pronounced temporal and spatial pattern in the distribution of the fungi in the roots, which was identifiable by both morphological (Merryweather & Fitter 1998b) and molecular (Helgason *et al.* 1999) approaches. Strikingly, both techniques were able to identify seven taxa in the roots, three in the genus *Glomus*, three in *Acaulospora* and one in *Scutellospora*.

These multiple infections and the high degree of non-specificity that arbuscular mycorrhizal symbioses apparently exhibit pose a challenge. Since all the fungal symbionts cannot be identical in their ability to provide phosphate to the host plant and in their demand for carbon from the plant, it might be assumed that there would be one 'best' partner among them, from the plant standpoint. Selection should therefore act to increase the probability of that fungus–plant pair and decrease colonization by the other fungi. In other words, how do ineffective symbionts persist, a question that Douglas (1998) has recently addressed. She offers three solutions: first, that 'best friend' symbioses are open to invasion by ineffective symbionts, presumably because the plant may be unable to discriminate among them. These invaders are effectively cheats, a possibility envisaged by Smith and Smith (1996). Detecting cheats is empirically very difficult, however, as shown below.

Second, there is the possibility that environmental conditions determine symbiont effectiveness. We know so little about the basic biology and ecology of most soil microbes, and especially the hard-to-culture ones such as arbuscular mycorrhizal fungi, that this must remain merely a highly likely explanation. Again, it is one that is hard to apply experimentally. Finally, Douglas (1998) suggests that many symbionts may have restricted distributions (in time or space) and therefore the probability of host and symbiont meeting is small, leading to variation in the pattern of colonization. This situation would be in marked contrast to that seen for freshwater protists (Finlay *et al.* 1999).

A further possibility is that the guild of symbionts is multifunctional (Newsham *et al.* 1995a). Although phosphate transport is usually given as the principal benefit that plants get from the arbuscular mycorrhizal symbiosis, there is clear evidence that other functions, notably protection from pathogenic fungi, can be more important in some cases. Newsham *et al.* (1995b) showed that the annual grass *Vulpia ciliata* did not experience greater phosphate uptake rates when mycorrhizal, but was protected from attack by *Fusarium oxysporum*, which could reduce fecundity by up to 40% (Fig. 5.5). Other benefits that have been demonstrated include drought resistance, micronutrient uptake and altered palatability (see Newsham *et al.* 1995a for review), and others may yet be found. Multifunctionality has large implications for the maintenance of diversity in plant–fungus symbioses.

All of these potential explanations for the persistence of multiple symbioses such as are found in arbuscular mycorrhizas are hard to demonstrate experimentally. It is relatively simple to demonstrate that they could be true, or even that they are in specific circumstances, but generalizing from there to the symbiosis as a whole is more tricky. Demonstrating cheating is especially hard. A cheat is a fungus that obtains benefits from the plant (i.e. a supply of carbon) without offering the plant benefits. In multiple colonizations, a fungus that takes relatively more carbon than it gives phosphorus compared to another fungus might also be regarded as cheating, and being on the evolutionary road to a full cheat. How can one detect a cheat? It is possible to demonstrate at a given time, under given conditions for a particular host, that a fungus offers no benefit. However, if Douglas's (1998) second explanation is true, there may be other conditions or other times in the life cycle when that fungus did offer benefit. If benefit is defined in terms of phosphorus transport, then the problem of multifunctionality also arises: benefit must then be measured, as it always should be, in terms of fitness, for example as a change in fecundity (Shumway & Koide 1994; Newsham *et al.* 1995b).

The potential sensitivity of the fungal symbionts to environmental conditions poses a special problem, because we know so little about their basic biology and ecology. When it was held that arbuscular mycorrhizal fungi had very low diversity and displayed no specificity, Law and Lewis (1983) explained this on the grounds

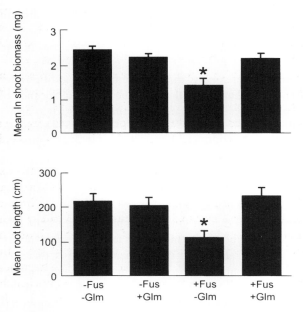

Figure 5.5 The effects of a factorial combination of *Fusarium oxysporum* (Fus) and an AM fungus, *Glomus* sp. (Glm), on root length and shoot biomass of *Vulpia ciliata* ssp. *ambigua* plants. Seedlings were grown in the laboratory, then transplanted into a natural population of *V. ciliata* ssp. *ambigua* and sampled after 62 days. Bars show standard error, and asterisks denote a significant difference ($P < 0.05$). (Reproduced with permission from Newsham et al. 1995b.)

that the fungus occupied a stable environment in the root, in comparison to ectomycorrhizal fungi, for example, which live outside the root. However, all mycorrhizal fungi have an extensive extra-radical mycelium that encounters a variable environment. Does this mean that arbuscular mycorrhizal fungi, the Glomales, show intertaxon differentiation in response to environmental factors, just as their plant partners do? Certainly not all fungi are found in all soils: acid soils have fungal taxa distinct from base-rich soils (Wang et al. 1993). There are some biogeographic patterns: *Gigaspora* species have never been recorded from Europe. There are seasonal patterns: in Pretty Wood, our own study site, *Scutellospora dipurpurescens* is most active in winter, while *Glomus* and *Acaulospora* spp. develop in spring and summer. There is an urgent need to undertake basic research on the biology of these fungi: how do different species respond to a range of environmental factors? Only with such information can we understand the potential for niche differentiation among the extra-radical mycelia to explain local diversity patterns in mycorrhizal communities.

To further complicate the situation, the genetic status of arbuscular mycorrhizal fungi remains obscure. As in all true fungi, the nuclei are haploid. The spores are produced asexually and are very large, up to 1 mm in diameter, containing thousands of nuclei (Burggraaf & Beringer 1989). These spores contain multiple sequences for some genes (Sanders *et al.* 1995; Lloyd-MacGilp *et al.* 1996), and such cultures may be heterokaryotic (Zézé *et al.* 1997). Giovannetti *et al.* (1999) have recently demonstrated fusion between hyphae of *Glomus mosseae*, and this could obviously represent a mechanism of genetic exchange. Feldmann (1998) recently showed that effective and ineffective strains of a single fungal species, *Glomus etunicatum*, could be selected within three multiplication cycles from a single spore. What other genetic distinctions can be selected for in this way, including possible host preference, remains unknown.

Is there then specificity? If there is, then the current list of 154 described taxa must be a gross underestimate, and indeed most mycorrhizal taxonomists and field workers would agree that this was so. Hoeksema (1999) offered three suggestions as to why arbuscular mycorrhizal fungi might show less specificity than ectomycorrhizal fungi, but only offered support to one of these: that the host plants of arbuscular mycorrhizal fungi are less predictable in time and space than those of ectomycorrhizal fungi. There is little evidence to support this hypothesis, especially since arbuscular mycorrhizal associations are so much more ancient than ectomycorrhizal ones (Simon *et al.* 1993; Fitter & Moyersoen 1996). Indeed it may well be unnecessary to look for an explanation, if the phenomenon turns out to be illusory, and the rarer, hard-to-culture specific Glomalean fungi are discovered.

Linkages

Although there may be a degree of specificity in mycorrhizal associations, there are indubitably large numbers of mycorrhizal fungi that are not selective as to host plant. Indeed, in types of mycorrhiza other than arbuscular mycorrhizas, it is now apparent that the same fungus can form different types of mycorrhiza with the roots of distinct host plant species. For example, one morphotype of ectomycorrhiza on the roots of *Picea abies* is formed by a close relative of *Hymenoscyphus ericae*, the fungus that forms the morphologically totally different ericoid mycorrhiza in the roots of Ericales (Vrålstad *et al.* 2000). Quite distinct hosts (a spruce tree and a heather) can therefore be linked by a single mycelium. Equally, those achlorophyllous and therefore non-photosynthetic plants that are not parasitic directly on other plants, are all obligately mycorrhizal. They fall into at least six families, indicating that the syndrome has evolved repeatedly (Leake 1994). In most cases the fungal partner is mycorrhizal with other plants, and the achlorophyllous plant is an apparent epiparasite on that normal mycorrhiza (McKendrick *et al.* 2000). In all these cases therefore there is linkage between unrelated plants in a community and net carbon flow from a chlorophyllous to the achlorophyllous plant.

More generally, however, it has been known for some time that such linkages may occur in all types of mycorrhizal association, and indeed arbuscular fungal hyphae have been observed running from roots of the plants of one species to another (Newman *et al.* 1994). Several workers demonstrated that if $^{14}CO_2$ were supplied to one plant, that ^{14}C could be detected in the roots, and in some cases the shoots of neighbouring plants (reviewed by Newman 1988). However, all these studies suffered from the fact that it was never possible to quantify the amount of carbon transferred, because the specific activity of the source plant was not fixed. There was also no way of determining whether this was genuine carbon transfer, or merely one side of a reciprocal movement of carbon among the members of a network linked by mycorrhizal hyphae. Even so, much speculation was engendered by these findings, to the effect that they implied that a traditional view of plant communities as being dominated by antagonistic interactions such as competition, should be replaced by one in which mutually supportive interactions were the norm. It was also suggested (Read *et al.* 1985) that the survival of seedlings might be promoted by the transfer of carbon to them from the 'universal mycelium'.

These speculations were given much weight by an important paper by Simard *et al.* (1997). They used a novel dual-labelling technique to measure reciprocal carbon transport between two ectomycorrhizal species (*Pseudotsuga menziesii* and *Betula papyrifera*) in which one plant was fed $^{14}CO_2$ and the other $^{13}CO_2$. The respective labels could then be measured in the partner. They also incorporated a control plant, *Thuja plicata*, which forms arbuscular mycorrhizas and therefore could not enter the common mycorrhizal network of the other two species. Their results demonstrated that carbon moved in both directions. In each case transferred carbon entered the shoots of the receiver, so that there was no doubt that it was transferred from fungus to plant, which is of course the opposite direction to that normally expected in mycorrhizas.

However, there are important caveats to be noted (Robinson & Fitter 1999). First, their results show that transfer from *B. papyrifera* to *P. menziesii* was much greater than in the other direction, which implies that the benefits of the association to the plant species may have been very unequal. Second, and critically, they did not unequivocally show that carbon moved through the mycorrhizal network although that has clearly been demonstrated in other situations. Strikingly carbon transport to the arbuscular mycorrhizal *T. plicata* was nearly 20% of that between *P. menziesii* and *B. papyrifera*. This carbon cannot have moved through the network and must have been lost to soil and then picked up by the roots or other fungal hyphae. They were not, therefore able to eliminate the possibility that the ectomycorrhizal fungi were merely effective scavengers of root exudates, which they are known to be (Finlay *et al.* 1992).

This point is of key importance because Simard *et al.* (1997) showed carbon transfer to the shoots of receiver plants, implying a reversal of the normal direction of carbon movement in mycorrhizas. Although this reverse transfer obviously happens in some mycorrhizas, for example those involving achlorophyllous plants, if it occurs more widely that has enormous implications for the determination of costs and benefits of mycorrhizas and their consequent impacts on plant performance and community structure. In arbuscular mycorrhizas, quantitatively significant carbon transfer to shoots of receiver plants has never been shown, and the small amounts found there almost certainly derive from photosynthetically re-fixed CO_2 respired by roots and soil microbes. Where ^{13}C has been used (which can be used to quantify transfer), all carbon remained in the roots. Fitter *et al.* (1998) showed that the extent of transfer from one plant to the roots of another was a positive function of fungal vesicle (storage structures) abundance and a negative function of hyphal abundance in the roots. They interpreted this from a 'mycocentric' viewpoint: transfer represents the normal behaviour of a mycelial fungus moving resources to parts of the mycelium where the storage function is predominant from those where carbon acquisition is occurring. The latter are likely to be young colonization units in roots where hyphae are growing between the plant cells. They also attempted to force transfer to shoots by defoliating the receiver plants and measuring ^{13}C in the regrowth: there was none. Why therefore did Simard *et al.* (1997) find transfer to shoots? There are two possibilities: first, some of the transferred carbon may have been taken up directly by the roots (as that acquired by *Thuja* probably was), in which case transfer is unsurprising. Secondly, the carbon may have been transferred as low molecular weight nitrogen-containing compounds, a phenomenon that is important in ectomycorrhizas (Chalot & Brun 1998) but as yet unknown in arbuscular mycorrhizas. In either case, it is apparent that carbon transport from plant to plant by arbuscular mycorrhizal fungal networks, which colonize two-thirds of land plants, is unlikely to be an important mechanism.

This mycocentric view of the mycorrhizal networks is perhaps less appealing to plant ecologists than one which sees them as passive channels through which plants can direct resources to vulnerable members of the community, but it has the merit of recognizing the fungi as organisms in their own right and as foci for natural selection. However, the complexity of having to consider two sets of interacting and interlinked organisms offers exciting dynamics.

Mycorrhizal networks and plant community structure

Although we are far from understanding the details of the linkages that mycorrhizal mycelia make in plant communities, it is apparent most plants do enter these common mycelial networks. It has been suggested that such networks will radically alter inter-plant interactions. For example, transfer of carbon from adult plants could ensure the survival of seedlings in an otherwise competitively harsh environment. The evidence discussed above, however, suggests that such an outcome is unlikely to occur, even if it were evolutionarily plausible. An alternative proposal is that seedlings that tap into the 'universal mycelium' can obtain a free symbiosis; in other words, they would obtain the benefits of mycorrhizal symbiosis (e.g. phosphorus uptake) without the costs because the fungus would obtain its carbon elsewhere. This is superficially a more plausible hypothesis, since it is clear that the fungus does move carbon around its mycelium and so gives the appearance of transferring carbon between plants. However, there is no evidence that the transmission of benefits to the host (e.g. phosphorus transfer) is regulated in this manner, and the concept assumes that the fungus's behaviour is determined by the demands of the plant rather than its own.

More importantly, networks are likely to have a profound influence on the feedbacks between plants and soil microbes. Bever (1999; Bever et al. 1997) has developed a model of plant community structure which explicitly incorporates the role of soil microbes. Initially, Bever et al. (1997) concentrated on the general community of microbes in soil surrounding plants, and produced a spatially explicit approach in which these rhizosphere microbes favoured either the existing plant species (producing positive feedback) or a different species (negative feedback). Experimentally, they found that negative feedback was more common. Bever (1999) developed this approach to cover mutualisms, in a situation where there were multiple symbionts. Even though all the interactions should be positive, it is possible to generate negative feedback if, for example, plant species A promotes the growth of fungus X more than of fungus Y, while fungus X promotes the growth of plant B more than plant A (Fig. 5.6). In other words, it is relative fitness benefit that matters. Depending on the relative values of the four possible interactions (in this 2 + 2 situation), several potential outcomes emerge. One possibility is specificity: if A and X promote each other, as do B and Y, then both combinations should survive, but only those.

If the plants, however, are connected to a number of distinct mycelial networks, the outcomes will be more complex. If all the plants connected to a particular mycelium are providing it with carbon, then they will in effect form a guild that all promote the performance of that fungus. If the fungi themselves are differentiating in a habitat with respect to environmental factors, as seems likely, then these guilds will themselves reflect the same environmental pattern, enhancing niche differentiation and promoting diversity. However, even within a guild of plants linked to a

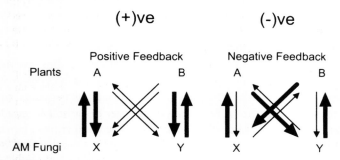

Figure 5.6 Feedback due to changes in composition of the community of soil mutualists. The direction of benefit delivered between two plant species, A and B, and their fungal mutualists, X and Y, are indicated by the arrows, with the thickness of the arrow indicating the magnitude of benefit. Plant A can increase the frequency of fungus X relative to fungus Y, which will then increase the rate of growth of plant A relative to plant B (resulting in positive feedback). Alternatively, plant A can decrease the frequency of fungus X relative to fungus Y, which will then decrease the rate of growth of plant A relative to plant B (resulting in negative feedback). (Reproduced with permission from Bever et al. 1997.)

common mycelium, it is unlikely that all obtain the same degree of benefit; indeed, they may be obtaining qualitatively distinct benefits from the same mycelium. To the plants, therefore, the common mycelial network is a club with variable subscription fee and a range of potential membership benefits; to the fungus, the plants are the potential club members whose subscriptions keep the club afloat.

Acknowledgements

I am grateful to the numerous colleagues who were responsible for much of the work described here, to the Natural Environment Research Council and the Biotechnology and Biological Sciences Research Council for funding, to James Merryweather for compiling the data in Table 5.1, and to Angela Hodge and Peter Young for valuable comments on a draft.

References

Albrecht, C., Geurts, R. & Bisseling, T. (1999) Legume nodulation and mycorrhizae formation: two extremes in host specificity meet. *EMBO Journal* **18**, 281–288.

Bever, J.D. (1999) Dynamics within mutualism and the maintenance of diversity: inference from a model of interguild frequency dependence. *Ecology Letters* **2**, 52–61.

Bever, J.D., Morton, J.B., Antonovics, J. & Schultz, P.A. (1996) Host-dependent sporulation and species diversity of arbuscular mycorrhizal fungi in a mown grassland. *Journal of Ecology* **84**, 71–82.

Bever, J.D., Westover, K.M. & Antonovics, J. (1997) Incorporating the soil community into plant population dynamics: the utility of the feedback approach. *Journal of Ecology* **85**, 561–573.

Blaszkowski, J. (1989) The occurrence of Endogonaceae in Poland. *Agriculture, Ecosystem and Environment* **29**, 45–50.

Bloemers, G.F., Hodda, M., Lambshead, P.J.D., Lawton, J.H. & Wanless, F.R. (1997) The effects of forest disturbance on diversity of tropical soil nematodes. *Oecologia* **111**, 575–582.

Burdon, J.J. (1987). *Diseases and Plant Population Biology*. Cambridge University Press, New York, MA.

Burggraaf, A.J.P. & Beringer, J.E. (1989) Absence of nuclear DNA synthesis in vesicular-arbuscular mycorrhizal fungi during *in vitro* development. *New Phytologist* **111**, 25–33.

Chalot, M. & Brun, A. (1998) Physiology of organic nitrogen acquisition by ectomycorrhizal fungi and ectomycorrhizas. *FEMS Microbiology Reviews* **22**, 21–44.

Chesson, P.L. (1985) Coexistence of competitors in spatially and temporally varying environments: a look at the combined effects of different soils of variability. *Theoretical Population Biology* **28**, 263–287.

Clapp, J.P., Young, J.P.W., Merryweather, J.W. & Fitter, A.H. (1995) Diversity of fungal symbionts in arbuscular mycorrhizas from a natural community. *New Phytologist* **130**, 259–265.

Cuenca, G. & Meneses, E. (1996) Diversity patterns of arbuscular mycorrhizal fungi associated with cacao in Venezuela. *Plant and Soil* **183**, 315–322.

Cuenca, G., De Andrade, Z. & Escalante, G. (1998) Diversity of glomalean spores from natural, disturbed and revegetated communities growing on nutrient-poor tropical soils. *Soil Biology and Biochemistry* **30**, 711–719.

Douglas, A.E. (1998) Host benefit and the evolution of specialization in symbiosis. *Heredity* **81**, 599–603.

Eom, A., Hartnett, D.C. & Wilson, G.W.T. (2000) Host plant species effects on arbuscular mycorrhizal fungal communities in tallgrass prairie. *Oecologia* **122**, 435–444.

Feldmann, F. (1998) The strain-inherent variability of arbuscular mycorrhizal effectiveness: II.

Effectiveness of single spores. *Symbiosis* **25**, 131–143.

Finlay, B.J. & Clarke, K.J. (1999) Apparent global ubiquity of species in the protist genus Paraphysomonas. *Protist* **150**, 419–430.

Finlay, R.D., Frostegard, A. & Sonnerfeldt, A.M. (1992) Utilization of organic and inorganic nitrogen sources by ectomycorrhizal fungi in pure culture and in symbiosis with *Pinus contorta* Dougl. ex Loud. *New Phytologist* **120**, 105–115.

Finlay, B.J., Esteban, G.F., Olmo, J.L. & Tyler, P.A. (1999) Global distribution of free-living microbial species. *Ecography* **22**, 138–144.

Fitter, A.H. (1987) Spatial and temporal separation of activity in plant communities: pre-requisite or consequence of co-existence? In: *Organisation of Communities: Past and Present* (eds P.S. Giller & J. Gee), pp. 119–139. 26th Symposium of the British Ecological Society. Blackwell Scientific Publications, Oxford.

Fitter, A.H. & Moyersoen, B. (1996) Evolutionary trends in root–microbe symbioses. *Philosophical Transactions of the Royal Society of London, Series B* **351**, 1367–1375.

Fitter, A.H., Graves, J.D., Watkins, N.K., Robinson, D. & Scrimgeour, C.M. (1998) Carbon transfer between plants and its control in networks of arbuscular mycorrhizas. *Functional Ecology* **12**, 406–412.

Giovannetti, M., Azzolini, D. & Citernesi, A.S. (1999) Anastomosis formation and nuclear and protoplasmic exchange in arbuscular mycorrhizal fungi. *Applied and Environmental Microbiology* **65**, 5571–5575.

Grime, J.P., MacKey, J.M., Hillier, S.H. & Read, D.J. (1987) Floristic diversity in a model system using experimental microcosms. *Nature* **328**, 420–422.

Guadarrama, P. & Alvarez Sanchez, F.J. (1999) Abundance of arbuscular mycorrhizal fungi in different environments in a tropical rain forest. *Mycorrhiza* **8**, 267–270.

Hamel, C., Dalpé, Y., Lapierre, C., Simard, R.R. & Smith, D.L. (1994) Composition of the vesicular-arbuscular mycorrhizal fungi population in an old meadow as affected by pH, phosphorus and soil disturbance. *Agriculture, Ecosystems and Environment* **49**, 223–231.

Hartnett, D.C. & Wilson, G.W.T. (1999) Mycorrhizae influence plant community structure and diversity in tallgrass prairie. *Ecology* **80**, 1187–1195.

Helgason, T., Daniell, T.J., Husband, R., Fitter, A.H. & Young, J.P.W. (1998) Ploughing up the wood-wide web? *Nature* **394**, 431–432.

Helgason, T., Fitter, A.H. & Young, J.P.W. (1999) Molecular diversity of colonising *Hyacinthoides non-scripta* (bluebell) in a semi-natural woodland. *Molecular Ecology* **8**, 659–666.

Hettrick, B.A.D. & Bloom, J. (1983) Vesicular-arbuscular mycorrhizal fungi associated with native tall grass prairie and cultivated winter wheat. *Canadian Journal of Botany* **61**, 2140–2146.

Hoeksema, J.D. (1999) Investigating the disparity in host specificity between AM and EM fungi: lessons from theory and better-studied systems. *Oikos* **84**, 327–332.

Holah, J.C. & Alexander, H.M. (1999) Soil pathogenic fungi have the potential to affect the co-existence of two tallgrass prairie species. *Journal of Ecology* **87**, 598–608.

Janzen, D.H. (1970) Herbivores and the number of tree species in tropical forests. *American Naturalist* **104**, 501–508.

Johnson, N.C., Zak, D.R., Tilman, D. & Pfleger, F.L. (1991) Dynamics of vesicular-arbuscular mycorrhizae during old field succession. *Oecologia* **86**, 349–358.

Kennedy, A.C. (1999) Bacterial diversity in agroecosystems. *Agriculture, Ecosystems and Environment* **74** (1–3), 65–76.

Koske, R.E. (1987) Distribution of VA mycorrhizal fungi along a latitudinal temperature gradient. *Mycologia* **79**, 55–68.

Koske, R.E. & Gemma, J.N. (1997) Mycorrhizae and succession in plantings of beachgrass in sand dunes. *American Journal of Botany* **84**, 118–130.

Koske, R.E. & Halvorson, W.L. (1981) Ecological studies of vesicular-arbuscular mycorrhizae in a barrier sand dune. *Canadian Journal of Botany* **59**, 1413–1422.

Koske, R.E. & Tews, L.L. (1987) Vesicular-arbuscular mycorrhizal fungi of Wisconsin sandy soils. *Mycologia* **79**, 901–905.

Koske, R.E., Gemma, J.N. & Jackson, N. (1997) Mycorrhizal fungi associated with three species of turfgrass. *Canadian Journal of Botany* **75**, 320–332.

Law, R. & Lewis, D.H. (1983) Biotic environments

and the maintenance of sex—some evidence from mutualistic symbioses. *Biological Journal of the Linnean Society* **20**, 249–276.

Leake, J.R. (1994) The biology of mycoheterotrophic saprophytic plants. *New Phytologist* **127**, 171–216.

Lloyd-MacGilp, S.A., Chambers, S.M., Dodd, J.C., Fitter, A.H., Walker, C. & Young, J.P.W. (1996) Diversity of the ribosomal internal transcribed spacers within and among isolates of *Glomus mosseae* and related fungi. *New Phytologist* **133**, 103–112.

Massicotte, H.B., Molina, R., Tackaberry, L.E., Smith, J.E. & Amaranthus, M.P. (1999) Diversity and host specificity of ectomycorrhizal fungi retrieved from three adjacent forest sites by five host species. *Canadian Journal of Botany* **77**, 1053–1076.

McKendrick, S.L., Leake, J.R., Taylor, D.L. & Read, D.J. (2000) Symbiotic germination and development of myco-heterotrophic plants in nature: ontogeny of *Corallorhiza trifida* and characterization of its mycorrhizal fungi. *New Phytologist* **145**, 523–537.

Merryweather, J. & Fitter, A.H. (1998a) The arbuscular mycorrhizal fungi of *Hyacinthoides non-scripta* I. Diversity of fungal taxa. *New Phytologist* **138**, 117–129.

Merryweather, J. & Fitter, A.H. (1998b) The arbuscular mycorrhizal fungi of *Hyacinthoides non-scripta* II. Seasonal and spatial patters of fungal populations. *New Phytologist* **138**, 131–142.

Miller, S.P. & Bever, J.D. (1999) Distribution of arbuscular mycorrhizal fungi in stands of the wetland grass *Panicum hemitomon* along a wide hydrologic gradient. *Oecologia* **119**, 586–592.

Miller, D.D., Domoto, P.A. & Walker, C. (1985) Mycorrhizal fungi at eighteen apple rootstock plantings in the United States. *New Phytologist* **100**, 379–391.

Mills, K.E. & Bever, J.D. (1998) Maintenance of diversity within plant communities: Soil pathogens as agents of negative feedback. *Ecology* **79**, 1595–1601.

Morton, J.B., Bentivenga, S.P. & Bever, J.D. (1995) Discovery, measurement and interpretation of diversity in symbiotic endomycorrhizal fungi (Glomales, Zygomycetes). *Canadian Journal of Botany* **73**, 25–32.

Newman, E.I. (1988) Mycorrhizal links between plants: their functioning and ecological significance. *Advance in Ecological Research* **18**, 243–271.

Newman, E.I., Devoy, C.L.N., Easen, N.J. & Fowles, K.J. (1994) Plant species that can be linked by VA mycorrhizal fungi. *New Phytologist* **126**, 691–693.

Newsham, K.K., Fitter, A.H. & Watkinson, A.R. (1995a) Multi-functionality and biodiversity in arbuscular mycorrhizas. *Trends in Ecology and Evolution* **10**, 407–411.

Newsham, K.K., Fitter, A.H. & Watkinson, A.R. (1995b) Arbuscular mycorrhizas protect an annual grass from root pathogenic fungi in the field. *Journal of Ecology* **83**, 991–1000.

Newsham, K.K., Watkinson, A.R., West, A.H. & Fitter, A.H. (1995c) Symbiotic fungi determine plant community structure: changes in a lichen-rich community induced by fungicide application. *Functional Ecology* **9**, 442–447.

Pacala, S.W. & Crawley, M.J. (1992) Herbivores and plant diversity. *American Naturalist* **140**, 243–260.

Packer, A. & Clay, K. (2000) Soil pathogens and spatial patterns of seedling mortality in a temperate tree. *Nature* **404**, 278–281.

Read, D.J., Francis, R. & Finlay, R.D. (1985) Mycorrhizal mycelia and nutrient cycling in plant communities. In: *Ecological Interactions in the Soil* (eds A.H. Fitter, D.J. Read & D. Atkinson), pp. 103–130. Blackwell Scientific Publications, Oxford.

Robinson, D. & Fitter, A.H. (1999) The magnitude and control of carbon transfer between plants linked by a common mycorrhizal network. *Journal of Experimental Botany* **50**, 9–13.

Saito, M. & Vargas, R. (1991) Vesicular-arbuscular mycorrhizal fungi in some humus rich Ando soils of Japan. *Soil Micro-organisms* **38**, 3–15.

Sanders, I.R. & Fitter, A.H. (1992) Evidence for differential responses between host-fungus combinations of vesicular-arbuscular mycorrhizas from a grassland. *Mycological Research* **96**, 415–419.

Sanders, I.R., Alt, M., Groppe, K., Boller, T. & Wiemken, A. (1995) Identification of ribosomal DNA polymorphisms among and within spores of Glomales: application to studies on genetic diversity of arbuscular mycorrhizal fungal communities. *New Phytologist* **130**, 419–427.

Shumway, D.L. & Koide, R.T. (1994) Reproductive responses to mycorrhizal colonisation of *Abutilon theophrasti* Medic. plants grown for two generations in the field. *New Phytologist* **128**, 219–224.

Simard, S.W., Perry, D.A., Jones, M.D., Myrold, D.D., Durall, D.M. & Molina, R. (1997) Net transfer of carbon between ectomycorrhizal tree species in the field. *Nature* **388**, 579–582.

Simon, L., Bousquet, J., Levesque, R.C. & Lalonde, M. (1993) Origin at diversification of endomycorrhizal fungi and coincidence with vascular plants. *Nature* **363**, 67–69.

Smith, F.A. & Smith, S.E. (1996) Mutualism and parasitism: biodiversity in function and structure in the 'arbuscular' (VA) mycorrhizal symbiosis. *Advances in Botanical Research* **22**, 1–43.

Stürmer, S.L. & Bellei, M.M. (1994) Composition and seasonal variation of spore populations of arbuscular mycorrhizal fungi in sand dunes on the island of Santa Catarina, Brazil. *Canadian Journal of Botany* **72**, 359–363.

Stutz, J.C. & Morton, J.B. (1996) Sucessive pot cultures reveal high species richness of arbuscular endomycorrhizal fungi in arid ecosystems. *Journal of Botany* **74**, 1883–1889.

Tews, L.L. & Koske, R.E. (1986) Toward a sampling strategy for vesicular-arbuscular mycorrhizas. *Transactions of the British Mycological Society* **87**, 353–358.

Thomas, A.S. (1960) Changes in vegetation since the advent of myxomatosis. *Journal of Ecology* **48**, 287–306.

Tiedje, J.M., Assuming-Brempong, S., Nusslein, K., Marsh, T.L. & Flynn, S.J. (1999) Opening the black box of soil microbial diversity. *Applied Soil Ecology* **13**, 109–122.

Torsvik, V., Sorheim, R. & Goksøyr, J. (1996) Total bacterial diversity in soil and sediment communities: a review. *Journal of Industrial Microbiology* **17**, 170–178.

Trappe, J.M. (1987) Phylogenic and ecological aspects of mycotrophy in the angiosperms from an evolutionary standpoint. In: *Ecophysiology of VA Mycorrhizal Plants* (ed. G.R. Safir), pp. 5–25. CRC Press, Boca Raton, FL.

Van der Heijden, M.G.A., Klironomas, J.N., Ursic, M. *et al.* (1998) Mycorrhizal fungal diversity determines plant biodiversity, ecosystem variability and productivity. *Nature* **396**, 69–72.

Van der Heijden, M.G.A., Klironomos, J.N., Ursic, M. *et al.* (1999) 'Sampling effect', a problem in biodiversity manipulation? A reply to David A. Wardle. *Oikos* **87**, 408–410.

van der Putten, W.H., Van Dijk, C. & Peters, B.A.M. (1993) Plant-specific soil-borne diseases contribute to succession in foredune vegetation. *Nature* **362**, 53–56.

Vrålstad, T., Fossheim, T. & Schumacher, T. (2000) *Piceirhiza bicolorata*—the ectomycorrhizal expression of the *Hymenoscyphus ericae* aggregate? *New Phytologist* **145**, 549–563.

Walker, C., Mize, C.W. & McNabb, H.S. (1982) Populations of endogonaceous fungi at two locations in central Iowa. *Canadian Journal of Botany* **60**, 2518–2529.

Wang, G.M., Stribley, D.P., Tinker, P.B. & Walker, C. (1993) Effects of pH on arbuscular mycorrhiza. I. Field observations on the long-term liming experiments at Rothamsted and Woburn. *New Phytologist* **124**, 465–472.

Wardle, D.A. (1999) Is "sampling effect" a problem for experiments investigating biodiversity–ecosystem function relationships? *Oikos* **87**, 403–407.

Weste, G. & Ashton, D.H. (1994) Regeneration and survival of indigenous dry sclerophyll species in the Brisbane Ranges, Victoria, after *Phytophthora cinnamomi* epidemic. *Australian Journal of Botany* **42**, 239–253.

Westover, K.M., Kennedy, A.C. & Kelly, S.E. (1997) Patterns of rhizosphere microbial community structure associated with co-occurring plant species. *Journal of Ecology* **85**, 863–873.

Zézé, A., Sulistyowati, E., Ophel-Keller, K., Barker, S. & Smith, S.E. (1997) Intersporal genetic variation of *Gigaspora margarita*, a vesicular-arbuscular mycorrhizal fungus revealed by M13 minisatellite-primed PCR. *Applied and Environmental Microbiology* **63**, 676–678.

Chapter 6
Global change and the linkages between physiological ecology and ecosystem ecology

J. R. Ehleringer,* T. E. Cerling† and L. B. Flanagan‡

A physiological basis for many ecosystem-scale patterns

Plant physiological ecology has focused historically on describing the basis of adaptation between organisms and their environment (Mooney & Chabot 1985; Mooney *et al.* 1987; Larcher 1995; Lambers *et al.* 1998). While the roots of physiological ecology are in both plant physiology and plant geography, the last several decades have seen the emergence of strong linkages between physiological ecology and ecosystem ecology, and also in the globalization of ecological thought (Mooney & Drake 1986; Ehleringer & Field 1993; Mooney 1998; Mooney *et al.* 1999; Mooney & Hobbs 2000).

Not all of the controls over processes at the physiological scale are directly relevant to processes at the ecosystem scale. Whereas the fluxes of water and CO_2 between the ecosystem and the atmosphere are often predictable knowing many of the basic physiological controls over leaf-level gas exchange (Canadell *et al.* 2000), this need not be the case for all metabolic processes. Aggregation properties at higher levels can result in additional controls over fluxes that are sometimes not predictable without also knowing how elements aggregate into a forest stand or entire ecosystem (Ehleringer & Field 1993). Yet frequently by identifying the mechanistic basis of plant performance, we can gain insight into controls and components of fluxes in ecosystems, the invasibility of species into ecosystems, sensitivity of individual species and plant communities to changes in resource availability, and biotic interactions among different species.

Changes are taking place in our environment and across the face of this planet. Global changes are occurring in many ways: atmosphere, land-use, and now also climate (IPCC 1996; Mooney *et al.* 1999; Huang *et al.* 2000). Consider these undebatable changes which have accelerated since the early 1950s: human population increase, conversion of lands into cultivation, fertilizer production, atmospheric greenhouse gases, and species extinctions. Humans are dominating and changing this planet in a way that no organism has since the first microbes began to produce

---

**Department of Biology, University of Utah, 257 South 1400 East, Salt Lake City, Utah 84112, USA;*
†*Department of Geology and Geophysics, University of Utah, Salt Lake City, Utah 84112, USA;*
‡*Department of Biological Sciences, University of Lethbridge, 4401 University Drive, Lethbridge, Alberta, T1K 3M4, Canada*

oxygen 2–3 billion years ago. The consequences of these global changes are realized as loss of habitat, changes in species abundances and distributions, biological invasions, and the changes in dynamics and functioning of ecosystems. While it is easy to identify man as the driver for these changes, it is frequently less easy to predict exactly how plants, populations, and ecological systems will respond. Understanding the physiology and biochemistry of plants, however, does serve as a basis for scaling responses from tissue to whole organism to ecosystem scales (Ehleringer & Field 1993). Similarly, top-down approaches are useful in constraining our interpretations.

This chapter focuses on three global-change examples of how an understanding of physiological processes has led to new insights into controls over the functioning of terrestrial ecosystems. We focus on examples where physiological properties of the individual organisms contribute to the functioning of the ecosystem in direct ways. In the first example, a *palaeoecological* study, we examine how shifts in atmospheric CO_2 through history have influenced the abundances of different photosynthetic pathways, leading to changes in the animal species occurring across these landscapes. The second example, a *current-ecological* study, describes how measurements of the carbon isotope ratio of plant tissue and atmospheric CO_2 are used to study metabolic properties and partition CO_2 exchange processes within ecosystems and between the ocean and terrestrial ecosystems. The temporal and spatial integration that is provided by measurements of stable isotope ratios helps to improve understanding of controls on ecosystem function. In the last, a *future-climate change* example, we explore the physiological basis for the sensitivity of arid-land ecosystems to anticipated changes in the distribution of monsoonal precipitation. Here we explore the consequences of morphological and physiological differences in rooting patterns to water uptake at the ecosystem level.

Atmospheric CO_2 and C_3–C_4 ecosystems

The functioning of ecosystems in the geological past probably overlaps considerably with how they function today, except where climatic and atmospheric conditions today contrast with conditions in the past. Understanding the past and how ecosystems have changed through time may provide clues to the future. Atmospheric CO_2 is a global parameter that has varied in the past and has had an impact on the world's ecosystems. The rise in atmospheric CO_2 today is a direct consequence of fossil fuel consumption, cement production, and forest burning associated with land-use changes. Humans have played a significant role in increasing atmospheric CO_2 since the dawn of the Industrial Revolution. Over the past 420 000 years, atmospheric CO_2 had varied between approximately 180 and 280 p.p.m. (Petit *et al.* 1999). Humans and many other species evolved in a low-CO_2 world. Yet today the Earth's atmosphere is changing rapidly. Human activities now exert a significant impact on the atmosphere and thereby also on the Earth's climate system (Vitousek *et al.* 1997). With today's CO_2 levels more than 30% greater than values three centuries ago, global warming has received the greatest attention. Yet there are other direct effects of elevated CO_2 on ecosystems, not the least of which is that physiological processes, such as photosynthesis and photorespiration; the rates of these processes vary directly with atmospheric CO_2 levels.

At the biochemical level, ribulose 1,5-bisphosphate carboxylase-oxygenase (rubisco) catalyses the reduction of CO_2 with ribulose bisphosphate to form two molecules of phosphoglyceric acid, a three-carbon molecule. This initial photosynthetic reaction forms the basis of C_3 photosynthesis and the eventual production of sugars and other carbohydrates. However, rubisco also has an oxygenase activity in which O_2 is substituted for CO_2, leading to the formation of one molecule each of phosphoglyceric acid and phosphoglycolate. An eventual product of this oxygenase activity is CO_2, which leads to the oxygenase activity being referred to as photorespiration. In spite of the high selectivity for CO_2 by rubisco, oxygenase activity increases as the CO_2/O_2 ratio decreases. Oxygenase activity also increases faster with temperature increase than does carboxylase activity. As a consequence photorespiratory rates increase as atmospheric CO_2 decreases and/or habitat temperatures increase.

An evolutionary solution to the dilemma of increased photorespiration is C_4 photosynthesis, a modification in which the C_3-photosynthesis cycle is now restricted to specific cells within the leaf interior (often bundle-sheath cells). These bundle-sheath cells are now physically surrounded by mesophyll cells with high phosphoenolpyruvate (PEP)-carboxylase activity. PEP carboxylase combines PEP and CO_2 to form oxaloacetate, a four-carbon molecule, as the initial photosynthetic product. Since PEP carboxylase activity is so much higher than Rubisco activity, the [CO_2] environment of Rubisco in C_4 plants is high, resulting in a high CO_2/O_2 ratio and the virtual elimination of photorespiration. The [CO_2] inside C_4 leaves at the site of Rubisco activity is thought to be in excess of 2000 p.p.m. (Sage & Monson 1999), which makes the CO_2 concentration similar to what C_3 plants would have experienced back in the Cretaceous! Yet the C_4 cycle is more energetically expensive, because it requires additional adenosine triphosphate (ATP) to regenerate PEP from pyruvate in order to maintain the cycle. Modelling results indicate that the intrinsic differences in the C_3/C_4 photosynthetic pathways have implications that scale to the ecosystem level, influencing the relative distributions of C_3 and C_4 plants, and also influencing the mammalian herbivores that feed on these plants.

Ehleringer *et al.* (1997) modelled the trade-offs between photorespiratory carbon losses in C_3 photosynthesis vs. increased energetic costs in C_4 plants as a function of the environmental [CO_2] and temperature conditions. That analysis, known as the quantum-yield model, predicted clear boundaries in the environmental conditions favouring the presence of C_3 vs. C_4 photosynthesis (Fig. 6.1). Atmospheric CO_2 concentration is predicted to have a strong influence on the relative competitive abilities of C_4 plants. For the modern atmosphere of about 365 p.p.m. CO_2 the C_3/C_4 crossover temperature for conditions favourable to C_4 photosynthesis was about 22–24°C. This predicted C_3/C_4 crossover is consistent with the global distributions of grasses and sedges today (Ehleringer *et al.* 1997; Epstein

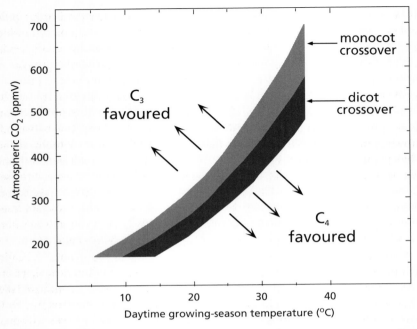

Figure 6.1 The combinations of atmospheric CO_2 and growing season temperatures that favour C_3 plants vs. C_4 plants. The light grey region represents the transition zone for C_3/C_4 monocot species and black region represent the transition zone for C_3/C_4 dicots. (Model is based on Ehleringer et al. 1997.)

et al. 1997; Tieszen et al. 1997; Sage & Monson 1999). A more detailed discussion of the evolution of C_4 photosynthesis is found in Sage and Monson (1999).

At lower CO_2 concentrations, such as were present during full glacial conditions, the crossover temperature is much lower (about 15°C). The emerging palaeo-ecological literature is revealing that C_4-dominated ecosystems were much more expansive during the glacial periods than today (Ficken et al. 1998; Street-Perrott et al. 1998; Huang et al. 1999). On the other hand, the quantum-yield model predicts that when CO_2 concentrations exceed about 500 p.p.m. the window of favourable temperatures for C_4 photosynthesis is set very high (at about 40°C), implying that few environments of Earth would favour C_4 photosynthesis. Therefore, this modelling exercise suggests that the 'C_4-world' is possible only when atmospheric CO_2 concentrations are low.

Carbon isotope ratios can be used as indicators of the relative abundances of C_3 vs. C_4 vegetation types through geological time (Cerling & Quade 1990; Cerling et al. 1997). The difference in the initial carboxylation reaction of C_3 and C_4 photosynthesis results in a large difference in their $^{13}C/^{12}C$ ratios (Farquhar et al. 1989). C_3 plants have carbon isotope ratios ranging from about −22 to −30‰

(though rarely as low as −35‰), whereas C_4 plants have carbon isotope ratios ranging from −11 to −15‰. Carbon isotope values form the main basis by which we can understand the spread of C_4 ecosystems in the geological record. Pollen morphology is not distinct enough to distinguish between C_3 and C_4 grasses, so pollen records cannot be used to determine the relative abundance of C_4 grasses in fossil floral records. In addition, because grasses are generally found in regions that are highly oxidizing, the fossil record of grasses is extremely poor. However, animal tissues preserve information about diet choices, and so the distinction between C_3 browsers and C_4 grazers is recorded in the fossil record in the form of tooth enamel, a bioapatite phase that is very resistant to recrystallization during diagenesis. Tooth enamel is enriched by about 14‰ compared to diet, so that C_3 hyper-browsers have carbon isotope ratios about −12‰ whereas C_4 hyper-grazers have carbon isotope ratios values about 0‰ (Cerling & Harris 1999). The high dietary selectively of grazers enhances the C_4 signal. However, some animals (including grazers) eat mixtures of C_3 and C_4 plants, resulting in intermediate $\delta^{13}C$ values.

Figure 6.2 shows the isotopic distinction between C_3 and C_4 plants, indicating that modern mammals consume either mostly C_3 or C_4 plants in their diets, and that before 8 million years ago there were no mammals with a C_4-dominated diet. Therefore, before 8 million years ago the global vegetation was dominated by C_3 photosynthesis (Cerling et al. 1997). By 6 million years ago, however, mammals with C_4-dominated diets are found in Asia, Africa, North America and South America. Equids dispersed from North America into the Old World by 10.5 million years ago and have C_3-dominated diets in North America, Asia and Africa until 8 million years ago. By 6 million years ago, however, they changed to a C_4-dominated diet in Pakistan, Africa and southern North America (Fig. 6.3). Equids in Europe and northern North America maintained a C_3-dominated diet, presumably C_3 grasses. The period at the end of the Miocene was one of global faunal change, with major faunal turnover in many parts of the world (Fig. 6.4). In many regions this time had been interpreted to be a major change in flora from a forested or closed habitat to a more open habitat. We now know it to be the time when C_4 grasses first underwent global expansion and established the significant presence of C_4 photosynthesis in low to intermediate latitudes, where they currently make up 30% or more of the net primary productivity.

The C_3/C_4 quantum-yield crossover model has important implications for evolution during the last several million years. Major changes occurred in the evolution of mammals, especially grazing mammals coincident with the expansion of C_4 ecosystems. Humans evolved in this low-CO_2 world and there are indications that a C_4 diet was associated with an ancestral hominid (Sponheimer & Lee-Thorp 1999). At present, we can document that for the last 400 000 years, the Earth has been alternately in Glacial or Interglacial conditions, with characteristic CO_2 concentrations of approximately 180 and 280 p.p.m., respectively (Petit et al. 1999). Today, humankind's thirst for cheap energy has resulted in fossil fuel burning which has increased atmospheric CO_2 to levels that should influence the competitive interactions between C_3 and C_4 grasses. If the predictions of the quantum-yield model

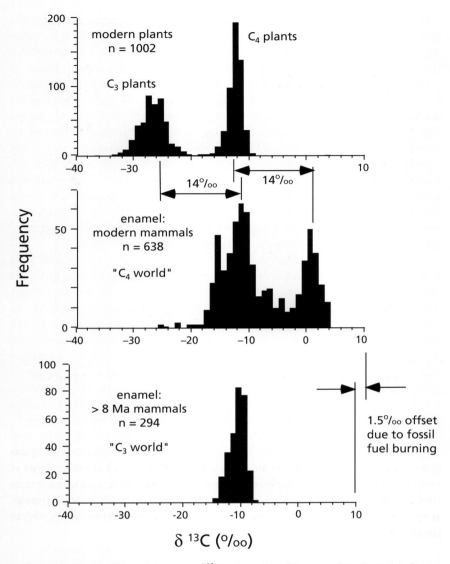

Figure 6.2 Frequency histograms of the $\delta^{13}C$ values of modern C_3 and C_4 plants (a), the tooth enamel of modern mammals (b), and the histogram of enamel from fossil tooth enamel for mammals that are older than 8 million years (c). (Adapted with permission from Cerling *et al.* 1997.)

hold up and stand the test of time, this physiological hypothesis predicts that C_4 taxa in a high-CO_2 world should eventually exhibit a decreased competitive ability relative to C_3 taxa in those areas where both photosynthetic types coexist today. Paired species comparisons may not always reflect a simple, uniform CO_2-

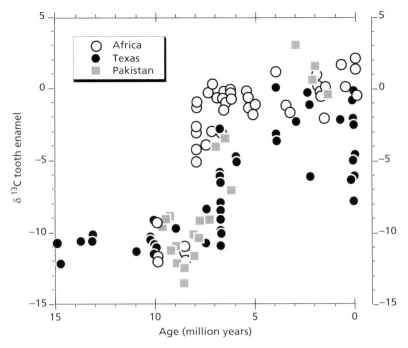

Figure 6.3 The $\delta^{13}C$ values of tooth enamel from fossil equids that grazed in grasslands in Kenya, Texas, and Pakistan between 15 million years ago and the present. (Adapted with permission from Cerling & Quade 1990.)

dependent competitive transition, because other physiological or phenological factors may come into play (Wand et al. 1999). Yet if overall the hypothesis is correct, then it is likely that both invertebrate and vertebrate herbivorous taxa in low-latitude regions will be exposed to new diets in the next century that will result in different digestibilities and therefore also influence the competitive ability of the fauna in these ecosystems.

Carbon isotopes in ecosystem and global carbon cycle studies

Linkages between ecophysiology and ecosystem ecology are important for understanding many of the global changes taking place today. Currently there are uncertainties about some specifics of the global carbon budget and the role that terrestrial ecosystems play in that budget (Fan et al. 1998; Lloyd 1999; Battle et al. 2000; Schimel et al. 2000). Atmospheric studies have been very important in illustrating the need to better understand the processes that control CO_2 exchange in terrestrial ecosystems. High-precision measurements of the concentration and stable isotope ratios of atmospheric CO_2, in combination with atmospheric

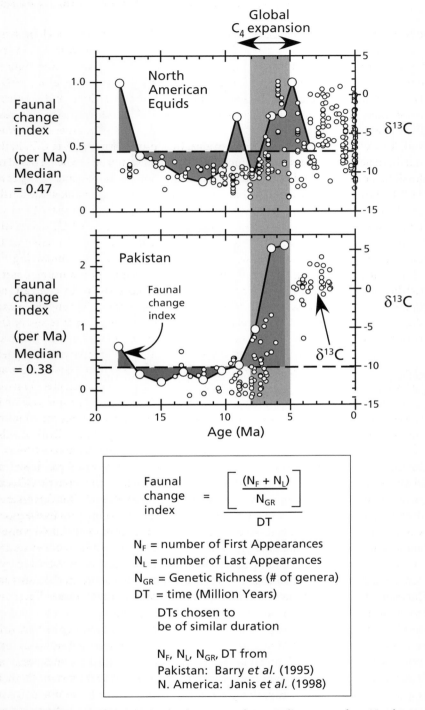

Figure 6.4 Changes in the faunal index for groups of mammalian grazers from North America and Pakistan between 20 million years ago and present.

transport models, have been used to infer CO_2 uptake and release on a global basis (Ciais *et al.* 1995; Battle *et al.* 2000). For example, there is a strong north–south gradient in the concentration of atmospheric CO_2, with higher concentrations observed in the northern hemisphere (Trolier *et al.* 1996; Fung *et al.* 1997). However, the difference in CO_2 concentration between northern and southern hemispheres is smaller than predicted, based on well-known emissions of fossil fuels. These CO_2-concentration data have been used to infer that there is a strong sink for CO_2 in northern temperate regions. In addition, atmospheric CO_2 in the northern hemisphere has higher amounts of ^{13}C than would be predicted based on fossil fuel emissions and observed CO_2 concentration measurements (Fung *et al.* 1997). These carbon-isotope data have been used to infer that the sink in the northern hemisphere is primarily associated with carbon uptake in terrestrial ecosystems (Battle *et al.* 2000). This argument is based on the fact that C_3 plants (the dominant plant type in northern temperate terrestrial ecosystems) preferentially take up ^{12}C during photosynthesis and discriminate against CO_2 containing ^{13}C (Farquhar *et al.* 1989). As a result, ^{13}C reaches higher levels in the atmosphere as it is left behind during net carbon uptake in terrestrial ecosystems. In contrast, net uptake of CO_2 from the atmosphere by the oceans does not result in any significant discrimination against ^{13}C (Ciais *et al.* 1995; Fung *et al.* 1997). Other recent analyses of atmospheric CO_2 concentration and isotopic measurements have shown large year-to-year changes in the net uptake of carbon by the terrestrial biosphere (Ciais *et al.* 1995; Francey *et al.* 1995; Battle *et al.* 2000). Some of these changes are probably associated with El Niño and other large-scale climate anomalies (Braswell *et al.* 1997). However, the mechanisms responsible for net carbon sequestration in terrestrial ecosystems and the exact spatial location of the ecosystems contributing most to the global terrestrial sink remain controversial (Lloyd 1999). Evidence also exists for changes in the seasonality of carbon exchange in terrestrial ecosystems. It appears that spring occurs earlier now in the northern hemisphere, based on changes in the timing and amplitude of the seasonal cycle of atmospheric CO_2 and on satellite observations (Keeling *et al.* 1996; Myneni *et al.* 1997; Randerson *et al.* 1998). All these results and interpretations about important changes in the global carbon cycle have come from atmospheric studies. Yet the basis of this response is physiological and ecological. We need further ecosystem studies to better understand the processes contributing to carbon uptake and release in terrestrial ecosystems. A major research challenge is to more accurately define the mechanisms and location of terrestrial sinks for atmospheric CO_2, and how these sinks will respond to changes in climate, land use and management practices.

The study of stable isotope fractionation during carbon cycling in terrestrial ecosystems provides a useful way to make the link between physiological and ecological processes on the ground and the atmospheric measurements described above (Flanagan & Ehleringer 1998; Bowling *et al.* 1999; Ehleringer *et al.* 2000). In C_3 plants there are two primary processes that cause carbon isotope ratios to change during photosynthesis, diffusional fractionation and enzymatic fractionation by rubisco. The carbon isotope ratio of plant organic material depends on the relative influence of diffusional and enzymatic fractionation which is controlled by the ratio of leaf intercellular CO_2 (c_i) and atmospheric CO_2 (c_a). This c_i/c_a ratio is important because it is a function of photosynthetic capacity and stomatal conductance. Changes in c_i/c_a and leaf carbon isotope ratio are a function of changes in either photosynthetic capacity or stomatal conductance or both (Farquhar et al. 1989).

Therefore, in addition to indicating photosynthetic pathway, carbon isotope discrimination against ^{13}C provides much information about the physiological and ecological functioning of plants in ecosystems. The carbon isotope ratio of C_3 plant tissue provides an assimilation-weighted average of c_i/c_a, a parameter that will vary as changes in environmental conditions induce variation in photosynthesis rate and stomatal conductance. Information about processes integrated over the life of the leaf are recorded by the ^{13}C content of leaf tissues and this allows subtle ecological differences in water-use efficiency and light-use efficiency to be distinguished within and among species in an ecosystem (Pearcy & Pfitsch 1991; Ehleringer et al. 1993; Berry et al. 1997; Buchmann et al. 1997a,b). Ehleringer (1994) and Brooks et al. (1997a) have illustrated distinct differences in ^{13}C among life forms or functional groups in different ecosystems. Variation in plant ^{13}C values appears to be related to plant water- and light-use strategies and to longevity of leaves in different plant functional groups. The carbon isotope ratio of plants averaged over the community can reflect environmental variation among different plant communities along rainfall, light or other environmental gradients (Ehleringer & Cooper 1988; Stewart et al. 1995). Ecosystem processes, such as the proportion of respired CO_2 that is re-fixed by photosynthesis before leaving a forest canopy, can also be studied using measurements of leaf carbon isotope ratio (Brooks et al. 1997b; Buchmann et al. 1997b, 1998).

The power of the stable isotope measurements on plant leaf tissues is that we can obtain time-integrated data on plant metabolism, and this is difficult or impossible to do with other physiological approaches because of limitation in equipment measurement resolution, or because of the time requirement for manually collecting many physiological measurements. In addition to the temporal integration of metabolic activities, stable isotope measurements of atmospheric CO_2 can be used to provide spatial integration of physiological information. At the ecosystem level, carbon isotope discrimination that occurs during net ecosystem carbon uptake should reflect the photosynthesis-weighted average of discrimination in all photosynthetically active plants in the ecosystem. Because there is no fractionation during mitochondrial respiration, the carbon isotope ratio of respired CO_2 reflects the composition of the substrate used for respiration. The isotope ratio of respired CO_2 can be measured at a range of spatial scales from individual ecosystem components (e.g. soil or leaf) to the entire ecosystem, to the landscape level using Keeling plot analyses (Flanagan & Ehleringer 1998).

Measurements of the isotope ratio of respired CO_2 reflect the spatially and temporally integrated ^{13}C content of organic matter within an ecosystem. A Keeling-plot analysis provides one means of quantifying the carbon isotope ratio of CO_2

leaving different terrestrial ecosystems (Fig. 6.5). The Keeling plot is basically the analysis of a mixing of two pools of CO_2. In effect, at any point in time the bulk atmosphere is considered to have a single fixed CO_2 concentration with a specific $\delta^{13}C$ value. Respiration from the plant, animal and microbial components increases the CO_2 concentration of this bulk air and alters its $\delta^{13}C$ value. The Keeling plot is a representation of this mixing process, obtained by sampling the air within canopies during periods of respiration. The intercept of a Keeling plot ($\delta^{13}C$ of CO_2 vs. $1/CO_2$) is the flux-weighted mean $\delta^{13}C$ value of the source (i.e. plants, animals and microbes). The basis for differences in the Keeling-plot intercepts is not only differences in the proportions of C_3 and C_4 photosynthesis, but also differences in physiological constraints associated with C_3 photosynthesis (Farquhar et al. 1989).

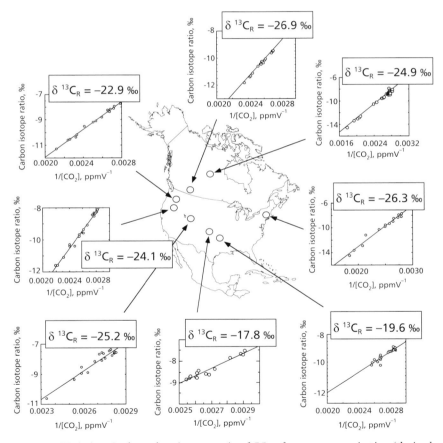

Figure 6.5 Variations in the carbon isotope ratio of CO_2 of ecosystem respiration (derived from Keeling plots) for different ecosystems across North America. (Data are from Ehleringer & Cook 1998; Flanagan et al. 1996; J. Berry, J. Ehleringer & L. Flanagan, unpublished.)

Ehleringer and Cook (1998) have used isotopic approaches to characterize how CO_2 respired from ecosystems along a precipitation gradient changes between wet and dry periods (Fig. 6.6). These data reveal shifts to lighter ^{13}C values of respired CO_2 as water stress is relieved. This same physiological pattern is commonly seen at the leaf level as stomata open to allow greater gas exchange under higher plant water potentials (Farquhar et al. 1989; Ehleringer et al. 1993).

There is much interest in making whole-ecosystem measurements of carbon isotope discrimination for several important reasons. First, knowledge of average photosynthetic discrimination at the global level, weighted by ecosystem type and productivity, is an important input to studies that partition the net uptake of anthropogenic CO_2 emissions between the ocean and terrestrial biosphere (Fung et al. 1997). Secondly, whole-ecosystem discrimination measurements could provide information about the canopy weighted value of c_i/c_a, and therefore the spatially integrated water-use efficiency of the canopy. Thirdly, in ecosystems with significant proportions of C_3 and C_4 plants, measurements of whole-ecosystem

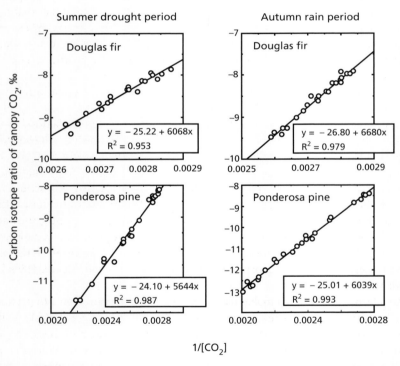

Figure 6.6 Variations in the carbon isotope ratio of CO_2 of ecosystem respiration (derived from Keeling plots) for Douglas fir and ponderosa pine ecosystems in central Oregon during the summer drought period and later in the autumn after rains have commenced. (Modified with permission from Ehleringer & Cook 1998.)

discrimination can be used to partition the contribution of the different photosynthetic pathways to total ecosystem productivity (Miranda *et al.* 1997). Fourthly, simultaneous measurements of ecosystem CO_2 fluxes using eddy covariance and isotope fluxes can allow the partitioning of net ecosystem CO_2 exchange into the component fluxes, gross photosynthesis and total ecosystem respiration (Yakir & Wang 1996; Bowling *et al.* 2001; Yakir & Sternberg 2000). We provide below further discussion and examples of a few of these important applications of ecosystem-level carbon isotope discrimination.

Many atmospheric inversion studies use a global average value of 18‰ for the carbon isotope discrimination that occurs during net uptake of CO_2 by terrestrial photosynthesis (Ciais *et al.* 1995; Battle *et al.* 2000). Fung *et al.* (1997) have shown how sensitive the results of the terrestrial–ocean partitioning exercise can be to variation in the terrestrial photosynthetic discrimination parameter. We know from empirical and modelling studies that carbon isotope values should vary greatly among ecosystems (e.g. Fig. 6.5), with significant change along latitudinal and altitudinal gradients (Körner *et al.* 1988, 1991; Lloyd & Farquhar 1994; Ciais *et al.* 1995; Fung *et al.* 1997). However, there are few measurements of isotope discrimination occurring in association with ecosystem CO_2 flux studies (Lloyd *et al.* 1996; Flanagan *et al.* 1996, 1997; Yakir & Wang 1996; Bakwin *et al.* 1998). This physiologically based, ecosystem information is needed in order to test global model discrimination calculations and improve confidence in the terrestrial–ocean ecosystem partitioning studies.

Ecosystem carbon budgets are controlled by the balance between carbon uptake during photosynthesis and carbon loss during respiration. Net CO_2 exchange of an ecosystem can now be measured with eddy covariance techniques (Baldocchi *et al.* 1988; Wofsy *et al.* 1993; Aubinet *et al.* 2000). However, a relatively small change in either photosynthesis or respiration can influence whether an ecosystem is a net source or sink for CO_2 (Goulden *et al.* 1998; Valentini *et al.* 2000). We would like to have direct, separate information on the rate of ecosystem photosynthesis and respiration, because the environmental controls on the two processes are quite different. In eddy covariance studies, total ecosystem respiration rate is often calculated from the net CO_2 exchange measurements made at night when photosynthesis is not active. However, this procedure is subject to a variety of important errors and often underestimates total ecosystem respiration rates (Goulden *et al.* 1996; Lee 1998; Aubinet *et al.* 2000).

An alternative approach makes use of stable isotope measurements to partition net ecosystem CO_2 exchange into its component fluxes (Bowling *et al.* 2001). In this approach simultaneous measurements of the $^{13}C/^{12}C$ ratio of atmospheric CO_2 are made while eddy covariance techniques are used to measure net ecosystem CO_2 exchange. Figure 6.7 shows an example of the results of flux partitioning using stable isotopes, for an active corn crop during August. The calculated rate of total ecosystem respiration at the start of the light period is much higher than would have been estimated from the night-time eddy covariance measurements made

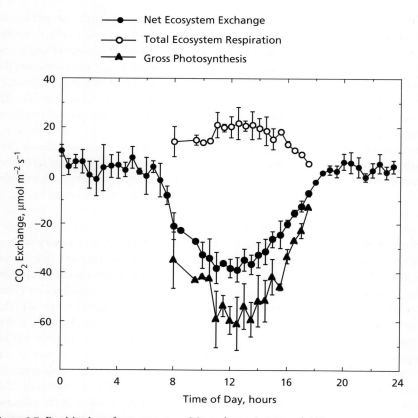

Figure 6.7 Partitioning of net ecosystem CO_2 exchange in a corn field into net photosynthesis and ecosystem respiration components, based on the approaches developed by Bowling *et al.* (2001). (From L.B. Flanagan, unpublished.)

under low turbulence. This stable isotope approach shows good promise for providing more mechanistic insights into the processes that control the carbon budget of an entire ecosystem.

A novel approach proposed by M.H. Conte and J.C. Weber (unpublished) for temporally and spatially integrating measurements of ecosystem carbon isotope discrimination is the isotopic analysis of plant leaf waxes collected in aerosol samples. A variety of wax compounds are produced to provide a protective covering on the surface of plant leaves. These waxes can be removed from the plant by physical disturbance (wind) and transported in the atmosphere as micron-sized particles. The carbon isotope composition of the leaf wax is dependent on fractionation during photosynthesis and secondary fractionation that occurs during the synthesis of the wax (wax and lipid molecules are generally depleted in ^{13}C relative to bulk plant tissue, Farquhar *et al.* 1989). Changes in the isotopic composition of the wax should reflect variation in photosynthetic discrimination against ^{13}C near the time of wax synthesis. Therefore, it is likely that the leaf-wax carbon isotope ratio will record daily to weekly changes in photosynthetic discrimination and can be used to study seasonal variation in ecosystem metabolism. M.H. Conte and J.C. Weber (unpublished) have shown strong (5–6‰) seasonal variation in the carbon isotope ratio of leaf waxes purified from aerosol samples collected in Bermuda. The Bermuda aerosol samples primarily reflect material produced and released from vegetation on the North American continent, and so represent a continental scale integration of ecosystem discrimination. This novel technique shows great promise for future analyses of ecosystem metabolism at a range of temporal and spatial scales. The interpretation of the data is based on physiological principles, yet the implications of these results are at the landscape and continental scales.

Precipitation dynamics and the functioning of an arid-land ecosystem

In response to global changes occurring today, species composition and the patterns of the cycling of carbon and other nutrients may be altered in the future. Here there are strong linkages between ecophysiology and ecosystem ecology. Arid-land ecosystems respond strongly to the patterns in which resources are supplied, most importantly the variability and timing of rainfall events (Noy-Meir 1973; Ehleringer et al. 1998). This dependency implies a strong relationship between physiological status and ecosystem productivity. In fact, primary productivity at the ecosystem scale is a strong linear function of cumulative precipitation (Le Houérou 1984; Noy-Meir 1985; Gutierrez & Whitford 1987; Gutierrez et al. 1988; Sala et al. 1988; Ludwig et al. 1989; Ehleringer et al. 1998), although the efficiency of production tends to be lower for summer-rain events than for winter-growth periods. Relevant to this, Reynolds et al. (1999) showed that interannual variation in summer precipitation in a desert ecosystem of western North America had less impact on the productivity of some desert perennials than changes in the amounts of winter precipitation. However, other woody and herbaceous species appear more sensitive to shifts in summer precipitation, with the northern distributions of many perennial species tied to the limits of summer monsoonal moisture inputs (Shreve & Wiggins 1964; Ehleringer & Phillips 1996). Arid-land ecosystems of south-western North American experience high year-to-year variability in precipitation, with El Niño and La Niña events representing two extremes of a moisture-input scale. These extreme-year types can have significant and persistent effects. For example, Brown et al. (1997) have attributed the recent expansion of woody perennials in the central Sonoran Desert to a series of unusually wet winters.

Throughout many North American deserts, precipitation distribution is biseasonal with the fraction of summer/winter precipitation following latitudinal gradients (Fig. 6.8). Winter precipitation falls at a time when the activity of many perennials is minimal or restricted by cold temperatures. Thus, wintertime moisture tends to percolate to deeper soil layers, with moisture from summer thunderstorms, in contrast, penetrating only the upper soil layers (Fig. 6.9). In effect, this

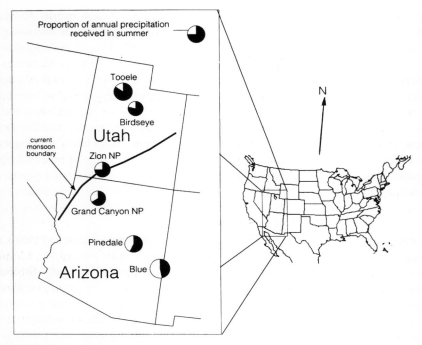

Figure 6.8 Changes in the fraction of total annual precipitation that falls in the summer along a geographical gradient in western North America. All sites average approximately 400 mm annual precipitation and are occupied by a *Juniperus–Pinus–Quercus* woodland. NP, National Park. (Modified with permission from Williams & Ehleringer 2000.)

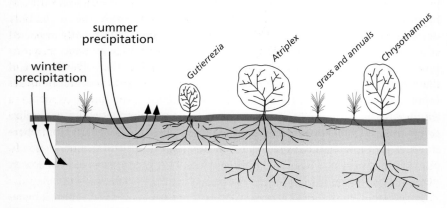

Figure 6.9 In arid-land ecosystems of western North America, moisture from winter-precipitation events tends to percolate into deeper soil layers, whereas moisture from summer-precipitation events tends to penetrate only into the upper soil layers. The bimodal precipitation patterns tends to create two distinct soil layers, each charged primarily by different precipitation regimes.

phenomenon creates a two-layer moisture distribution and forms the basis for distinguishing water relations responses among different species within the vegetation (Ehleringer et al. 1999). Some plant species are able to access soil moisture from both surface and deep soil layers (Ehleringer et al. 1991). Yet Shreve and Wiggins (1964) noted that in warmer arid-land sites, this bimodal precipitation pattern was associated with distinct and non-overlapping annual and herbaceous floras. At a larger scale, the northern distributions of many arid-land woody taxa are correlated with these summer precipitation limits. For example, the distribution of the turbinella live oak (*Quercus turbinella*) stops abruptly at the monsoon boundary, with this species occurring only in those habitats with summer rain. In contrast, the distribution of *Quercus gambelii* spans the monsoon boundary. Isotopic studies using naturally occurring differences in the abundances of deuterium or ^{18}O in winter- vs. summer-derived precipitation reveal a significant use of surface moisture by *Q. turbinella*, whereas *Q. gambelii* appears not to derive significant proportions of its water from moisture in upper soil layers (Phillips & Ehleringer 1995; Ehleringer & Phillips 1996). Similar northern distribution limits and summer-moisture dependencies appear to contribute to the sharp distribution boundaries of the alligator juniper *Juniperus deppeana* and pinyon pine *Pinus edulis* (Williams & Ehleringer 2000), whereas distributions of the related taxa *J. osteosperma* and *P. monophylla* appear not to be limited by summer-precipitation boundaries.

Because of the constraints water has on the gas-exchange activities of arid-land plants, changes in water availability may play a much more prominent role in the functioning of arid-land ecosystems in the near future than other global changes, such as increasing atmospheric CO_2 or increased dry nitrogen deposition (Ehleringer et al. 1999). Elevated CO_2 is expected to have an indirect effect and increase water-use efficiency (Field et al. 1997), but extensive evaluations of this hypothesis are not yet available. Ehleringer et al. (1999) showed that it was only during strong El Niño years that summer moisture inputs significantly impacted arid regions north of the normal monsoon boundary. Climate × change scenarios predict both warming and shifts in the monsoon boundary. Williams and Ehleringer (2000) observed that the responsiveness of trees to summer moisture input appeared related to the average monsoonal precipitation input in a threshold-type pattern (Fig. 6.10). In more northerly sites, trees did not utilize summer-derived moisture in the upper soil layer, whereas in southerly sites, where summer moisture inputs were greater, plants tend to utilize summer rains. In part, this difference between northern and southern populations may be due to differences in the intensity of individual rain events, but it is also likely that populations adjust their rooting habit to make better use of an increasingly reliable water source in the shallow soil. Herbaceous perennials should be even better adapted to exploit near-surface soil resources and should, in general, have a competitive advantage over most of the woody perennials on the wet end of the monsoonal gradient.

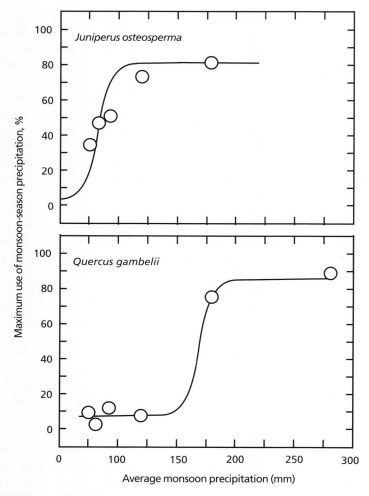

Figure 6.10 The maximum use of moisture derived by summer precipitation events by juniper and oak in arid woodland along the geographical gradient shown in Fig. 6.8. (Modified with permission from Williams & Ehleringer 2000.)

Most precipitation events in the arid lands are typically less than 10 mm (Ehleringer 1994), although during El Niño years the precipitation event size is larger (Cayan & Webb 1992). One adaptation to arid-land conditions is often thought to be the capacity to utilize small precipitation events before the water is lost through surface evaporation (Sala & Lauenroth 1982). Yet isotopic water-source observations indicate niche differentiation with respect to soil moisture use in arid-land and semiarid woodland ecosystems (Ehleringer *et al.* 1991; Flanagan *et al.* 1992; Donovan & Ehleringer 1994; Phillips & Ehleringer 1995). Not all species respond equally to summer moisture inputs, suggesting that some vegetation elements will be more sensitive to shifts in the ratio of winter- vs. summer-precipitation inputs. When large precipitation events (≥25 mm) which occur on average only once a summer were applied to trees in the arid woodlands or to shrubs in the desert regions of the Intermountain West, there was often a limited response by many of these species to utilize summer rain (Lin *et al.* 1996; Williams & Ehleringer 2000). While annuals and grass species located within the current monsoon boundary tend to fully respond to summer-rain events, woody species respond in a much more limited fashion. Given the strong general relationships between precipitation input and productivity, it is possible that there are ecotypic differences among woody plants in their relative capacities of roots to utilize surface moisture. Shifts in the amount of summer-rain input, whether through climate cycles or through global climate shifts, are likely to influence the productivities of different life form elements in arid-land ecosystems. The sensitivity of vegetation structure in these ecosystems is likely to be a function of how rapidly the changes occur in climate relative to the genetic changes.

We are beginning to understand trade-offs associated with pulse utilization and pulse frequency (Schwinning & Ehleringer 2001). As an extreme example, allocation of a large fraction of root biomass to shallow layers increases the capacity to take up water during a pulse, yet also lowers the rate of water uptake between pulses, when moisture is available only in deeper soil layers. Similarly, a high leaf-to-stem ratio enables rapid water uptake during a pulse, but increases the risk of hydraulic failure between pulses (Sperry *et al.* 1998; Sperry 2000). Conversely, a low leaf-to-stem ratio reduces potential transpiration but enables plants to maintain a more moderate water status throughout the drought period. The key to avoidance of low water potentials that may induce hydraulic failure is to reduce transpirational surface area. The costs and benefits of specific allocation strategies vary with the intensity and the duration of pulses, with pulse frequency and with the availability of alternative water sources, such as water stored in deeper soil layers left over from winter precipitation. While our understanding of these physiological trade-offs is incomplete, it is clear that physiological properties and morphological traits do shape the nature of plant responses to moisture input at the whole-plant and ecosystem scales. Consequently, it may be possible to predict the directions of vegetation shifts within arid-land ecosystems if we can better understand how drought cycles and monsoon patterns will shift in an elevated-CO_2 world.

Summary

Physiological ecology has a rich tradition investigating mechanisms of adaptation between organisms and their environment. Increasingly, our understanding of plant physiological ecology has proved valuable for addressing larger-scale questions related to ecosystem dynamics and biosphere–atmosphere fluxes. Nowhere is this linkage more apparent than in global change studies, where the focus is on determining how ecosystems are responding to atmospheric, hydrologic and land-use changes. The examples discussed in this chapter demonstrate the natural melding of physiological ecology and ecosystem ecology: a *palaeoecological example*—C_3/C_4 photosynthesis in response to atmospheric CO_2 and its impacts on productivity and animal diversity; a *current-ecological* example—stable isotope ratios in CO_2 fluxes between the biosphere and the atmosphere that allow determination of carbon sources and sinks in different ecosystems and under land-use change activities; and a *future-climate change* example—water-resource partitioning, competition and ecosystem dynamics in arid ecosystems in response to monsoon-boundary shifts.

References

Aubinet, M., Grelle, A., Ibran, A. *et al.* (2000) Estimates of the annual net carbon and water exchange of forests: the EUROFLUX methodology. *Advances in Ecological Research* **30**, 113–175.

Bakwin, P.S., Tans, P.P., White, J.W.C. & Andres, R.J. (1998) Determination of the isotopic ($^{13}C/^{12}C$) discrimination by terrestrial biology from a global network of observations. *Global Biogeochemical Cycles* **12**, 555–562.

Baldocchi, D.D., Hicks, B.B. & Meyers, T.P. (1988) Measuring biosphere-atmosphere exchanges of biologically related gasses with micrometeorological methods. *Ecology* **69**, 1331–1340.

Barry, J.C., Morgan, M.E., Flynn, L.J. *et al.* (1995) Patterns of faunal turnover and diversity in the Neogene Siwaliks of northern Pakistan. *Palaeogeography Palaeoclimatology Palaeoecology* **115**, 209–226.

Battle, M., Bender, M.L., Tans, P.P. *et al.* (2000) Global carbon sinks and their variability inferred from atmospheric O_2 and $\delta^{13}C$. *Science* **287**, 2467–2470.

Berry, S.C., Varney, G.T. & Flanagan, L.B. (1997) Leaf $\delta^{13}C$ in *Pinus resinosa* trees and understory plants: variation associated with light and CO_2 gradients. *Oecologia* **109**, 499–506.

Bowling, D.R., Baldocchi, D.D. & Monson, R.K. (1999) Dynamics of isotopic exchange of carbon dioxide in a Tennessee deciduous forest. *Global Biogeochemical Cycles* **13**, 903–922.

Bowling, D.R., Tans, P.P. & Monson, R.K. (2001) Partitioning net ecosystem carbon exchange with isotopic fluxes of CO_2. *Global Change Biology* in press.

Braswell, B.H., Schimel, D.S., Linder, E. & Moore, B. (1997) The response of global terrestrial ecosystems to interannual temperature variability. *Science* **278**, 870–872.

Brooks, J.R., Flanagan, L.B., Buchmann, N. & Ehleringer, J.R. (1997a) Carbon isotope composition of boreal plants: functional grouping of life forms. *Oecologia* **110**, 301–311.

Brooks, J.R., Flanagan, L.B., Varney, G.T. & Ehleringer, J.R. (1997b) Vertical gradients in photosynthetic gas exchange characteristics and refixation of respired CO_2 within boreal forest canopies. *Tree Physiology* **17**, 1–12.

Brown, J.H., Valone, T.J. & Curtin, C.G. (1997) Reorganization of an arid ecosystem in response to recent climate change. *Proceedings of the National Academy of Science of the USA* **94**, 9729–9733.

Buchmann, N., Brooks, J.R., Flanagan, L.B. & Ehleringer, J.R. (1998) Carbon isotope discrimination of terrestrial ecosystems. In: *Stable Isotopes, Integration of Biological, Ecological, and Geochemical Processes* (ed. H. Griffiths), pp. 203–221. BIOS Scientific Publishers, Oxford.

Buchmann, N., Guehl, J.-M., Barigah, T. & Ehleringer, J.R. (1997a) Interseasonal comparison of CO_2 concentrations, isotopic composition, and carbon cycling in an Amazonian rainforest (French Guiana). *Oecologia* **110**, 120–131.

Buchmann, N., Kao, W. & Ehleringer, J.R. (1997b)

Influence of stand structure on carbon-13 of vegetation, soils, and canopy air within deciduous and evergreen forests of Utah, United States. *Oecologia* **110**, 109–119.

Canadell, J., Mooney, H.A., Mooney, D. *et al.* (2000) Carbon metabolism of the terrestrial biosphere: a multi-technique approach for improved understanding. *Ecosystems* **3**, 115–130.

Cayan, D.R. & Webb, H. (1992) El Niño/Southern Oscillation and streamflow in the western United States. In: *El Niño Historical and Paleoclimatic Aspects of the Southern Oscillation* (eds H.F. Diaz & V. Markgraf), pp. 26–69. Cambridge University Press, Cambridge.

Cerling, T.E. & Harris, J.M. (1999) Carbon isotope fractionation between diet and bioapatite in ungulate mammals and implications for ecological and paleoecological studies. *Oecologia* **120**, 347–363.

Cerling, T.E. & Quade, J. (1990) Global ecologic and climatic change during the neogene: stable isotopic evidence from soils. *Chemical Geology* **84**, 164–165.

Cerling, T.E., Harris, J.M., MacFadden, B.J. *et al.* (1997) Global vegetation change through the Miocene–Pleistocene boundary. *Nature* **389**, 153–158.

Ciais, P., Tans, P.P., White, J.W.C. *et al.* (1995) Partitioning of ocean and land uptake of CO_2 as inferred by $\delta^{13}C$ measurements from the NOAA Climate Monitoring and Diagnostics Laboratory Global Air Sampling Network. *Journal of Geophysical Research* **100**, 5051–5070.

Donovan, L.A. & Ehleringer, J.R. (1994) Water stress and use of summer precipitation in a Great Basin shrub community. *Functional Ecology* **8**, 289–297.

Ehleringer, J.R. (1994) Variation in gas exchange characteristics among desert plants. In: *Ecophysiology of Photosynthesis* (eds E.-D. Schulze & M.M. Caldwell), pp. 361–392. Springer-Verlag, New York.

Ehleringer, J.R. & Cook, C.S. (1998) Carbon and oxygen isotope ratios of ecosystem respiration along an Oregon conifer transect: preliminary observations based upon small-flask sampling. *Tree Physiology* **18**, 513–519.

Ehleringer, J.R. & Cooper, T.A. (1988) Correlations between carbon isotope ratio and microhabitat in desert plants. *Oecologia* **76**, 562–566.

Ehleringer, J.R. & Field, C.B., eds (1993). *Scaling Physiological Processes: Leaf to Globe*. Academic Press, San Diego.

Ehleringer, J.R. & Phillips, S.L. (1996) Ecophysiological factors contributing to the distributions of several *Quercus* species in the Intermountain West. *Annals of Forestry Science* **53**, 291–302.

Ehleringer, J.R., Phillips, S.L., Schuster, W.F.S. & Sandquist, D.R. (1991) Differential utilization of summer rains by desert plants, implications for competition and climate change. *Oecologia* **88**, 430–434.

Ehleringer, J.R., Hall, A.E. & Farquhar, G.D., eds (1993). *Stable Isotopes and Plant Carbon/Water Relations*. Academic Press, San Diego.

Ehleringer, J.R., Cerling, T.E. & Helliker, B.R. (1997) C_4 photosynthesis, atmospheric CO_2, and climate. *Oecologia* **112**, 285–299.

Ehleringer, J.R., Evans, R.D. & Williams, D. (1998) Assessing sensitivity to change in desert ecosystems—a stable isotope approach. In: *Stable Isotopes, Integration of Biological, Ecological, and Geochemical Processes* (ed. H. Griffiths), pp. 223–237. BIOS Scientific Publishers, Oxford.

Ehleringer, J.R., Schwinning, S. & Gebauer, R.L. (1999) Water use in arid land ecosystems. In: *Advances in Plant Physiological Ecology* (ed. M.C. Press), pp. 347–365. Blackwell Science, Oxford.

Ehleringer, J.R., Buchmann, N. & Flanagan, L.B. (2000) Carbon isotope ratios in below-ground carbon cycle processes. *Ecological Applications* **10**, 412–422.

Epstein, H.E., Lauenroth, W.K., Burke, I.C. & Coffin, D.P. (1997) Productivity patterns of C_3 and C_4 functional types in the U.S. Great Plains. *Ecology* **78**, 722–731.

Fan, S., Gloor, M., Mahlman, J. *et al.* (1998) A large terrestrial carbon sink in North America implied by atmospheric and oceanic carbon dioxide data and models. *Science* **282**, 442–446.

Farquhar, G.D., Ehleringer, J.R. & Hubick, K.T. (1989) Carbon isotope discrimination and photosynthesis. *Annual Review of Plant Physiology and Molecular Biology* **40**, 503–537.

Ficken, K.J., Street-Perrott, F.A., Perrott, R.A., Swain, D.L., Olago, D.O. & Eglington, G. (1998) Glacial/interglacial variations in carbon cycling revealed by molecular and isotope stratigraphy of Lake Nkunga, Mt. Kenya, East Africa. *Organic Geochemistry* **29**, 1701–1719.

Field, C.B., Lund, C.P., Chiariello, N.R. & Mortimer, B.E. (1997) CO_2 effects on the water budget of grassland microcosm communities. *Global Change Biology* **3**, 197–206.

Flanagan, L.B. & Ehleringer, J.R. (1998) Ecosystem–atmosphere CO_2 exchange: interpreting signals of change using stable isotope ratios. *Trends in Ecology and Evolution* **13**, 10–14.

Flanagan, L.B., Brooks, J.R., Varney, G.T., Berry, S.C. & Ehleringer, J.R. (1996) Carbon isotope discrimination during photosynthesis and the isotope ratio of respired CO_2 in boreal forest ecosystems. *Global Biogeochemical Cycles* **10**, 629–640.

Flanagan, L.B., Brooks, J.R., Varney, G.T. & Ehleringer, J.R. (1997) Discrimination against $C^{18}O^{16}O$ during photosynthesis and the oxygen isotope ratio of respired CO_2 in boreal forest ecosystems. *Global Biogeochemical Cycles* **11**, 83–98.

Flanagan, L.B., Ehleringer, J.R. & Marshall, J.D. (1992) Differential uptake of summer precipitation and groundwater among co-occurring trees and shrubs in the southwestern United States. *Plant Cell and Environment* **15**, 831–836.

Francey, R.J., Tans, P.P., Allison, C.E., Enting, I.G., White, J.W.C. & Trolier, M. (1995) Changes in oceanic and terrestrial uptake since 1982. *Nature* **373**, 326–330.

Fung, I., Field, C.B., Berry, J.A. *et al.* (1997) Carbon-13 exchanges between the atmosphere and the biosphere. *Global Biogeochemical Cycles* **11**, 507–533.

Goulden, M.L., Munger, J.W., Fan, S.-M., Daube, B.C. & Wofsy, S.C. (1996) Exchange of carbon dioxide by a deciduous forest: response to interannual climate variability. *Science* **271**, 1576–1578.

Goulden, M.L., Wofsy, S.C., Harden, J.W. *et al.* (1998) Sensitivity of boreal forest carbon balance to soil thaw. *Science* **279**, 214–217.

Gutierrez, J.R. & Whitford, W.G. (1987) Responses of Chihuahuan Desert herbaceous annuals to rainfall augmentation. *Journal of Arid Environments* **12**, 127–139.

Gutierrez, J.R., DaSilva, O.A., Pagani, M.I., Weems, D. & Whitford, W.G. (1988) Effects of different patterns of supplemental water and nitrogen fertilization on productivity and composition of Chihuahuan Desert annual plants. *American Midland Naturalist* **119**, 336–343.

Huang, Y., Freeman. K.H., Eglington, T.I. & Street-Perrott, F.A. (1999) $\delta^{13}C$ analyses of individual lignin phenols in Quaternary lake sediments: a novel proxy for deciphering past terrestrial vegetation changes. *Geology*, **27**, 471–474.

Huang, S., Pollack, H.N. & Shen, P.Y. (2000) Temperature trends over the past five centuries reconstructed from borehole temperatures. *Nature* **403**, 756–758.

IPCC (1996) *Climate Change 1995. The Science of Climate Change*. Contribution of Working Group I to the Second Assessment Report of the Intergovernmental Panel on Climate Change. Cambridge University Press, Cambridge.

Janis, C.M., Scott, K.M. & Jacobs, L.L. (1998). *Evolution of Tertiary Mammals of North America*, Vol. 1. *Terrestrial Carnivores, Ungulates, and Ungulatelike Mammals*. Cambridge University Press, Cambridge.

Keeling, C.D., Chin, J.F.S. & Whorf, T.P. (1996) Increased activity of northern vegetation inferred from atmospheric CO_2 measurements. *Nature* **382**, 146–149.

Körner, Ch., Farquhar, G.D. & Roksandic, Z. (1988) A global survey of carbon isotope discrimination in plants from high altitude. *Oecologia* **74**, 623–632.

Körner, Ch., Farquhar, G.D. & Wong, S.C. (1991) Carbon isotope discrimination by plants follows latitudinal and altitudinal trends. *Oecologia* **88**, 30–40.

Lambers, H., Chapin, F.S. III & Pons, T.L. (1998) *Plant Physiological Ecology*. Springer-Verlag, Heidelberg.

Larcher, W. (1995). *Plant Physiological Ecology*. Springer-Verlag, Heidelberg.

Le Houérou, H.N. (1984) Rain use efficiency: a unifying concept in arid-land ecology. *Journal of Arid Environments* **7**, 213–247.

Lee, X. (1998) On micrometeorological observations of surface–air exchange over tall vegetation. *Agricultural and Forest Meteorology* **91**, 39–49.

Lin, G., Phillips, S.L. & Ehleringer, J.R. (1996)

Monsoonal precipitation responses of shrubs in a cold desert community on the Colorado Plateau. *Oecologia* **106**, 8–17.

Lloyd, J. (1999) Current perspectives on the terrestrial carbon cycle. *Tellus* **51B**, 336–342.

Lloyd, J. & Farquhar, G.D. (1994) ^{13}C discrimination during CO_2 assimilation by the terrestrial biosphere. *Oecologia* **99**, 201–215.

Lloyd, J., Krujit, B., Hollinger, D.Y. et al. (1996) Vegetation effects on the isotopic composition of atmospheric CO_2 at local and regional scales: theoretical aspects and a comparison between rain forest in Amazonia and a boreal forest in Siberia. *Australian Journal of Plant Physiology* **23**, 371–399.

Ludwig, J.A., Whitford, W.G. & Cornelius, J.M. (1989) Effects of water, nitrogen and sulfur amendments on cover, density, and size of Chihuahuan Desert ephemerals. *Journal of Arid Environments* **16**, 35–42.

Miranda, A.C., Miranda, H.S., Lloyd, J. et al. (1997) Fluxes of carbon, water and energy over Brazilian cerrado: an analysis using eddy covariance and stable isotopes. *Plant Cell and Environment* **20**, 315–328.

Mooney, H.A. (1998). *The Globalization of Ecological Thought*. Ecology Institute, Oldendorf.

Mooney, H.A. & Chabot, B.F., eds (1985) *Physiological Ecology of North American Plant Communities*. Chapman & Hall, New York.

Mooney, H.A. & Drake, J.A., eds (1986) *Ecology of Biological Invasions of North America and Hawaii*. Springer-Verlag, New York.

Mooney, H.A. & Hobbs, R.J. (2000) *Invasive Species in a Changing World*. Island Press, New York.

Mooney, H.A., Pearcy, R.W. & Ehleringer, J.R. (1987) Plant physiological ecology today. *Bioscience* **37**, 18–20.

Mooney, H.A., Canadell, J., Chapin III, F.S. et al. (1999) Ecosystem physiology responses to global change. In: *The Terrestrial Biosphere and Global Change: Implications for Natural and Managed Ecosystems* (eds B.H. Walker, W. Steffen, J. Canadell & J. Ingram), pp. 141–189. Cambridge University Press, Cambridge.

Myneni, R.B., Keeling, C.D., Tucker, C.J., Asrars, G. & Nemani, R.R. (1997) Increased plant growth in the northern high latitudes from 1981 to 1991. *Nature* **386**, 698–702.

Noy-Meir, I. (1973) Desert ecosystems, environment and producers. *Annual Reviews of Ecology and Systematics* **4**, 25–41.

Noy-Meir, I. (1985) Desert ecosystem structure and function. In: *Ecosystems of the World* (eds M. Evenari, I. Noy-Meir & D. Goodall), pp. 92–103. Elsevier, Amsterdam.

Pearcy, R.W. & Pfitsch, W.A. (1991) Influence of sunflecks on the $\delta^{13}C$ of *Adenocaulon bicolor* plants occurring in contrasting forest understory microsites. *Oecologia* **86**, 457–462.

Petit, J.R., Jouzel, J., Raynaud, D. et al. (1999) Climate and atmospheric history of the past 420,000 years from the Vostok ice core, Antarctica. *Nature* **399**, 429–436.

Phillips, S.L. & Ehleringer, J.R. (1995) Limited uptake of summer precipitation by bigtooth maple (*Acer grandidentatum* Nutt) and Gambel's oak (*Quercus gambelii* Nutt). *Trees* **9**, 214–219.

Randerson, J.T., Thompson, M.V. & Field, C.B. (1998) Linking ^{13}C-based estimates of land and ocean sinks with predictions of carbon storage from CO_2 fertilization of plant growth. *Tellus* **51B**, 668–678.

Reynolds, J.F., Virginia, R.A., Kemp, P.R., de Soyza, A.G. & Tremmel, D.C. (1999) Impact of drought on desert shrubs: effects on seasonality and degree of resource island development. *Ecological Monographs* **69**, 69–106.

Sage, R.F. & Monson, R.K., eds (1999) C_4 *Plant Biology*. Academic Press, San Diego.

Sala, O.E. & Lauenroth, W.K. (1982) Small rainfall events: an ecological role in semiarid regions. *Oecologia* **53**, 301–304.

Sala, O.E., Parton, W.J., Joyce, L.A. & Lauenroth, W.K. (1988) Primary production of the central grassland region of the United States. *Ecology* **69**, 40–45.

Schimel, D.S., Melillo, J., Tian, H. et al. (2000) Contribution of increasing CO_2 and climate to carbon storage by ecosystems in the United States. *Science* **287**, 2004–2006.

Schwinning, S. & Ehleringer, J.R. (2001) Water use tradeoffs and optimal adaptations to pulse-driven arid ecosystems. *Journal of Ecology* **89**, in press.

Shreve, F. & Wiggins, I.R. (1964) *Vegetation and Flora of the Sonoran Desert*. Stanford University Press, Palo Alto.

Sperry, J.S. (2000) Hydraulic constraints on plant gas exchange. *Agricultural and Forest Meteorology* **104**, 13–23.

Sperry, J.S., Campbell, G.S. & Alder, N. (1998) Hydraulic limitation of flux and pressure in the soil–plant continuum: results from a model. *Plant Cell and Environment* **21**, 347–359.

Sponheimer, M. & Lee-Thorp, J.A. (1999) Isotopic evidence for the diet of an early hominid, *Australopithecus africanus*. *Science* **283**, 368–370.

Stewart, G.R., Turnbull, M.H., Schmidt, S. & Werskine, P.D. (1995) $\delta^{13}C$ natural abundance in plant communities along a rainfall gradient: a biological integrator of water availability. *Australian Journal of Plant Physiology* **22**, 51–55.

Street-Perrott, F.A., Huang, Y., Perrott, R.A. & Eglington, G. (1998) Carbon isotopes in lake sediments and peats of the last glacial age: implications for the global carbon cycle. In: *Stable Isotopes* (ed. H. Griffiths), pp. 381–396. BIOS Scientific, Oxford.

Tieszen, L.L., Reed, B.C., Bliss, N.B., Wylie, B.K. & DeJong, D.D. (1997) NDVI, C_3 and C_4 production, and distribution in Great Plains grassland land cover classes. *Ecological Applications* **7**, 59–78.

Trolier, M., White, J.W.C., Tans, P.P., Masarie, K.A. & Gemery, P.A. (1996) Monitoring the isotopic composition of atmospheric CO_2: measurements from the NOAA global air sampling network. *Journal of Geophysical Research* **101**, 25 897–25 916.

Valentini, R., Matteucci, G., Dolman, A.J. *et al.* (2000) Respiration as the main determinant of carbon balance in European forests. *Nature* **404**, 861–865.

Vitousek, P.M., Mooney, H.A., Lubchenco, J. & Melillo, J.M. (1997) Human domination of Earth's ecosystems. *Science* **277**, 494–499.

Wand, S.J.E., Midgley, G.F., Jones, M.H. & Curtis, P.S. (1999) Responses of wild C_4 and C_3 grass (Poaceae) species to elevated atmospheric CO_2 concentration: a meta-analytic test of current theories and perceptions. *Global Change Biology* **5**, 723–741.

Williams, D.G. & Ehleringer, J.R. (2000) Intra- and interspecific variation for summer precipitation use in pinyon-juniper woodlands. *Ecological Monographs* **70**, 517–537.

Wofsy, S.C., Goulden, M.L., Munger, J.W. *et al.* (1993) Net exchange of CO_2 in a mid-latitude forest. *Science* **260**, 1314–1317.

Yakir, D. & Sternberg, L.S.L. (2000) The use of stable isotopes to study ecosystem gas exchange. *Oecologia* **123**, 297–311.

Yakir, D. & Wang, X.-F. (1996) Fluxes of CO_2 and water between terrestrial vegetation and the atmosphere estimated from isotope measurements. *Nature* **380**, 515–417.

Chapter 7
Biodiversity, ecosystem processes and climate change

J. H. Lawton

Introduction

We know that climate change will have a major impact on biodiversity, both regionally and locally. It is also obvious that climate change will alter ecosystem processes—nature's 'goods and services'. A third area of theoretical and experimental work focuses on how changes in local biodiversity impact upon ecosystem processes. The task ahead is to link these three areas into a coherent body of knowledge, and in particular for ecologists to rise to the challenge of predicting the responses of communities and ecosystems to global environmental change.

Climate affects ecosystem processes directly. Low temperatures, for instance, reduce decomposition rates. Climate also indirectly affects ecosystem processes because it has a profound effect on the zonation of vegetation across the face of the Earth (the major biomes; Whittaker 1975), and different types of vegetation may differ in their potential productivity, rates of litter decomposition, etc. Various measures of energy input to terrestrial ecosystems, typically some combination of solar energy input and rainfall (Rosenzweig 1995; Francis & Currie 1998) are also strongly correlated with species richness, although the mechanism is poorly understood (Srivastava & Lawton 1998; Lawton 2000). We are therefore dealing with a complex web of interconnected variables (Fig. 7.1), and I do not have the space to discuss all of them here. I propose to say very little, for example, about the direct effects of climate on ecosystem processes. Instead I have chosen to focus on what I personally regard as some of the most interesting parts of the linkages in Fig. 7.1, namely how climate change will alter the species richness and species composition of ecological communities, and the effects of local species richness on ecosystem processes. To keep within reasonable bounds, I have ignored evolutionary responses to change (see Chapters 1 and 2). Readers must decide for themselves whether this is reasonable. To hedge my bets I have added evolution in response to climate change to the list of research challenges at the end of the chapter.

My examples are primarily from terrestrial and freshwater systems, because those are the ones with which I am most familiar. Climate is not the only part of the

NERC Centre for Population Biology, Imperial College, Silwood Park, Ascot SL5 7PY, UK; and Natural Environment Research Council, Polaris House, Swindon, SN2 1EU. E-mail: HQPO@nerc.ac.uk

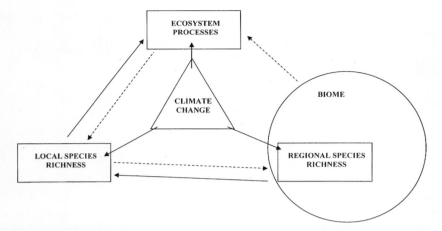

Figure 7.1 Overview of some of the complex web of relationships linking biodiversity, ecosystem processes and climate change. Climate change will alter regional species richness, and in the long run will lead to shifts in entire biomes. In the shorter term, climate change will alter local species richness. Ecosystem processes are influenced by climate, by vegetation type (biome) and by species richness; ecosystem processes themselves can also impact on species richness. The chapter focuses on relationships indicated by the solid arrows.

Earth system being abused by human actions. But again, to keep within my allotted space, I will merely, and only very occasionally, touch on other key aspects of global environmental change.

If we were able to fast-forward the video, and watch a small patch of habitat respond to Earth's changing climate, we would first observe accelerating shifts in the strength and nature of the interactions among the pool of original inhabitants. Then, with increasing rapidity, the pool itself starts to change, as species become locally extinct and new ones invade, tearing apart existing webs of interactions and creating new ones. Wholesale shifts in species' geographical ranges inevitably follow. It is these complex changes, and the accompanying changes in ecosystem processes, that we seek to understand and, if possible, to predict.

There is no magic bullet that will improve the ability of ecologists to predict the impacts of climate change. At the present time some of the problems that confront us seem intractable (see below). The challenge, however, is to reduce current uncertainty, without trying to solve all the problems at once. To do this we need to use a series of approaches, ranging from laboratory experiments in controlled environment facilities (CEFs), through large-scale field manipulations of whole systems, to the search for, and exploitation of, macroecological patterns. Each approach has its own advantages and disadvantages. I use these different approaches to tackle a range of problems. In the order in which they appear these are:

1 the relationship between biodiversity and ecosystem processes, drawing upon large-scale field manipulation experiments;
2 the challenge posed by the individualistic nature of species responses to environmental change;
3 the disadvantages of highly controlled, reductionist experiments on the impacts of change on communities and ecosystems, and the need for more whole-system manipulations, both in the laboratory (using CEFs) and in the field;
4 changes in the geographical ranges of species in response to climate change, not least the need to recognize that ranges are determined by more than physiological tolerance—I illustrate this with work in a CEF;
5 the use of macroecological patterns to make predictions about the impacts of change, particularly local vs. regional species richness, and the energy–diversity relationship;
6 I close with some general comments on the advantages and disadvantages of work in CEFs, and lay out what I see as future priorities in the emerging research field linking biodiversity, ecosystem processes and climate change.

Biodiversity and ecosystem processes

Human impacts are altering the species composition and species richness of ecological communities all over the world. Sometimes we enrich local diversity, but mainly we reduce it. What are the effects of reducing species richness on ecosystem processes? To keep the argument within reasonable bounds I focus here on plant species richness, and on plant biomass production (other ecosystem processes and other trophic levels are briefly considered in Lawton 2000). There is now a growing theoretical and empirical literature on this problem, summarized by Allison (1999), Tilman (1999; Chapter 9, this volume) and Lawton (2000).

Consider a reasonably large species pool (thereby avoiding strong stochastic effects). On average the most likely theoretical relationship between loss of species and ecosystem processes is a roughly linear decline in productivity with declining species richness plotted on a log scale (Fig. 7.2b). (The more usual, but less transparent way, of plotting this same relationship is in Fig. 7.2(a), which shows productivity increasing asymptotically with increasing species richness plotted on an arithmetic scale.) The fundamental underpinning mechanism is that species differ in their ecologies, i.e. no two species have identical niches. Niche differences translate into differences in productivity via three processes (see Chapter 9). In summary these are: (i) that species-poor systems 'sample' less total niche space (the 'sampling effect'); (ii) they have reduced functional diversity (the 'niche-differentiation effect' *sensu stricto*); and (iii) reduced opportunities for positive (mutualistic) interactions between species (see Hector *et al.* (1999), Tilman (1999; Chapter 9, this volume) and Lawton (2000) for further discussion).

Two key points follow from this theoretical framework. First, there is currently some debate about the relative importance of biodiversity attributes other than species richness in determining ecosystem processes (e.g. plant species richness vs.

LOCAL, WITHIN SITE EFFECTS

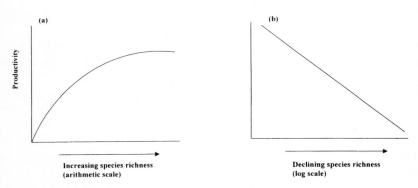

Figure 7.2 Expected relationship between local plant species richness and primary productivity (see text, and Chapter 9). The relationship is normally plotted with species richness on an arithmetic axis, increasing from left to right (a). The alternative plot (b) has species richness *decreasing* from left to right, on a logarithmic scale; this is the form used to illustrate the outcome of the BIODEPTH experiment in Fig. 7.4.

plant functional types and plant species identities) (Grime 1997; Hooper & Vitousek 1997; Tilman 1997; Tilman *et al.* 1997; Wardle *et al.* 1997a,b; Allison 1999). The debate is artificial. If niche differences between species are small, then ecosystem processes influenced by these niche differences will be little affected by loss of species. By these same arguments, deliberately selecting species to be as different as possible (by selecting different functional types) is likely to have a big effect on ecosystem processes. Because functional types are arbitrary divisions of continuous niche space, deciding whether species richness or functional types has a bigger impact on ecosystem processes is to arbitrarily divide a continuum. By the same logic, it follows that if only a few species (or functional groups) are involved in the analyses, and/or if there are major changes in dominants with diversity, the effects of species identities (rather than species- or functional-group richness) will appear to be paramount.

A second, and in my view even more fundamental area of confusion in the literature, involves the impacts of changes in biodiversity on ecosystem processes within one area (*within-site effects*), and comparisons between different areas (*across-site effects*) (Fig. 7.3). So far I have been talking only about within-site effects, where the theoretical problem being addressed is about loss of species at one locality. Across-site comparisons test a fundamentally different problem, because they simultaneously examine changes in species richness and differences between environments. We will see a real example of these differences in a moment. It is important to distinguish between, and to know the relative magnitudes of, within-site species-richness effects, and across-site environmental effects on

BIODIVERSITY, ECOSYSTEMS AND CLIMATE CHANGE

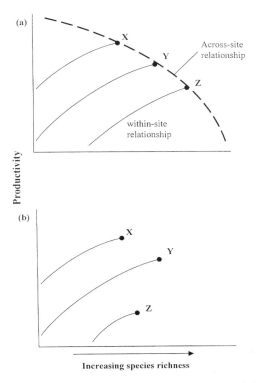

Figure 7.3 The difference between within-site and across-site effects of changes in species richness on ecosystem processes (Lawton 2000). This model imagines that there are three different sites (X, Y and Z) that differ in their natural species richness (Z > Y > X in (a), and Y > Z > X in (b)) and in their environments. In example (a), ecosystem processes (e.g. primary production) decline with increasing natural species richness across sites (dashed line), because of differences in weather, soil fertility, etc. But within each site (solid lines), declining species richness leads to a decline in ecosystem processes, exactly as in Fig. 7.2(a). In example (b) sites differ in productivity, and productivity again declines within sites as species are lost from the system, but there is no systematic change in productivity with natural species richness across sites (mirroring the results from the BIODEPTH experiment in Fig. 7.4).

ecosystem processes. For example, by substituting space for time, contemporary differences in ecosystem processes in comparable ecosystems at different sites give us some feel for how ecosystem processes might change at one place over time as the climate changes, and how such changes may be modulated by associated changes in species richness and species composition.

An example of these theoretical considerations: the BIODEPTH experiment

BIODEPTH (*Bio*diversity and *E*cological *P*rocesses in *T*errestrial *H*erbaceous ecosystems) is a pan-European field experiment (Hector *et al.* 1999). There are eight field sites (Umeå in Sweden; Sheffield and Silwood in the UK; Cork in Ireland; Bayreuth in Germany; Zurich in Switzerland; Lisbon in Portugal; and Lesbos in Greece). At all the field sites, colleagues and I set up replicated model ecosystems with five levels of plant species richness, from monocultures to polycultures with a maximum richness per plot characteristic of the natural vegetation for each site. The sites embrace a wide range of environmental conditions.

We found we could fit one statistical model to data on plant productivity across all sites (Fig. 7.4). Sites differed from one another in overall productivity, because of inherent differences in soil fertility, the weather and the length of the growing season. However, because there was not even a hint of a significant site–treatment interaction, the overall result was that at all eight sites plant biomass increased linearly with the log of plant species richness (Fig. 7.4a) as envisaged in Fig. 7.2. As well as the within-site effects of plant species richness on biomass production, there were also significant effects of plant functional group (Fig. 7.4b) and plant species identity, particularly the presence of the legume *Trifolium pratense*. Together, location, richness, and species identity (composition) explained approximately 28%, 18% and 39%, respectively, of the total variation in productivity. The richness (biodiversity) effect was predominantly due to species richness (15%), with a small, additional contribution (3%) from the richness of functional types. (This last result is unexpected, given my earlier comments. One explanation is that our *a priori* groupings into functional types reflect smaller niche differences between functional types than in reality exist between species within these groups.)

Plant diversity and ecosystem processes in a changing world

Summarizing, in the BIODEPTH experiment biodiversity (species- and functional-group richness) was not the most important determinant of local productivity. Changes in species identity had the largest effect on local productivity, followed by changes in the environment (including, but not exclusively, changes in climate). But loosing species, on average, always reduced productivity, with the same underlying pattern at every site. These results give us some feel for how changes in climate, and changes in species identity and total richness, may alter local ecosystem processes in a rapidly changing world. The proportion of change allocated to each of the main drivers is context dependent, and since BIODEPTH is unique in its geographical scale, we must await other experiments to see if environment, species identity and species richness influence ecosystem processes in quantitatively similar ways in other ecosystems and continents.

In the present context the BIODEPTH experiment must also be interpreted cautiously for other reasons. Not everybody interprets the results in the way that I have done. For instance, this and similar experiments (see Chapter 9) tell us what happens 'on average', based as they are on experimental assemblages built without favouring particular species by drawing combinations at random from the local pool. In the real world of climate change, the species lost from ecosystems are unlikely to be a random sample from the local set, and the loss of particularly

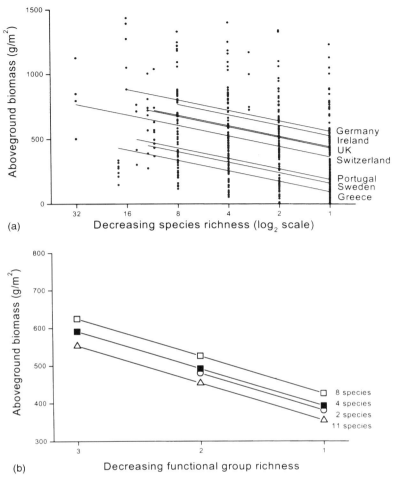

Figure 7.4 (a) Productivity declines with loss of plant species across eight sites in seven European countries in the BIODEPTH experiment (see text) (Hector *et al.* 1999). Points are total above-ground biomass for individual plots. Fitted lines are slopes from a multiple regression model with species richness on a $\log_2$ scale. Because there was no statistically significant site–treatment interaction, all sites are fitted by the same underlying slope, but differ in intercept. This figure shows real data for the hypothetical models shown in Figs 7.2(b) and 7.3. Across sites in the BIODEPTH experiment there is no relationship between biodiversity and productivity (look at the left-hand ends of the fitted regression lines, where maximum local diversity and average maximum productivity for each site form a cloud of eight points and are clearly uncorrelated). (b) Controlling for plant species richness, productivity in the BIODEPTH plots also declines with reductions in the number of plant 'functional effect groups' present. The loss of a functional group has the same proportional effect on above-ground biomass irrespective of the number of species (2–11) involved (see text). (Reproduced with permission from *Science*.)

vulnerable species in nature may not mimic what happens in the experiments particularly closely.

Consider some of the possibilities. Walker et al. (1999) have suggested that in changing environments, as formerly dominant plant species decline or are lost from the community, ecosystem processes may be maintained by functionally equivalent, previously rare species that 'swap ranks' and substitute for the lost dominants. This model requires a pool of rare, apparently redundant species to reside in most communities. However, it is just these locally rare species that may be most vulnerable to local and regional extinction (Peters & Darling 1985; Lawton 1993, 1995b). It is clear that BIODEPTH does no more than put a toe into some rather deep water.

In the longer run, of course, none of these arguments will be relevant. As global environmental change speeds up, as species distributions change substantially and communities collapse and reform, even a halving or doubling of ecosystem productivity due to changes in species richness and composition of an existing species pool may be small beer compared with the effects of changing climate, and the wholesale destruction and resorting of species assemblages.

The individualistic nature of species responses to change

As we grapple with what the natural world might be like 100 years from now, it would be enormously helpful if we could predict which species, particularly currently dominant plant species (e.g. see Chapter 8) will decline, disappear, shift their range (how quickly, and where to?) and so on. Unfortunately, given present knowledge, such predictions look like an impossibly complex task.

Without question, one of the greatest barriers to predicting how biodiversity will respond to global change is the apparently individualistic nature of species responses. By individualistic I mean that every species does something different. Sometimes the responses seem downright idiosyncratic—individualistic to the point of being quirky, even illogical (Lawton 2000). Closely related species may respond quite differently to change, whilst unrelated taxa may respond in broadly similar (but never identical) ways, and we do not understand why. Examples are many (see also Chapters 11 and 13). They include changes in the phenology of flowering by higher plants (Fitter et al. 1995), and the laying dates of birds (Crick & Sparks 1999) in response to changes in temperature; alterations in the performance of insect herbivores in response to CO_2-induced changes in their host plants (Bezemer & Jones 1998); the differential growth responses of plant species to increasing temperature (Harte & Shaw 1995) or CO_2 enrichment (Körner & Bazzaz 1996; Mooney et al. 1999); and on much longer time scales, the disaggregation and reassembly of species during glacials and interglacials.

The Quaternary record shows that migration has been the usual, long-term response of organisms to climate change (Huntley 1991, 1994; Bazzaz 1996; Shugart 1998; Chapter 10, this volume). Extensive data from a variety of taxa show that species respond individualistically, moving at different rates and in different directions (Coope 1978, 1995; Davis 1986; Foster *et al.* 1990; Elias 1991; Huntley 1991, 1994; Clark 1993; Buzas & Culver 1994; Graham *et al.* 1996; Jablonski & Sepkoski 1996; Shugart 1998). In the longer run, we therefore expect human-induced climate change to lead to large-scale readjustment and reorganization of the species interactions within, and the composition of, ecological communities across the globe. Some species have already started to move (Kozár & Dávid 1986; Pollard *et al.* 1995; Parmesan 1996; Thomas & Lennon 1999), with different species moving at different rates, and not all in the same direction (Parmesan *et al.* 1999). Pounds *et al.* (1999) describe recent altitudinal shifts in the ranges of cloud-forest birds, mammals and amphibians as 'profound and unpredictable'.

The challenge of understanding individualistic responses

The challenge is to find ways of gathering species into functional groups—sets of species with similar responses to change (Hunt *et al.* 1993; Woodward & Cramer 1996; Grime *et al.* 1997; Smith *et al.* 1997; Catovsky 1998; Walker *et al.* 1999; Chapter 8, this volume) that have some predictive power. It is responses to change that matter, not grouping by simple ecological roles of the type discussed earlier, and illustrated in Fig. 7.4(b) (nitrogen fixers, grasses and so on; Catovsky 1998; Walker *et al.* 1999; Chapter 8, this volume). Catovsky (1998) suggests the terms 'functional *response* groups' for sets of species that respond in similar ways to change, and 'functional *effect* groups' for groups based on functional roles. Previous attempts to group plants into functional effect groups have 'little predictive value' for responses to change (Mooney *et al.* 1999).

There is also the further challenge that responses to one kind of change (say increasing temperature) may not group in the same way as responses to another type of change (say reductions in rainfall), although one might hope that evolutionary constraints and life-history trade-offs would prevent a complete free-for-all. As the data come in, mapping the responses of species onto modern, reliable phylogenies may reveal order in the midst of apparent chaos. Turning apparently idiosyncratic species responses into something we can understand and predict would be a very significant step towards reducing uncertainty about the impacts of climate change on biodiversity and ecosystem processes.

Highly controlled, reductionist experiments on small parts of a system

There is a long tradition in community ecology of doing highly controlled, reductionist experiments to discover the 'local rules of engagement'—the nature and strength of the interactions between particular species (rarely more than a few), and between species and their environment (Lawton 1999, 2000). I do not believe that the approach is particularly helpful confronted with the need to predict the responses of local biodiversity and ecosystem processes to global change.

The main problem with reductionism is that it almost inevitably fails to incorporate many of the key drivers and longer-term feedbacks that exist in complex ecological systems. (Mooney *et al.* 1991, 1999; Carpenter *et al.* 1995; Körner 1995;

Chapter 11, this volume; Weiner 1996; Chapter 17, this volume). Accordingly, predictions about responses to change may be seriously in error because they fail to incorporate the effects of slow, long-term changes in the big drivers. Using mathematical models to scale up from small experiments to the big picture (e.g. Shugart 1998) is not a solution if the models also fail to incorporate the correct feedbacks (Pace 1993; Bolker et al. 1995).

Manipulation of whole systems

Given this state of affairs, the only realistic solution is to experimentally manipulate whole systems, with all the key feedbacks in place, for long enough to detect community-level responses. Then, and only then, will it be worthwhile concentrating on underlying details, once we know what the big changes really are.

This conclusion is not new (Mooney et al. 1991, 1999; Körner 1995; Chapter 11, this volume). But whole-system manipulation can be expensive, and it takes a long time to obtain answers. The 1991 paper by Mooney and colleagues is visionary and depressing. Mooney et al. are concerned with predicting ecosystem responses to elevated CO_2 concentrations and rising global temperatures. They lay out a vision of coordinated, large-scale manipulation studies of both CO_2 and temperature over a range of different types of ecosystems spanning all six of Earth's main biomes (because different types of ecosystems will probably have different responses). They recognize that, to be meaningful, experiments will need to run for at least a decade. They are level-headed enough to realize that 'the size and number of experiments that are needed is unprecedented in ecology'. The paper is depressing because a decade later the vision is nowhere near realized. There is progress, but compared with the magnitude of the challenge, it is glacial.

Worse, hardly anybody is manipulating more than one aspect of change (Hättenschwiler & Körner (1998) and Stocker et al. (1999) are notable recent exceptions), despite the urgency of understanding interactions between temperature and CO_2, temperature and rainfall, biodiversity loss and CO_2, biodiversity loss and temperature, and so on. If we are really serious about reducing uncertainty we have no choice but to study combinations of the big drivers. It will be very difficult and expensive, but trivial compared with the cost of a small war.

Manipulating whole systems is not a panacea. Such experiments deal with the middle-term impacts of change, before wholesale migration of species and local extinctions tear existing communities apart. It is also possible that every system will again turn out to be different, but my intuition says otherwise, if only because the idiosyncratic responses of individual species may be lost in a game of statistical averages when we focus on system-level properties (see also Chapter 11).

Changes in distribution

Over the next 50–100 years, changes in the rules of engagement in local communities formed from a given species pool will gradually but inexorably be swamped by wholesale changes in species distributions (see Chapter 13). We already know (see above) that communities are not going to migrate as nice coherent packages, with all their species and interactions intact. Range changes will tear apart existing species assemblages and create new ones (Peters & Darling 1985), transforming local and regional biodiversity. Estimates vary, but anywhere between 16 and 55% of the Earth's land surface could eventually undergo a change to a different biome with a doubling of atmospheric CO_2 (Neilson 1993).

How should we tackle the huge challenges posed by changes in geographical range? The first step is to recognize the difficulties we face in making useful predictions.

'Climate envelopes' as the null hypothesis

Current attempts to predict the future distributions of species focus on climate mapping (Rogers & Randolph 1993; Huntley 1994; Porter 1995; Sutherst *et al.* 1995; Jeffree & Jeffree 1996; Shugart 1998). The approach is based on the attractively simple idea that species ranges will simply follow shifts in their characteristic 'climate envelopes'. As Pacala & Hurtt (1993), Mack (1996) and Shugart (1998) amongst others have pointed out, this is tantamount to assuming that fundamental, physiologically based niches are sufficient to predict future distributions. A more cautious view would be that climate envelopes predict potential ranges (Huntley 1994) and as such constitute the biological null hypothesis. In many cases this null hypothesis will be wrong (Pacala & Hurtt 1993), for well-known reasons (Davis *et al.* 1998a,b; Lawton 2000).

In no particular order, the null hypothesis will be wrong because: (i) current ranges are not solely determined by climate, but may be strongly influenced by species interactions (Fig. 7.5) (see also Chapter 10); (ii) in the real world there are barriers to dispersal (Peters & Darling 1985; Thomas *et al.* 1992; Thomas & Jones 1993); (iii) some species may not be able to migrate fast enough to stay within tolerable limits, and will die out (Huntley 1991, 1994); (iv) new ranges and patterns of population abundance are unlikely to mirror old ranges very closely unless metapopulation processes (Hanski 1998) remain the same in different locations, which seems unlikely; and (v) future worlds will not merely have existing climates in different places, but will expose organisms to new combinations of climate and photoperiod etc. (Graham & Grimm 1990; Huntley 1991; Geber & Dawson 1993).

Evidence that the null hypothesis is wrong

For all these reasons it seems extremely unlikely that climate envelopes will make the right predictions about future distributions, but there is uncertainty about how wrong this approach might be. If climate envelopes get future distributions 'about right' (say with an error of 10 or 20%), this may be sufficient for many purposes. Common sense suggests that sometimes climate-space mapping will work and sometimes it will not. Here is a major research challenge aimed at reducing uncertainty about the long-term impacts of climate change on biodiversity.

There already exists experimental evidence that it is easy to make mistakes when

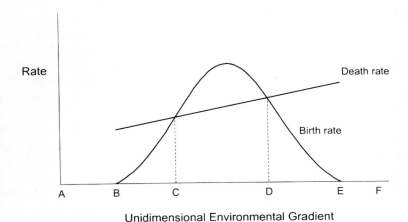

Figure 7.5 A hypothetical species has a positive birth rate over the range of conditions B–E, which define its fundamental niche. It is unable to reproduce under conditions A–B and E–F. The species experiences an average death rate due to competitors and enemies that increases across its range from left to right. Populations of the species will only persist in the region C–D where the birth rate exceeds the death rate; C–D is the realized niche of the species. Notice that in the absence of any further information it would be possible to fit a 'climate envelope' to the observed geographical range (C–D) of the species and hence conclude that its distribution is primarily determined by climate, when in fact it is jointly determined by climate and interspecific interactions. If the species range shifts because of climate change, the fitted climate envelope will only predict the new range correctly if the interspecific interactions generating the overall death rate remain identical between present and future ranges. In practice this seems rather unlikely. (From Lawton 2000, with the permission of the Ecology Institute).

predicting shifts in species ranges and abundances in a warming world (Davis *et al.* 1998a,b). These experiments used laboratory microcosms containing up to three species of *Drosophila* and a parasitoid, and mimicked long-term population dynamics along geographical-scale temperature clines. Even in this greatly simplified world, changes in the distributions and abundances of some of the species in a warming world were both unexpected and counter-intuitive. More experiments of this kind would be valuable.

Macroecology

Confronted with the challenge of reducing uncertainties about the impacts of climate change on biodiversity and ecosystem processes, particularly what will happen to local and regional biodiversity, I see powerful arguments for using macroecological patterns (Gaston & Blackburn 2000) to guide our thinking (Brown 1995; Maurer 1999; Lawton 2000). Here I will touch on two such patterns relevant to Fig. 7.1—the relationship between local and regional species richness, and between energy inputs to ecosystems and local and regional diversity.

Local vs. regional diversity in a changing world

Possible theoretical relationships between regional species richness (the pool) and local species richness (the community, local guild, assemblage, etc.) are illustrated in Fig. 7.6 (Cornell & Lawton 1992; Cornell & Karlson 1997; Srivastava 1999). To create Fig. 7.6 we need data on the species richness of a set of independent regional pools, and on the corresponding species richness of local communities drawn from these pools. Because every species cannot live everywhere and there are other strong environmental filters, with a few notable exceptions (e.g. Dawah *et al.* 1995) we do not expect every species in the regional pool to occur in every community (Zobel 1992). The simplest model of the relationship between regional and local species richness is then one of proportional sampling (Type I in Fig. 7.6a), with local richness directly proportional to, but less than, regional richness. At the other end of the continuum the graph of local vs. regional richness is a negatively accelerating curve to a plateau (Type II in Fig. 7.6a). Cramming a large literature into a very small space, in the real world Type I, and intermediate, weakly curvilinear systems outnumber strongly saturating Type II systems by a ratio of about 3:1 (Lawton 1999, 2000) from a sample of some 40 studies.

In the real world, of course, there is variation round the average Type I and Type II trends. Some of this variation (illustrated in Fig. 7.6a) is due to other significant and generally familiar ecological variables that vary from site to site within a geographical area. These local differences might be in productivity, nutrient availability, disturbance intensity, patch (island) size, the spatial arrangement and density of patches, habitat complexity (architecture), altitude, seasonality, and so on (Begon *et al.* 1990; Ricklefs & Schluter 1993; Huston 1994; Rosenzweig 1995).

Now imagine what might happen in a changing world over intermediate time scales. In other words, Fig. 7.6(b) is a cartoon of what might happen to a given regional pool of species as environments change, prior to wholesale invasion by new species. As global change speeds up, increasing local and regional rates of extinction are inevitable. The net effect will be to move existing regional species pools to the left in Fig. 7.6(b).

If this caricature has any validity, some interesting consequences follow. First, declines in the richness of extant regional species pools mean that assemblages will become more similar to one another across a region (β diversity declines). This is true for both Type 1 and Type II systems, but more so for Type II. To see this, recall that β diversity is proportional to the difference between the 1:1 line (everybody is everywhere), and the average trend line relating local to regional richness (Cornell & Lawton 1992). Thus, β diversity collapses more quickly in Type II than Type I systems in a changing world. But second, at least while they remain on the asymptote shown in Fig. 7.6(a), local richness may be maintained for longer in Type II compared with Type I systems, albeit that species composition must change and become more similar from site to site.

One can broadly see what this might mean for ecosystem processes from what has gone before. One insight in particular is worth emphasizing. For Type I and weakly curvilinear Type II systems (i.e. on current evidence the majority of

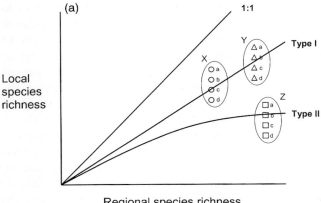

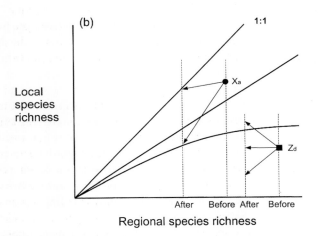

Figure 7.6 The relationship between regional and local species richness (Cornell & Lawton 1992). (a) Every species in the regional pool is unlikely to occur everywhere (the line labelled 1 : 1). Rather, local richness will usually be less than regional richness, but with two bounds defining the relationship. Type I systems show proportional sampling, in which local richness is, on average, a constant proportion of regional richness. Type II systems saturate with species, so that above a certain level of regional richness, on average local richness does not increase. Real systems may lie anywhere between Type I and Type II. Intermediate cases are referred to as 'weakly curvilinear' in the text. These are average trends. Within any one region, local ecological processes may cause variation round the trend. To illustrate this, X, Y and Z are three (hypothetical) regions, each with four local assemblages (a–d). If these local assemblages differ in habitat complexity, productivity, area, etc. (such that a > b > c > d), local richness will vary round the average trend lines as illustrated. (b) How reductions in regional and local species richness due to environmental change may move assemblages in (a). In general, species extinctions will move the richness

communities), reductions in regional species richness through, for example, habitat destruction leads inevitably to reductions in local richness, even in surviving habitat patches. A widespread misinterpretation of recent experimental work on biodiversity and ecosystem processes is that only a few tens of plant species are required to maintain ecosystem processes close to, or at, their maximum value. This interpretation is incorrect. To maintain a local assemblage of 20 plant species requires a regional species pool much larger than this. We must maintain regional biodiversity to sustain local ecosystem processes.

These very broad, even crude, predictions could also be modified in several ways. Rapid invasions of aliens could halt, or even reverse, changes in regional and local richness (Dukes & Mooney 1999). More subtly, even in the absence of invasions, significant changes in other local ecological conditions (patch size, habitat complexity, productivity, etc.) could modify, or even override the influence of changes in the regional pool, allowing some local assemblages to increase in richness even though regional diversity declines, and so on (Fig. 7.6b). Given that there are no time scales on Fig. 7.6(b) (it basically assumes that local and regional processes have had time to come to an approximate equilibrium) the potential for still more complex trajectories seems endless. But it is hard to avoid the conclusion that the general trends in indigenous regional and local species pools will be down and to the left, as we progressively simplify the Earth's biota (Lawton & May 1995; Hughes *et al.* 1997).

Energy and diversity

At biogeographic scales there are striking, generally positive (but sometimes more complex) relationships between energy inputs to ecological systems and species richness (Currie & Paquin 1987; Turner *et al.* 1987, 1988; Wright *et al.* 1993; Rosenzweig 1995; Francis & Currie 1998). The commonest metrics of energy input for terrestrial systems are potential or actual evapotranspiration, or other combinations of rainfall and/or solar energy that determine ecosystem productivity. Species richness over geographical scales increases with energy input in taxa as varied as trees, birds, mammals, amphibia, reptiles, fish, butterflies and moths. Hence if we can predict how climate will change in the future, particularly changes in solar energy input and rainfall, we can in principle predict changes in regional species richness simply by assuming that extant correlations will continue to work in the future (Kerr & Packer 1998). And if we can predict regional richness, we should be able to say something about changes in local richness, by the arguments laid out above.

---

of any particular point in (a) (e.g. Z_d, or X_a) down and to the left. But changes in local conditions (e.g. an increase in patch area or local productivity) could conceivably override the general trend and lead to an increase in local richness despite a reduction in the size of the regional pool (upper arrow from Z_d). Equally, unfavourable changes in local conditions could lead to greater losses than expected simply from changes in the regional pool (e.g. lower arrow from X_a). (From Lawton 2000, with the permission of the Ecology Institute.)

There are all kinds of reasons to disagree with this view. For example, species–energy relationships show a lot of unexplained variation. Second, we do not understand the mechanisms underlying the relationship (Srivastava & Lawton 1998) and it is dangerous to use correlations blindly. Simply assuming that present correlations will hold in the future is basically extrapolating climate spaces for many species at once, rather than one species at a time, and I have cautioned against this. But if we are to reduce uncertainty about the impacts of climate change on biodiversity and ecosystem processes, one way to do this would be to better understand the species–energy relationship. We do not know how long new species–energy relationships will take to emerge in a rapidly changing world, but it strikes me as informative to have a prediction about which direction ecosystem productivity and regional richness might be heading in a changing world, driven by changes in energy inputs to the biosphere.

Experimental model systems and controlled environment facilities

At the other end of the methodological spectrum from macroecology lie laboratory experiments with model systems housed in CEFs. I am an enthusiastic proponent of the use of CEFs for ecological research in general (see also Chapter 2) and for global change work in particular. I see their deployment as a major contribution to reducing uncertainty about the impacts of global change on communities and ecosystems, because (as advocated earlier) they allow us to study the responses of intact systems to change, and to delve into mechanisms with relative ease.

Not everybody shares this vision (e.g. Wise 1993; Carpenter et al. 1995; Carpenter 1999), and as with any tool they do some things well, other things badly, and some things not at all. In a thoughtful piece, Carpenter (1999) has argued that experiments in CEFs are a good way of teasing apart qualitative processes that we might expect to be a general feature of many systems. But if we are interested in the absolute magnitude of whole ecosystem processes and wish to compare fluxes between ecosystems, CEFs are less useful. I have argued the pros and cons of using CEFs elsewhere (Lawton 1995a, 1996, 1998, 2000), and will not rehearse the arguments here. The interested reader can make up his or her own mind by referring to the papers for and against.

CEFs housing model ecosystems vary enormously in complexity, from small beakers or Petri dishes of water swarming with bacteria, algae and protozoa, to large, extremely sophisticated facilities housing complex terrestrial ecosystems like the Ecotron at Silwood Park (Lawton et al. 1993; Lawton 1995a, 1996, 1998). In the last few years, experiments in CEFs have been used to examine several of the key relationships and processes sketched in Fig. 7.1. These include: (i) the first experimental study of the relationship between biodiversity and ecosystem processes involving changes in species richness across four trophic levels (Naeem et al. 1994); (ii) the demonstration, already referred to, that increases in temperature can lead to unexpected changes in the distribution and abundance of insect species (Davis et al. 1998a,b); (iii) startling differences in the sensitivity of different trophic levels to experimental warming (Petchey *et al.* 1999), illustrating how communities might 'fall apart' in a warming world; (iv) pioneering exploration of the codependency of ecosystem productivity on the species diversity of primary producers and decomposers (Naeem *et al.* 2000); (v) the demonstration that biodiversity regulates variability in ecosystem processes (McGrady-Steed *et al.* 1997); and (vi) studies of the complex effects of changes in biodiversity on the productivity of primary and secondary consumers embedded in complex soil food webs (Mikola & Setälä 1998).

These are valuable insights, none of which has been tested by whole-system manipulations in the field, either because of the time it would take to get an answer, the cost, shear logistical constraints, or some combination of all three! I predict that some of the most rapid advances in our understanding of the complex web of interactions shown in Fig. 7.1 over the next decade will continue to come from studies of model ecosystems housed in CEFs. They will provide only a first cut at the problems, but a first cut may be all we get to do before rapid climate change does the field manipulations for us.

Epilogue

No great new synthesis emerges from this brief review. But in the spirit of this symposium, we can celebrate how far ecology has come in understanding key parts of Fig. 7.1. I am heartened by the progress we have made in bringing together strands of science extending back half a century or more (the relationship between climate and the distribution of Earth's major biomes) with science that did not even exist a decade ago (the relationship between biodiversity and ecosystem processes).

Nevertheless, considerable challenges confront us. Those that emerge from this chapter include:

1 the need for more multisite experiments at continental scales, teasing apart the relative contributions of climate, soil and biodiversity (including species identity, species richness and functional-group richness) to ecosystem processes;
2 an urgent requirement to find ways of making sense of apparently individualistic, not to say idiosyncratic, species responses to environmental change (the problem of defining robust 'functional response groups');
3 the need for many more experiments manipulating whole systems, particularly the need for factorial experiments that simultaneously manipulate more than one of the 'big drivers' (biodiversity × climate change × CO_2, etc.);
4 greater use of work in CEFs, to tackle many of the problems that currently seem too difficult or too expensive to do in the field, for example experimental work on changes in geographical range;
5 greater exploitation of the power offered to us by macroecology—we need to understand macroecological patterns better (e.g. the energy–diversity relationship) and use them more frequently to make predictions about future directions of ecological change.

In this chapter I have deliberately ignored evolutionary responses. It is unclear, to me at least, whether this is a serious or a minor omission. Evolution in response to environmental change has the potential to greatly complicate matters and to increase uncertainty.

Some these issues currently look intractable (defining useful functional response groups, for instance), but for others we can see a clear way forward. Overall, I have no reason to doubt that ecologists will be any less successful over the next decade than they have been over the previous 30 years, as we rise to the challenge of advising society about the likely consequences of global change for biodiversity and ecosystem processes. But we have no more than a decade, and my real worries lie not with my science's ability to deliver the goods, but with society's ability to understand, and to act upon, the message.

Acknowledgements

I thank the BES and ESA for the invitation to take part in this joint symposium, two referees for helpful comments on the manuscript, and the Natural Environment Research Council for giving me time off from running it to be a scientist again for a few days!

References

Allison, G.W. (1999) The implications of experimental design for biodiversity manipulations. *American Naturalist* **153**, 26–45.

Bazzaz, F.A. (1996) Plants in Changing Environments. *Linking Physiological, Population, and Community Ecology.* Cambridge University Press, Cambridge.

Begon, M., Harper, J.L. & Townsend, C.R. (1990) *Ecology. Individuals, Populations and Communities.* Blackwell Scientific Publications, Boston, MA.

Bezemer, T.M. & Jones, T.H. (1998) Plant–herbivore interactions in elevated atmospheric CO_2: quantitative analyses and guild effects. *Oikos* **82**, 212–222.

Bolker, B.M., Pacala, S.W., Bazzaz, F.A., Canham, C.D. & Levin, S.A. (1995) Species diversity and ecosystem response to carbon dioxide fertilization: conclusions from a temperate forest model. *Global Change Biology* **1**, 373–381.

Brown, J.H. (1995) *Macroecology.* University of Chicago Press, Chicago.

Buzas, M.A. & Culver, S.J. (1994) Species pool and dynamics of marine paleocommunities. *Science* **264**, 1439–1441.

Carpenter, S.R. (1999) Microcosm experiments have limited relevance for community and ecosystem ecology: Reply. *Ecology* **80**, 1085–1088.

Carpenter, S.R., Chisholm, S.W., Krebs, C.J., Schindler, D.W. & Wright, R.F. (1995) Ecosystem experiments. *Science* **269**, 324–327.

Catovsky, S. (1998) Functional groups: clarifying our use of the term. *Bulletin of the Ecological Society of America* **79**, 126–127.

Clark, J.S. (1993) Paleoecological perspective on modeling broad-scale responses to global change. In: *Biotic Interactions and Global Change* (eds P.M. Kareiva, J.G. Kingsolver & R.B. Huey), pp. 315–332. Sinauer, Sunderland, MA.

Coope, G.R. (1978) Constancy of insect species versus inconstancy of Quaternary environments. In: *Diversity of Insect Faunas. Symposia of the Royal Entomological Society of London 9* (eds L.A. Mound & N. Waloff), pp. 176–187. Blackwell Scientific Publications, Oxford.

Coope, G.R. (1995) The effects of Quaternary climatic changes on insect populations: lessons from the past. In: *Insects in a Changing*

Environment (eds R. Harrington & N.E. Stork), pp. 29–48. Academic Press, London.

Cornell, H.V. & Karlson, R.H. (1997) Local and regional processes as controls of species richness. In: *Spatial Ecology. The Role of Space in Population Dynamics and Interspecific Interactions* (eds D. Tilman & P. Kareiva), pp. 250–268. Princeton University Press, Princeton.

Cornell, H.V. & Lawton, J.H. (1992) Species interactions, local and regional processes, and limits to the richness of ecological communities: a theoretical perspective. *Journal of Animal Ecology* **61**, 1–12.

Crick, H.Q.P. & Sparks, T.H. (1999) Climate change related to egg-laying trends. *Nature* **399**, 423–424.

Currie, D.J. & Paquin, V. (1987) Large-scale biogeographic patterns of species richness of trees. *Nature* **329**, 326–327.

Davis, M.B. (1986) Climatic instability, time lags, and community disequilibrium. In: *Community Ecology* (eds J. Diamond & T.J. Case), pp. 269–284. Harper & Row, New York.

Davis, A.J., Jenkinson, L.S., Lawton, J.H., Shorrocks, B. & Wood, S. (1998a) Making mistakes when predicting shifts in species range in response to global warming. *Nature* **391**, 783–786.

Davis, A.J., Lawton, J.H., Shorrocks, B. & Jenkinson, L. (1998b) Individualistic species responses invalidate simple physiological models of community dynamics under global environmental change. *Journal of Animal Ecology* **67**, 600–612.

Dawah, H.A., Hawkins, B.A. & Claridge, M.F. (1995) Structure of the parasitoid communities of grass-feeding chalcid wasps. *Journal of Animal Ecology* **64**, 708–720.

Dukes, J.S. & Mooney, H.A. (1999) Does global change increase the success of biological invaders? *Trends in Ecology and Evolution* **14**, 135–139.

Elias, S.A. (1991) Insects and climate change. *Bioscience* **41**, 552–559.

Fitter, A.H., Fitter, R.S.R., Harris, I.T.B. & Williamson, M.H. (1995) Relationships between first flowering date and temperature in the flora of a locality in central England. *Functional Ecology* **9**, 55–60.

Foster, D.R., Schoonmaker, P.K. & Pickett, S.T.A. (1990) Insights from paleoecology to community ecology. *Trends in Ecology and Evolution* **5**, 119–122.

Francis, A.P. & Currie, D.J. (1998) Global patterns of tree species richness in moist forests: another look. *Oikos* **81**, 598–602.

Gaston, K.J. & Blackburn, T.M. (2000) *Macroecology: Pattern and Process*. Blackwell Science, Oxford.

Geber, M.A. & Dawson, T.E. (1993) Evolutionary responses of plants to global change. In: *Biotic Interactions and Global Change* (eds P.M. Kareiva, J.G. Kingsolver & R.B. Huey), pp. 179–197. Sinauer, Sunderland, MA.

Graham, R.W. & Grimm, E.C. (1990) Effects of global climate change on the patterns of terrestrial biological communities. *Trends in Ecology and Evolution* **5**, 289–292.

Graham, R.W., Lundelius, E.L. Jr, Graham, M.A. *et al.* (1996) Spatial response of mammals to late Quaternary environmental fluctuations. *Science* **272**, 1601–1606.

Grime, J.P. (1997) Biodiversity and ecosystem function: the debate deepens. *Science* **277**, 1260–1261.

Grime, J.P., Thompson, K., Hunt, R. *et al.* (1997) Integrated screening validates primary axes of specialisation in plants. *Oikos* **79**, 259–281.

Hanski, I. (1998) Metapopulation dynamics. *Nature* **396**, 41–49.

Harte, J. & Shaw, R. (1995) Shifting dominance within a montane vegetation community: results of a climate-warming experiment. *Science* **267**, 876–880.

Hättenschwiler, S. & Körner, C. (1998) Biomass allocation and canopy development in spruce model ecosystems under elevated CO_2 and increasing N deposition. *Oecologia* **113**, 104–114.

Hector, A., Schmid, B., Beierkuhnlein, C. *et al.* (1999) Plant diversity and productivity experiments in European grasslands. *Science* **286**, 1123–1127.

Hooper, D.U. & Vitousek, P.M. (1997) The effects of plant composition and diversity on ecosystem processes. *Science* **277**, 1302–1305.

Hughes, J.B., Daily, G.C. & Ehrlich, P.R. (1997) Population diversity: Its extent and extinction. *Science* **278**, 689–692.

Hunt, R., Hand, D.W., Hannah, M.A. & Neal, A.M. (1993) Further response to CO_2-enrichment in

British herbaceous species. *Functional Ecology* 7, 661–668.

Huntley, B. (1991) How plants respond to climate change: migration rates, individualism and the consequences for plant communities. *Annals of Botany* 67 (Suppl. 1), 15–22.

Huntley, B. (1994) Plant species' response to climate change: implications for the conservation of European birds. *Ibis* 137 (Suppl. 127), 138.

Huston, M.A. (1994) Biological Diversity. *The Coexistence of Species on Changing Landscapes.* Cambridge University Press, Cambridge.

Jablonski, D. & Sepkoski, J.J. Jr (1996) Paleobiology, community ecology, and scales of ecological pattern. *Ecology* 77, 1367–1378.

Jeffree, C.E. & Jeffree, E.P. (1996) Redistribution of the potential geographical ranges of Mistletoe and Colorado Beetle in Europe in response to the temperature component of climate change. *Functional Ecology* 10, 562–577.

Kerr, J. & Packer, L. (1998) The impact of climate change on mammal diversity in Canada. *Environmental Monitoring and Assessment* 49, 263–270.

Körner, C. (1995) Towards a better experimental basis for upscaling plant responses to elevated CO_2 and climate warming. *Plant, Cell and Environment* 18, 1101–1110.

Körner, C. & Bazzaz, F.A. (1996) *Carbon Dioxide, Populations, and Communities.* Academic Press, San Diego.

Kozár, F. & Dávid, A.N. (1986) The unexpected northward migration of some species of insects in Central Europe and climate change. *Anzeiger fur Schädlingskunde, Pflanzenschutz und Umweltschutz* 58, 90–94.

Lawton, J.H. (1993) Range, population abundance and conservation. *Trends in Ecology and Evolution* 8, 409–413.

Lawton, J.H. (1995a) Ecological experiments with model systems. *Science* 269, 328–331.

Lawton, J.H. (1995b) Population dynamic principles. In: *Extinction Rates* (eds J.H. Lawton & R.M. May), pp. 147–163. Oxford University Press, Oxford.

Lawton, J.H. (1996) The Ecotron facility at Silwood Park: the value of 'big bottle' experiments. *Ecology* 77, 665–669.

Lawton, J.H. (1998) Ecological experiments with model systems. The Ecotron facility in context. In: *Experimental Ecology. Issues and Perspectives* (eds W.J. Resetarits, Jr & J. Bernardo), pp. 170–182. Oxford University Press, Oxford.

Lawton, J.H. (1999) Are there general laws in ecology? *Oikos* 84, 177–192.

Lawton, J.H. (2000) *Community Ecology in a Changing World.* Oldendorf/Luhe: Ecology Institute.

Lawton, J.H. & May, R.M. (1995) *Extinction Rates.* Oxford University Press, Oxford.

Lawton, J.H., Naeem, S., Woodfin, R.M. et al. (1993) The Ecotron: a controlled environmental facility for the investigation of population and ecosystem processes. *Philosophical Transactions of the Royal Society of London, Series B* 341, 181–194.

Mack, R.N. (1996) Predicting the identity and fate of plant invaders: emergent and emerging approaches. *Biological Conservation* 78, 107–121.

Maurer, B.A. (1999) *Untangling Ecological Complexity.* University of Chicago Press, Chicago.

McGrady-Steed, J., Harris, P.M. & Morin, P.J. (1997) Biodiversity regulates ecosystem predictability. *Nature* 390, 162–165.

Mikola, J. & Setälä, H. (1998) Relating species diversity to ecosystem functioning: mechanistic background and experimental approach with a decomposer food web. *Oikos* 83, 180–194.

Mooney, H.A., Drake, B.G., Luxmoore, R.J., Oechel, W.C. & Pitelka, L.F. (1991) Predicting ecosystem responses to elevated CO_2 concentrations. *Bioscience* 41, 96–104.

Mooney, H.A., Canadell, J., Chapin III, F.S. et al. (1999) Ecosystem physiology responses to global change. In: *The Terrestrial Biosphere and Global Change* (eds B.H. Walker, W.L. Stephan, J. Canadell & J.S.I. Ingram), pp. 141–189. Cambridge University Press, London.

Naeem, S., Thompson, L.J., Lawler, S.P., Lawton, J.H. & Woodfin, R.M. (1994) Declining biodiversity can alter the performance of ecosystems. *Nature* 368, 734–737.

Naeem, S., Hahn, D. & Schuurman, G. (2000) Producer-decomposer co-dependency influences biodiversity effects. *Nature* 403, 762–764.

Neilson, R.P. (1993) Transient ecotone response to

climatic change: some conceptual and modelling approaches. *Ecological Applications* **3**, 385–395.

Pacala, S.W. & Hurtt, G.C. (1993) Terrestrial vegetation and climate change: integrating models and experiments. In: *Biotic Interactions and Global Change* (eds P.M. Kareiva, J.G. Kingsolver & R.B. Huey), pp. 57–74. Sinauer, Sunderland, MA.

Pace, M.L. (1993) Forecasting ecological responses to global change: The need for large-scale comparative studies. In: *Biotic Interactions and Global Change* (eds P.M. Kareiva, J.G. Kingsolver & R.B. Huey), pp. 356–363. Sinauer, Sunderland, MA.

Parmesan, C. (1996) Climate and species' range. *Nature* **382**, 765–766.

Parmesan, C., Ryrholm, N., Stefanescu, C. et al. (1999) Poleward shifts in geographic ranges of butterfly species associated with regional warming. *Nature* **399**, 579–583.

Petchey, O., McPhearson, T., Casey, T. & Morin, P. (1999) Environmental warming alters food-web structure and ecosystem function. *Nature* **402**, 69–72.

Peters, R.L. & Darling, J.D. (1985) The greenhouse effect and nature reserves. *Bioscience* **35**, 707–717.

Pollard, E., Moss, D. & Yates, T.J. (1995) Population trends of common British butterflies at monitored sites. *Journal of Applied Ecology* **32**, 9–16.

Porter, J. (1995) The effects of climate change on the agricultural environment for crop insect pests with particular reference to the European corn borer and grain maize. In: *Insects in a Changing Environment* (eds R. Harrington & N.E. Stork), pp. 93–123. Academic Press, London.

Pounds, J.A., Fogden, M.P.L. & Campbell, J.H. (1999) Biological responses to climate change on a tropical mountain. *Nature* **398**, 611–615.

Ricklefs, R.E. & Schluter, D. (1993) Species Diversity in Ecological Communities. *Historical and Geographical Perspectives*. University of Chicago Press, Chicago.

Rogers, D.J. & Randolph, S.E. (1993) Distribution of tsetse and ticks in Africa: past, present and future. *Parasitology Today* **9**, 266–271.

Rosenzweig, M.L. (1995) *Species Diversity in Space and Time*. Cambridge University Press, Cambridge.

Shugart, H.H. (1998) *Terrestrial Ecosystems in Changing Environments*. Cambridge University Press, Cambridge.

Smith, T.M., Shugart, H.H. & Woodward, F.I. (1997) *Plant Functional Types: their Relevance to Ecosystem Properties and Global Change*. Cambridge University Press, Cambridge.

Srivastava, D.S. (1999) Using local-regional richness plots to test for species saturation: pitfalls and potentials. *Journal of Animal Ecology* **68**, 1–16.

Srivastava, D.S. & Lawton, J.H. (1998) Why more productive sites have more species: an experimental test of theory using tree-hole communities. *American Naturalist* **152**, 510–529.

Stocker, R., Körner, C., Schmid, B., Niklaus, P.A. & Leadley, P.W. (1999) A field study of the effects of elevated CO_2 and plant species diversity on ecosystem-level gas exchange in a planted calcareous grassland. *Global Change Biology* **5**, 95–105.

Sutherst, R.W., Maywald, G.F. & Skarratt, D.B. (1995) Predicting insect distributions in a changed climate. In: *Insects in a Changing Environment* (eds R. Harrington & N.E. Stork), pp. 59–91. Academic Press, London.

Thomas, C.D. & Jones, T.M. (1993) Partial recovery of a skipper butterfly (*Hesperia comma*) from population refuges: lessons for conservation in a fragmented landscape. *Journal of Animal Ecology* **62**, 472–481.

Thomas, C.D. & Lennon, J.J. (1999) Birds extend their ranges northward. *Nature* **399**, 213.

Thomas, C.D., Thomas, J.A. & Warren, M.S. (1992) Distribution of occupied and vacant butterfly habitats in fragmented landscapes. *Oecologia* **92**, 563–567.

Tilman, D. (1997) Distinguishing between the effects of species diversity and species composition. *Oikos* **80**, 185.

Tilman, D. (1999) The ecological consequences of changes in biodiversity: a search for general principles. *Ecology* **80**, 1455–1474.

Tilman, D., Knops, J., Wedin, D., Reich, P., Ritchie, M. & Siemann, E. (1997) The influence of functional diversity and composition on ecosystem processes. *Science* **277**, 1300–1305.

Turner, J.R.G., Gatehouse, C.M. & Corey, C.A.

(1987) Does solar energy control organic diversity? Butterflies, moths and the British climate. *Oikos* **48**, 195–205.

Turner, J.R.G., Lennon, J.J. & Lawrenson, J.A. (1988) British bird species distributions and the energy theory. *Nature* **335**, 539–541.

Walker, B., Kinzig, A. & Langridge, J. (1999) Plant attribute diversity, resilience, and ecosystem function: the nature and significance of dominant and minor species. *Ecosystems* **2**, 95–113.

Wardle, D.A., Bonner, K.I. & Nicholson, K.S. (1997a) Biodiversity and plant litter: experimental evidence which does not support the view that enhanced species richness improves ecosystem function. *Oikos* **79**, 247–258.

Wardle, D.A., Zackrisson, O., Hörnberg, G. & Gallet, C. (1997b) The influence of island area on ecosystem properties. *Science* **277**, 1296–1299.

Weiner, J. (1996) Problems in predicting the ecological effects of elevated CO_2. In: *Carbon Dioxide, Populations, and Communities* (eds C. Körner & F.A. Bazzaz), pp. 431–441. Academic Press, San Diego.

Whittaker, R.H. (1975) *Communities and Ecosystems*. Macmillan, New York.

Wise, D.H. (1993) *Spiders in Ecological Webs*. Cambridge University Press, Cambridge.

Woodward, F.I. & Cramer, W. (1996) *Plant Functional Types and Climate Change*. Special Features in Vegetation Science 12. Opulus Press, Uppsala.

Wright, D.H., Currie, D.J. & Maurer, B.A. (1993) Energy supply and patterns of species richness on local and regional scales. In: *Species Diversity in Ecological Communities. Historical and Geographical Perspectives* (eds R.E. Ricklefs & D. Schluter), pp. 66–74. University of Chicago Press, Chicago.

Zobel, M. (1992) Plant species coexistence—the role of historical, evolutionary and ecological factors. *Oikos* **65**, 314–320.

Chapter 8
Plant functional types, communities and ecosystems

J. P. Grime

Introduction

In genetics, taxonomy, evolutionary biology and biomedical science it is widely accepted that the current rapid rates of progress in these various fields of research depend in large measure upon a 'scaling down' of methods and models to the molecular level. In the midst of this exciting revolution within the biologies the relatively new science of ecology continues to seek its identity and to define its research objectives. Although this paper begins with an illustration of the use of reductionist methods in ecology, my purpose is to suggest that although molecular insights into the ecology of organisms are interesting and useful, the success or failure of our emerging science is likely to rest ultimately on our ability to understand large-scale processes. In particular, we need to know what determines the structure, biodiversity and dynamics of ecosystems and causes their properties to change from place to place, with the passage of time and under various human pressures. This suggests that in contrast to what is taking place in much of the remainder of biology, the challenge for ecologists is to 'scale up'.

How is this objective to be achieved? Can we now gain an understanding of large-scale processes sufficiently rapidly to allow effective diagnosis and intervention where ecosystems are degraded as a consequence of impacts such as pollution episodes or over-intensive exploitation? As the end of the last century approached, ecologists from several subdisciplines responded to this challenge. Later in this paper comment is made on the fascinating interaction that is taking place as ecophysiologists and population biologists converge on the ecosystem and bring their contrasted perspectives, skills and priorities to bear upon it. First, however, I shall try to justify the assertion that much of the essential information about an ecosystem can be obtained from knowledge of the life-histories and resource dynamics of its dominant plants. No apology is necessary for this concentration on the plants; approximately 99% of the biomass of an ecosystem consists of vegetation and most of the dead organic material originates directly from the plants.

One of the implications of this plant-centred approach to the study of ecosys-

Unit of Comparative Plant Ecology, Department of Animal and Plant Sciences, University of Sheffield, Sheffield, S10 2TN

J. P. GRIME

tems is recognition of the need to identify and classify plants in terms of those functional characteristics most likely to influence ecosystem properties. Whereas classical Linnaen taxonomy (fortified by modern molecular insights) tells us about the origins of plants, the quite different purpose of an ecological classification is to group plants according to their functional similarities regardless of their evolutionary and taxonomic affiliations.

The hierarchy of plant functional types

Classifications of functional types differ with respect to the geographical scale at which they are applied, the criteria they use and the purpose for which they are designed. It is often useful to recognize a continuous hierarchy or nesting of functional types. Figure 8.1 classifies functional types arbitrarily into those with *global*, *regional* or *local* applicability.

At the global scale are systems of classification that include all plants and animals, and attempt to aggregate them into a small number of primary functional types (Macleod 1894; Ramenskii 1938; MacArthur & Wilson 1967; Rabotnov 1985). Here, generality is sought at the expense of precision, and the number of useful criteria is restricted by the need to work with attributes common to all organisms. An inevitable consequence of this wide focus is a strong dependence upon life histories and the capture and utilization of resources, tissue lifespan, defence and reproduction-interrelated subjects, the study of which forms an essential unifying theme in any modern and mechanistic approach to plant ecology. Functional types corresponding to the *regional* scale depicted in Fig. 8.1 have been recognized primarily by plant geographers and physiologists seeking to explain patchiness in species distributions and vegetation types across the land surface.

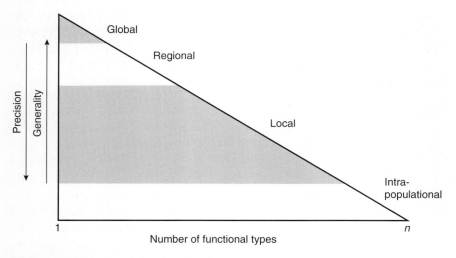

Figure 8.1 A hierarchy of plant functional types.

Much of this patchiness coincides with the world's major climatic zones and has resulted in classifications that focus upon variation in the primary mechanism of photosynthesis (C_3, C_4 or CAM) or which rely upon morphological traits suspected to confer a selective advantage under particular combinations of temperature and moisture supply (Raunkiaer 1934; Holdridge 1947; Box 1981). At a finer scale, the definitions of regional functional types may often bear a strong imprint of the underlying geology and soil types; particular importance has been attached to functional types associated with saline, calcareous, ultramific and acidic substrata. A variety of plant strategies are also evident with respect to the capacity of particular taxa to fix atmospheric nitrogen or to capture phosphorus in conditions where this element is in short supply.

At the *local* scale, it is possible to define functional types extremely narrowly, and at the base of the hierarchy of plant functional types we must consider the implications of the fine-scale variation in functional types that can exist below the level of the species and within local populations.

At the base of the hierarchy: an example

Species-rich perennial plant communities are composed mainly of strongly outbreeding species, and it is suspected that genetic variation can influence the persistence and relative abundance of individual populations (Antonovics 1976; Aarssen 1989; Burdon 1993; Miller & Fowler 1994; Prentice *et al.* 1995).

In 1997 following 3 years of systematic clonal propagation of large numbers of randomly selected cuttings from within a $10\,m \times 10\,m$ area of ancient species-rich limestone pasture at Cressbrookdale in North Derbyshire, UK, a microcosm experiment was initiated to examine the consequences of genetic impoverishment. The experiment utilizes a stock of material derived from cuttings from 16 randomly selected established individuals from each of 11 species (four grasses, three sedges, four forbs) all of which occur at high frequency in the Cressbrookdale turf. Three levels of genetic diversity have been imposed. In one treatment, every individual in each of the 11 species is unique. In a second treatment, each species is represented by four each of four randomly selected biotypes. The third treatment contains no genetic diversity in any of the 11 species; here each replicate consists of a unique combination of biotypes. In this experimental design, therefore, all microcosms are closely similar in appearance and contain communities with the same initially high level of species richness but with three contrasted levels of genetic diversity. Frequent point analysis is conducted to measure the rate of decline in species diversity. The objective is to test the prediction that as a consequence of lower variation in neighbourhood interactions the rate of loss of species evenness and diversity will be greatest in the genetically impoverished communities. It is interesting to note, however, that it is not possible to predict the identity of expanding and declining species in particular replicates; this component of the results is predicted to vary idiosyncratically according to the various combinations of biotypes in particular replicates.

By the end of the second growing season, effects of genetic impoverishment had begun to appear in the experiment and will be reported in detail elsewhere. In the communities lacking genetic diversity populations of five of the component species have begun to show higher variation in abundance between replicates in comparison with the genetically diverse populations. Repercussions at the community level are also apparent; the communities without genetic diversity have become more susceptible to disease and more variable in canopy structure (Fig. 8.2). Longer-term observation will be necessary to reveal whether these changes are a precursor to collapse of the genetically uniform communities.

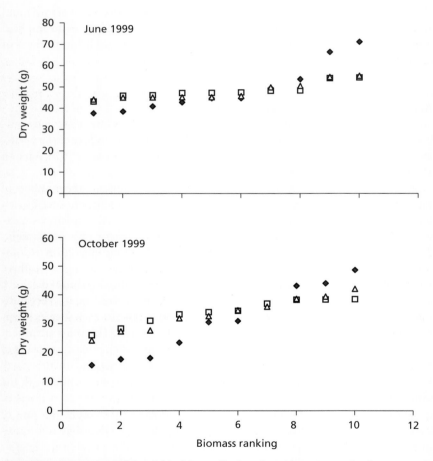

Figure 8.2 Comparison of the yield of shoot clippings from above 25 mm in plant communities with three levels of genetic impoverishment in all component populations. △: 16 biotypes per population, □: four biotypes per population ◆: one biotype per population (R.E. Booth and J.P. Grime, unpublished).

At the apex of the hierarchy

Early attempts to achieve functional classifications of terrestrial plants of world-wide applicability were based upon the architecture of the above-ground parts and through the efforts of Raunkiaer (1934) and his successors it has been possible to identify many consistent correlations between plant form and climate. A key factor explaining the popularity and success of classifications of plants by reference to architecture and climate has been the availability world-wide, as a byproduct of taxonomy, of the required information on plant form and structure.

A more difficult task is to recognize functional types that reflect the dynamic relationships of plants with other plants, with resources, herbivores, carnivores, decomposers, pathogens and symbionts, and with the disruptive effects of climatic extremes and human interference on vegetation development. Following the early ideas of Ramenskii (1938), it has been recognized that much of the variation in the adaptive responses of plants to this complex of interacting factors can be predicted and explained by recognizing the overwhelming importance of habitat productivity and the frequency and severity of biomass destruction (disturbance) in the evolution and current ecology of plants. When a templet comprising four types of plant habitat is constructed by combining the extremes of high and low productivity with high and low disturbance, it is evident that only three of the resulting contingencies are capable of supporting vegetation. Plants are excluded where low productivity coincides with frequent and severe destruction. It is proposed that the three remaining cells of the productivity:disturbance templet correspond to primary plant strategies (competitors (C), stress-tolerators (S) and ruderals (R)) of universal occurrence and with distinctive traits.

The defining characteristic of the competitor is the ability to rapidly monopolize resource capture by the spatially dynamic foraging of roots and shoots. Stress-tolerators are distinguished by the capacity of their long-lived tissues to resist herbivory and effects of environmental stress in conditions where growth is severely restricted by low rates of mineral nutrient supply. Ruderals are characterized by a short life history and the tendency to rapidly invest captured resources in the production of offspring.

The pervading nature of C, S and R axes of specialization in the established phase of plant life histories is a consequence of the occurrence of unavoidable and identifiable trade-offs in core aspects of plant function including mechanisms of resource capture, growth, storage, defence and reproduction. Contrary to the view of Grubb (1985) it is possible to reconcile the occurrence of both fundamental trade-offs and primary plant strategies with the individuality of plant ecologies. The constraints that predetermine channelling into C, S and R paths of specialization do not preclude fine-tuning of ecologies, for example through possession of particular types of juveniles (regenerative strategies), distinct phenologies, resistances to particular edaphic or climatic stresses, or association with symbionts.

J. P. GRIME

Testing predictions based on plant functional types

A protocol

In recent years a consensus has begun to develop with regard to the procedures necessary for the recognition of plant functional types and for tests of our ability to use them to interpret or predict the structure and properties of communities and ecosystems. Figure 8.3 summarizes a research protocol advocated (Grime 1993) as a general approach to this problem. The starting point is the accumulation of standardized information (*sensu* Clapham 1956; Hendry & Grime 1993) on the functional characteristics of large numbers of crops, weeds and native plant species of contrasted ecology. Examples of this approach include Grime and Hunt (1975), Jurado *et al.* (1991), Hunt *et al.* (1991), Keddy (1992), Diaz and Cabido (1997), and Diaz *et al.* (2000). The purpose of such screening is to document variation in basic attributes of plant morphology, physiology and biochemistry. This is followed by multivariate analyses to determine whether there are positive or negative associations between particular traits; when correlations between traits recur widely in screening operations conducted in many different floras we may suspect the existence of primary plant strategies. There is growing evidence that certain attributes used singly (Noble & Slatyer 1979; van der Valk 1981) or as sets (Grime *et al.* 1987a;

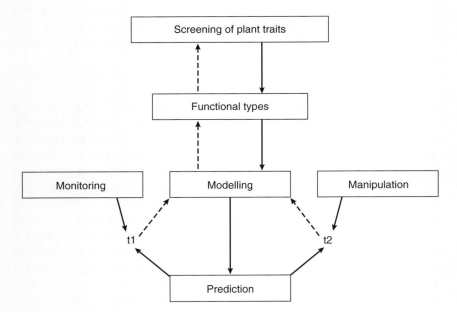

Figure 8.3 Protocol for development and testing of predictions of vegetation response to environmental change. Discrepancies revealed at t1 and t2 initiate further modelling cycles, each of which may necessitate refinement of the functional types or even additional screening.

Leishman & Westoby 1992) can provide a basis for interpreting individual plant responses to specific changes in climate, soils or management, or to predict the composition of plant communities and the relative abundance of component populations. Where sufficient and appropriate data are available to classify the majority of component species into functional types, it may be possible to predict primary and secondary successional changes and vegetation responses to specific scenarios of changed climate or land use.

In Fig. 8.3 two alternative methods are suggested by which predictions of community and ecosystem structure and properties based upon plant functional types may be tested and refined The first, illustrated in the left-hand side of the figure, involves comparison of model predictions against field data collected by surveys or long-term monitoring of permanent plots. This represents an efficient mechanism for recognizing errors and weaknesses in that, as indicated by the arrows in Fig. 8.3, discrepancies can stimulate not only changes in the model but also, where necessary, further data inputs from new screening procedures.

Monitoring studies on communities and ecosystems are few in number and in consequence alternative mechanisms of hypothesis testing (illustrated on the right-hand side of Fig. 8.3) must be used. These follow a logical pathway similar to that involving surveys and monitoring but rely upon manipulative experiments. Some of these experiments can be conducted on a small scale and involve synthesis of vegetation and ecosystems under controlled conditions (e.g. Grime *et al.* 1987b; Diaz *et al.* 1993; Lawton *et al.* 1993). Often, however, it is necessary to perform replicated manipulations at a comparatively large scale in natural environments and necessitating cooperation between ecologists and engineers (e.g. Thorpe *et al.* 1993; MacGillivray & Grime 1995; Grime *et al.* 2000).

Example 1: rarification and extinction
Under the pressures generated by rising human populations and changing patterns of land use, many plant species are experiencing decline and, in some cases, extinction. For the purposes of long-term conservation and management it is necessary, wherever possible, to interpret and predict rarification and extinction in terms of measurable functional shifts within regional or national floras and within individual plant communities. From recent research there is abundant evidence of our ability to achieve this objective.

Comparisons between various Western European countries with respect to the C–S–R profiles of expanding and declining plant species (Thompson 1994) has revealed consistent differences correlated with human population density. In countries with relatively few people, no functional differences can be detected between increasing and decreasing components of the national flora but where human density exceeds $100\,km^{-2}$ a polarization is evident between declining stress-tolerators and expanding C- or R-strategists (Fig. 8.4). It would appear, therefore, that human population density can provide a convenient surrogate for a set of landscape changes that above a critical threshold have the combined effect of causing a predictable functional shift.

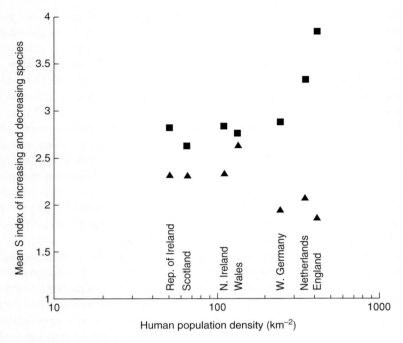

Figure 8.4 Relationship between mean S radius of increasing (▲) and decreasing (■) species and human population density in seven European countries. S radius of two groups not significantly different in Scotland, Northern Ireland or Wales. Two groups are significantly different in R of Ireland, England, W. Germany and the Netherlands.

By monitoring changes in the abundances of primary and secondary C–S–R strategies in herbaceous vegetation in permanent plots it is possible to detect functional shifts within plant communities and to discriminate between effects of eutrophication, dereliction and increased disturbance (Hodgson 1991; Bunce et al. 1999). The same technique also provides an early warning system for changes likely to threaten the survival of local populations of rare plant species (Thompson & Jones 1999; Wilson 1998).

Example 2: ecosystem resistance and resilience
All ecosystems are subject to impacts that affect their functioning and, in extreme form, may threaten their viability and survival. The forces responsible may be climatic (droughts, frost, fires, floods, windstorms) or biotic (grazing episodes, disease outbreaks, cultivation, pollution, urbanization). Building on Levitt's (1975) pioneering and largely physiological broad surveys of plant responses to environmental stresses, Westman (1978) identified two key aspects of the response of an individual plant or a stand of vegetation to an extreme event. Resistance was

defined as the ability of the plant biomass to resist displacement from control levels. Resilience was recognized as the speed and completeness of the subsequent return to control levels.

There are at least three main hypotheses that can be put forward to devise predictions of resistance and resilience in particular ecosystems. The first arises from an historical perspective; it is argued that responses to an extreme event will be strongly affected by past exposures to similar events (e.g. Sankaran & McNaughton 1999). This theory clearly depends upon the assumption that extreme events select species or genotypes with traits that confer resistance and/or resilience. Whether this occurs or not is likely to be affected by the frequency and severity of events and the nature and potency of the selection forces operating on component populations in the periods between extreme events.

The second hypothesis (MacArthur 1955; Elton 1958; Ehrlich & Ehrlich 1981; Tilman & Downing 1994) proposes that resistance and resilience will be higher in species-rich communities. This theory is based on the assumption that species-rich vegetation may contain a greater diversity in the genetic traits conferring tolerance and recovery. The weakness of this theory is that the same diversity that may ensure the presence of advantageous traits within a subset of the community of plants is likely to dictate that there are other subsets that lack resistance or resilience. Clearly, the highest resistance or resilience is likely to be observed where the requisite traits occur in all of the vegetation regardless of whether it is composed of many or a few species. Some monocultures are capable of remarkable resistance; an example is provided by the flood and fire history of the coastal redwoods of California (Stone & Vasey 1968).

A third set of predictions arises directly from the triangular model of primary plant strategies. Here the predictions are distinctive and specific in proposing: (i) a trade-off between the two very different sets of traits associated with resistance and resilience (Leps *et al.* 1982; MacGillivray & Grime 1995); and (ii) a predictable relationship between resistance and the particular set of traits identified with stress tolerance and between resilience and the sets of traits associated with ruderals, and to a lesser extent, competitors. The mechanistic and specific character of the predictions arising from CSR theory make them relatively easy to test by observation and experiment. However, in devising tests of these implications of the triangular model, four points of clarification are necessary:

1 The prediction of greater resistances of stress-tolerators to extreme events is founded on the assumption that during the evolution of the species exploiting chronically unproductive, mineral nutrient-limited conditions, selection has led to a long lifespan both in the individuals and in their vegetative tissues. An inevitable consequence of this longevity is increased experience of extreme events, particularly those arising from climatic fluctuations. However, the mechanisms promoting resistance in stress-tolerators are not confined to those related to greater past exposure to extreme events. Powerful selection forces are also likely to arise from the greater ecological and evolutionary penalties associated with low

resistance to damage in stress-tolerators; this is because the capacity for replacement of lost tissues is severely limited in these slow-growing plants. Following these arguments it may be expected that many stress-tolerators will exhibit resistance to several different kinds of extreme events, a prediction supported by the investigations of cotolerance of different climatic stresses documented by Levitt (1975). It does not follow, however, that all stress-tolerators will be highly resistant to all types of extreme events; the spectrum of resistances is likely to be strongly dependent upon the evolutionary history of the species or population concerned.

2 Qualifications are also necessary with respect to predictions of the resistances of stress-tolerators to mechanical damage. The majority of plants of unproductive habitats have tough foliage and are resistant to attack by generalist herbivores. It is also evident that the physical structure of the shoots of many stress-tolerators confers resistance to other forms of mechanical damage including trampling (Liddle 1975). However, circumstances arise in which the greater toughness of the foliage of stress-tolerators cannot be expected to confer a significant degree of differential resistance; examples include damage by more catastrophic forces such as fire, flood, wind, the attentions of large herbivores such as elephant and giraffe, or the operation of harvesting and mowing machines.

3 In comparison with the rather complex arguments surrounding the concept of resistance, the phenomenon of resilience is relatively straightforward both in theory (De Angelis 1980) and operationally. Regardless of the agency responsible for an extreme event, greater resilience is to be expected in ruderal and competitive strategists on the basis of their faster rates of resource capture and growth. It is also predictable that resilience will be most readily achieved by ruderals because they are composed of species with shorter life histories than competitors, and their communities are relatively unstructured and rapidly reassembled.

4 Although predictions of resistance and resilience using CSR theory can be applied rapidly across ecosystems of widely contrasted productivities and species composition, there is no justification for exclusive reliance on this approach. For the study of responses to particular types or intensities of perturbation there are opportunities for more refined analysis and prediction utilizing additional plant attributes. Reference to the regenerative strategies of plants may be particularly helpful; here an excellent example is the capacity to accumulate large banks of persistent seeds Where this trait is allied to the ruderal strategy, as in ephemeral vegetation situated near to the fluctuating water levels of lakes (van der Valk & Davis 1978; Furness & Hall 1981; Singer *et al.* 1996) very precise predictions of resilience can be made that refer not only to the restoration of biomass but also to the degree to which the original species composition of the community is reconstituted.

An early account of an attempt to examine the use of specified criteria as predictors of resistance and resilience is that of Leps *et al.* (1982) who examined the effects of the 1976 drought on two neighbouring grasslands similar in diversity but differing in productivity at a study site in (the former) Czechoslovakia. One of the grasslands was of recent origin and consisted of fast-growing ruderal and competitive strategists growing on a fertile soil. The second grassland was of greater antiquity and lower productivity and was composed of relatively slow-growing stress-tolerant species. The results of the investigation are consistent with predictions of resistance derived from CSR theory in that the depression of yield caused by the drought was greater in the more productive vegetation on fertile soil. Also in accordance with plant strategy theory, the rate of recovery (resilience) following the drought was higher in the vegetation composed of populations of ruderal and competitive strategists.

A similar study was reported by Tilman and Downing (1994) who compared the impacts of a summer drought on areas of grassland of contrasted productivity and species composition. In this case, however, the monitored vegetation consisted of experimental plots, some of which had been subjected to applications of nitrogenous fertilizer over the decade preceding the drought converting the original unproductive species-rich prairie community to a species-poor assemblage of robust, fast-growing perennials. In this investigation, resistance to the drought was higher in the unfertilized prairie community. As Givnish (1994) and Huston (1997) point out, the most likely explanation for the greater resistance to drought of the vegetation in the unfertilized plots was the lower water use and physiological resistance to moisture stress that would be expected in a natural prairie community as compared to a stand of productive vegetation on nitrogen-rich soil. On this basis, there appears to be a close correspondence between the study of Tilman and Downing (1994) and that of Leps *et al.* (1982); in both studies the impact of drought was strongly correlated with site productivity and vegetation responses were predictable from CSR theory.

Two further experimental studies have confirmed the usefulness of CSR theory as a predictor of resistance and resilience. The first, reported by MacGillivray & Grime (1995) compares the response of five different herbaceous communities to controlled applications of drought, late frost, and fire. The results for individual species calculated as weighted averages for each community reveal that both resistance and resilience are predictable from the functional characteristics of component plant species.

Recently, data have become available (Grime *et al.* 2000) from the first 5 years of a long-term experiment in which climate manipulations using non-invasive techniques have been applied to two calcareous grasslands closely similar in species richness but contrasted in productivity and functional composition. The first grassland situated at Wytham in Oxfordshire is representative of the rapidly expanding proportion of the English landscape that has been impacted by eutrophication and disturbance. It is of recent origin following abandonment of arable cultivation. The soil contains residues of mineral fertilizer, and the flora consists mainly of competitive or ruderal strategists. The second site, at Buxton in north Derbyshire, is an ancient pasture that has never been fertilized and is an isolated relic of the unproductive grasslands, dominated by long-lived, slow-growing perennial plants, whch extended over large areas of countryside prior to the introduction of intensive agriculture. The results based upon a principal components analysis (Fig. 8.5) summarize the extent of changes in plant species composition at both sites over a 5-year period (1994–98) in which treatments were applied to elevate winter temperature (+3°C), inflict summer drought (no rainfall in July and August) or supplement summer rainfall (to 20% above the long-term average from June to September). Because Wytham and Buxton have almost no plant species in common, it was possible by conducting the principal component analysis (PCA) analysis on the combined floristic data from the two sites to allow the first axis of the PCA analysis to effectively separate Wytham and Buxton permitting a direct comparison of treatment trajectories at the two sites using PCA axes 2 and 3. The results (Fig. 8.5) expose clearly the very contrasted responses of the vegetation at the two sites. Over 5 years of climate manipulation the composition of the stress-tolerant vegetation in the treated plots at Buxton varied hardly at all. At Wytham large changes were detected; PCA axis 3 reveals a successional path followed

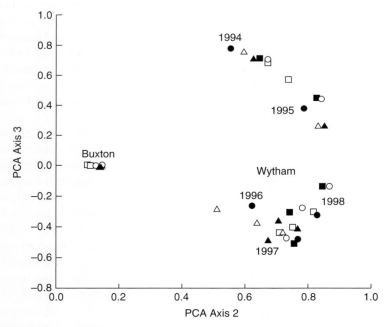

Figure 8.5 Principal component analysis of the combined floristic data from Buxton and Wytham for all treatments over 5 consecutive years. Proportion of variance accounted for by the PCA axes: Axis 1 (not shown), 46%; axis 2, 27%; axis 3, 12%. Experimental treatments. ■ Control, ● Summer drought, □ Supplementary summer rain, ▲ Elevated winter temperature (+3°C), △ Elevated winter temperature and summer drought, ○ Elevated winter temperature and supplementary summer rain.

regardless of treatment, and axis 2 reflects marked treatment effects superimposed on the successional trend.

Example 3: ecosystem productivity
The production of living matter in terrestrial ecosystems ultimately depends upon the supply, on an annual basis, of the essential resources for plant growth and the rates at which carbon, energy and mineral nutrients are captured by the vegetation and converted to plant tissue. Productivity is also affected by the rates at which plant parts are consumed by herbivores and pathogens or removed by harvesting; if a high proportion of the plant biomass is continuously destroyed there is a threshold of damage above which even fast-growing vegetation cannot sustain high rates of production. It is also worth noting that the controlling effects on productivity of plants themselves are not restricted to resource capture and growth. It has been demonstrated in theoretical models (Grime 1987; Aerts & van der Peijl 1993; Pastor & Cohen 1997) and in numerous experiments (e.g. Aber & Melillo 1982; Berendse *et al.* 1989; Wedin & Tilman 1990; Wardle *et al.* 1997) that the rate at which mineral nutrients (so often the limiting resource in natural ecosystems!) are recycled within an ecosystem is strongly affected by the functional characteristics of the dominant plant species. Whereas the short lifespan and weaker defences of the leaves and roots of competitor and ruderal strategists make them subjects of rapid decay and mineral nutrient release, the greater longevity and high carbon: mineral nutrients ratio of the tissues of stress-tolerators tend to reduce the rate of recycling by imposing long residence times for mineral nutrients in both living plant parts and litter (Grime *et al.* 1997; Aerts & Chapin 2000).

Against a theoretical background in which soil fertility, climate and the traits of dominant plants were widely suspected to be acting as the overriding controllers of ecosystem productivity, considerable interest and controversy was generated when, in the period 1994–99, several papers appeared purporting to demonstrate immediate benefits to ecosystem productivity arising from high species richness in experimental plant assemblages (Naeem *et al.* 1994; Tilman 1996, 1999; Tilman *et al.* 1996; Hector *et al.* 1999). In each case it was suggested that benefits arose in the species-rich mixtures from the presence of a wider range of morphologies and physiologies, generating complementary and more complete exploitation of resources. Interest in these publications extending beyond the realm of ecology was stimulated by commentaries (Karieva 1994, 1996) suggesting that studies of this kind provided a justification for the conservation of species-rich ecosystems.

Subsequently, doubts have been cast on the validity of the conclusions drawn by the authors of these papers and these have been reviewed in detail elsewhere (André *et al.* 1994; Aarssen 1997; Garnier *et al.* 1997; Grime 1997; Huston 1997; Hodgson *et al.* 1998; Huston *et al.* 2000; Wardle *et al.* 2000). It appears that, in these experiments, the higher productivies attributed to high species richness were in reality due to the presence in the more diverse communities of dominant species with traits (e.g. large size, rapid growth or capacity to fix atmospheric nitrogen) conducive to high productivity. Other experiments (Hooper & Vitousek 1997;

Tilman *et al.* 1997; Wardle *et al.* 1997) have failed to provide convincing support for effects of high species richness on ecosystem functions; the most parsimonious explanation (Grime 1987, 1997) for the data presented in all these papers is that the ecosystem properties examined were controlled by the functional traits of a relatively small number of species accounting for a high proportion of the total plant biomass.

On first inspection, synthesis of ecosystems from seed mixtures varying in richness appears to provide a direct and informative method by which to determine whether productivity is controlled by the characteristics of dominant species or by the species richness of vegetation. There are two main reasons why this approach has not so far resulted in definitive tests.

1 A crucial flaw in the experiment by Naeem *et al.* (1994) was the inclusion of species in the species-rich assemblages that were not represented in the species-poor mixtures. In an attempt to rectify this design fault, experiments such as those of Tilman *et al.* (1997) and Hector *et al.* (1999) have used species mixtures in which the composition of the seed mixture sown into individual replicate plots is based upon a random draw of species. However, as pointed out by Huston (1997), this procedure does not avoid the risk that the species-rich mixtures more often than the species-poor mixtures will contain species of high potential productivity. This, of course, arises as a simple consequence of the greater chance of representation of all types of species, including the most productive ones, in the species-rich mixtures. This sampling effect (Wardle 1999) leads to a confounding of the two possible explanations for any rise in productivity that is found to be associated with increasing the number of species in the seed mixture. Higher yield might be the result of greater species richness but it is also possible that this effect could be the result of including particular species of high potential productivity.

2 In order to differentiate between the two possible explanations (described under (1) above) for any rise in productivity associated with species richness of the seed mixture, it is imperative that data analysis and interpretation involves measurements of the numbers of species surviving and their relative abundance in the synthesized communities. In the experiments reported by Naeem *et al.* (1994), Tilman *et al.* (1996, 1997) and Hector *et al.* (1999), estimates of productivity are plotted against the numbers of species sown in the seed mixtures. In the absence of data on the actual membership of the synthesized communities and the composition of harvested biomass, it is not possible to validate the claims in all four of these published studies that benefits to productivity of species richness *per se* have been demonstrated.

In an attempt to circumvent some of the difficulties outlined in (1) and (2) and, in particular to avoid problems with the creation of communities from seed, R.E. Booth and J.P. Grime (unpublished) measured productivity in microcosms in which assemblages were created from vegetative transplants of 12 species removed from a small area of ancient calcareous pasture in north Derbyshire. By this method, the vagaries of seedling establishment were avoided and it was possible to

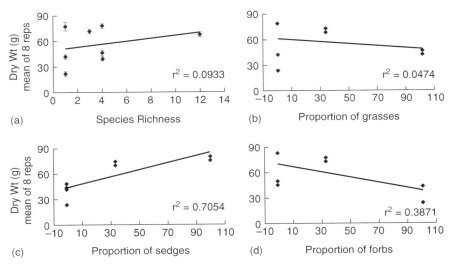

Figure 8.6 Relationships of plant species richness and functional composition to harvested shoot biomass in experimental plant communities. (a) Species richness; (b) grasses; (c) sedges; (d) forbs.

exercise close control of both the species richness and functional composition of the synthesized communities.

The results of this experiment (Fig. 8.6a) provided no evidence of beneficial effects of increasing species richness on productivity. However, interesting insights became available when the data were rearranged (Fig. 8.6b–d) to examine the relationship between productivity and the abundance of grasses, sedges and forbs in the experimental assemblages. Whereas no effect of increasing abundance on productivity was detected in the grasses and forbs, a substantial and statistically significant benefit was associated with the presence of sedges (Fig. 8.6c). In physiological terms, this effect is not hard to interpret; grassland sedges possess dauciform roots (Davies *et al.* 1973), specialized structures which are suspected to be capable of facilitating the mineralization and capture of phosphorus, the limiting element for plant growth in many calcareous ecosystems.

These experimental results have two main implications concerning the use of synthesized ecosystems to investigate the control of productivity. The first becomes obvious if we consider what would have happened if communities drawing upon the same 12 species but varying in species richness had been allowed to assemble from randomly selected seed mixtures following the experimental protocol of Naeem *et al.* (1994) or Tilman *et al.* (1996). It seems most likely that sedges would have prospered under the conditions of the experiment and their more frequent inclusion in the species-rich seed mixtures would have led inevitably to confounding of a beneficial 'sedge effect' with increasing species richness and rising productivity. Only careful analysis taking account of the species composition of harvested material would have prevented false description of this sedge effect as one arising from species richness.

The second implication of the results of the Booth and Grime (2001) experiment is that vegetation development in ancient, unproductive, calcareous pastures involves processes that, at least in the short term (2 years), were not expressed in the microcosms. A conclusion that can be drawn from Fig. 8.6 is that productivity in these ecosystems is strongly affected by the capacity of sedges to capture phosphorus and if we rely exclusively upon inferences based upon the microcosm assemblages we might suppose that natural selection in the field would drive the composition of such infertile grasslands towards a turf consisting exclusively of sedges. In reality ancient infertile calcareous pastures contain a large sedge component (Willis 1989) but there is also present a rich diversity of grasses, forbs, small shrubs and bryophytes. How can these insights from microcosm experiments and field observations be reconciled?

In addressing this paradox perhaps the most important first step we can take is to acknowledge the limitations of experiments. Here it is instructive to consider a comment from the world of quantum mechanics:

> In the beginning natural philosophers tried to understand the world around them . . . . They hit upon the idea of contriving artificially simple situations . . . . Experimental science was born. But experiment is a tool. The aim remains to understand the world. To restrict quantum mechanics to be exclusively about piddling laboratory operations is to betray the great enterprise. (Bell 1990)

The message from physics for ecosystem ecology is clear. Much can be learned, and perhaps some things can *only* be learned, from the tactic associated with the synthesis and study of simplified systems. However, the approach must not become an end in itself. The insights from synthetic ecosystems will need to be continuously confronted by field realities. The dangers in exclusive reliance on simplified ecosystems are only too apparent in the unsubstantiated claims for benefits of species richness to productivity contained in some recent publications.

When insights from the synthesis of ecosystems are viewed in a broader perspective it is not difficult to envisage why the selective advantage of sedges in a phosphorus-stressed ecosystem does not lead to exclusive dominance by these species. The inconclusive nature of competitive interactions has been well documented (Hillier *et al.* 1990) in circumstances where life histories are long, growth is slow, biomass is continuously lost to herbivores, and the quantity and vigour of the vegetation is low. In such conditions, the capacity of sedges to capture phosphorus is not translated into exclusive dominance of the vegetation because the mineral nutritional and grazing constraints on biomass accumulation prevent the dynamic foraging of shoots and roots and the accumulation of the density of vegetation necessary for competitive dominance. The inability of sedges to dominate unproductive grassland may be further explained by the seasonal delay in the leaf expansion exhibited by sedges (Grime et al. 1985); this dependence upon warmer temperatures is likely to provide numerous opportunities for incursions by less thermophilous grasses, forbs and bryophytes.

What can be concluded concerning the failure to establish a consistent relationship between species diversity and productivity? Three points deserve consideration.

1 Although the composition and species richness of plant communities is dependent upon the fitness of component plant populations, fitness itself does not invariably involve the assumption of traits that maximize dry-matter production. In conditions of high potential productivity and low incidence of vegetation disturbance fitness is likely to depend upon high competitive ability and the struggle for dominance between contending species, a process that will tend to drive up productivity. However, there is abundant evidence that on less fertile soils fitness may depend upon attributes that, at least in the short term, can have the effect of driving down ecosystem productivity.

2 Whereas some theoretical models suggest that increasing species richness may increase productivity, there is a very extensive literature indicating a unimodal relationship in which high botanical diversity occurs at rather low productivity (Grime 1973; Al-Mufti et al. 1977; Janssens et al. 1998; Grace 1999) and high production coincides with dominance by a small number of species.

3 Failure to detect *immediate* benefits of species richness on productivity or other ecosystem properties does not mean that losses of plant diversity should be ignored. The priority in the next phase of research on declining plant diversity should be to consider its long-term consequences for ecosystem assembly and function. Losses in species richness may be associated with less obvious impacts which operate through failures in filter and founder effects controlling the recruitment of dominant species (Grime 1998). A progressive loss of ecosystem functions may be predicted in circumstances where vegetation patch dynamics and ecosystem reassembly continue against the background of declining diversity in the pool of colonizing propagules. Effects on the recruitment of dominants, rather than the immediate consequences of declining richness *per se* deserve our curiosity and attention.

The debate about the relationship between species diversity and ecosystem function was prompted by animal ecologists testing theories originating from population biology. Their conclusions have been challenged by plant ecophysiologists and the discussion is now engaging the attention of an even wider range of subdisciplines within ecology. I submit that this is a benchmark debate in the sense that it is a forerunner of many more discussions that will be necessary as ecology matures into a coherent predictive science.

> It is at this stage that relationships within the family are tested and begin to determine how quickly the jigsaw is completed. At one extreme is the family in which overall progress is kept under review so that connections between

developing islands are established as early as possible allowing the completed picture to be visualized and the remaining gaps to be filled with minimal delay. Not unknown, however, is the family for which the jigsaw provides a longer and more enjoyable diversion in which the construction of each island becomes an absorbing activity in its own right with individual logic, rules and rivalries with neighbouring islands. (Grime 1985)

These recent exchanges between plant and animal ecologists are a healthy development and suggest that a constructive course can be followed as we move into the late stages of assembling the ecological jigsaw.

Acknowledgements

This paper is dedicated to W. H. Pearsall, A. R. Clapham and C. D. Pigott, the founders of plant ecology at Sheffield University, and to all past and present members of the Unit of Comparative Plant Ecology.

References

Aarssen, L.W. (1989) Competitive ability and species coexistence. a 'plant's-eye' view. *Oikos* **56**, 386–401.

Aarssen, L.W. (1997) High productivity in grassland ecosystems: effected by species diversity or productive species? *Oikos* **80**, 183–184.

Aber, J.D. & Melillo, J.M. (1982) Nitrogen immobilization in decaying hard wood leaf litter as a function of initial nitrogen and lignin content. *Canadian Journal of Botany* **60**, 2263–2269.

Aerts, R. & Chapin, F.S. III (2000) The mineral nutrition of wild plants revisited: a re-evaluation of processes and patterns. *Advances in Ecological Research* **30**, 1–67.

Aerts, R. & van der Peijl, L. (1993) A simple model to explain the dominance of low-productive perennials in nutrient-poor habitats. *Oikos* **66**, 144–147.

Al-Mufti, M.M., Sydes, C.L., Furness, S.B., Grime, J.P. & Band, S.R. (1977) A. quantitative analysis of shoot phenology and dominance in herbaceous vegetation. *Journal of Ecology* **65**, 759–791.

André, M., Brechignac, F. & Thibault, P. (1994) Biodiversity in model ecosystems. *Nature* **371**, 565.

Antonovics, J. (1976) The population genetics of mixtures. In: *Plant Relations in Pastures* (ed. J. R. Wilson), pp. 233–252. CSIRO, Melbourne.

Bell, J.S. (1990) Against 'measurement'. *Physics World* **3**, 33–40.

Berendse, F., Bobbink, R. & Rouwenhorst, G. (1989) A comparative study on nutrient cycling in wet heathland ecosytems II. Litter decomposition and nutrient mineralization. *Oecologia* **78**, 388–348.

Box, E.O. (1981) *Macroclimate and Plant Forms. An Introduction to Predictive Modelling in Phytogeography*. Kluwer, The Hague.

Bunce, R.G.H., Barr, C.J., Gillespie, M.K. et al. (1999) *Vegetation of the British countryside. The Countryside Vegetation System*. ECOFACT, 1 London DETR.

Burdon, J.J. (1993) The structure of pathogen populations in natural plant communities. *Annual Reviews of Phytopathology* **31**, 305–323.

Clapham, A.R. (1956) Autecological studies and the 'Biological Flora of the British Isles'. *Journal of Ecology* **44**, 1–11.

Davies, J., Briarty, L.G. & Rieley, J.O. (1973) Observations on the swollen lateral roots of the Cyperaceae. *New Phytologist* **72**, 167–174.

De Angelis, D.L. (1980) Energy flow, nutrient cycling and ecosystem resilience. *Ecology* **61**, 764–771.

Diaz, S. & Cabido, M. (1997) Plant functional types

and ecosystem function in relation to global change. *Journal of Vegetation Science* **8**, 463–474.

Diaz, S., Grime, J.P., Harris, J. & Macpherson, E. (1993) Evidence of a feedback mechanism limiting plant response to elevated carbon dioxide. *Nature* **364**, 616–617.

Ehrlich, P.R. & Ehrlich, A.H. (1981) . *Extinction. The Causes and Consequences of the Disappearance of Species.* Random House, London.

Elton, C.S. (1958) The reasons for conservation. *The Ecology of Invasions by Animals and Plants.* Chapman & Hall, London.

Furness, S.B. & Hall, R.H. (1981) An. explanation for the intermittent occurrence of *Physcomitrium sphaericum. Journal of Bryology* **11**, 733–742.

Garnier, E., Navas, M.L., Austin, M.P., Lilley, J.M. & Gifford, R.M. (1997) A problem for biodiversity-productivity studies: how to compare the productivity of multispecific plant mixtures to that of monocultures. *Acta Ecologica* **18**, 657–670.

Givnish, T.J. (1994) Does diversity beget stability? *Nature* **371**, 113–114.

Grace, J.B. (1999) The factors controlling species density in herbaceous plant communities. In: *Perspectives in Plant Ecology, Evolution and Systematics* **3**, 1–28. Urban and Fischer-Verlag.

Grime, J.P. (1973) Competitive exclusion in herbaceous vegetation. *Nature* **242**, 344–347.

Grime, J.P. (1985) Towards a functional description of vegetation. In: *Population Structure of Vegetation* (ed. J. White), pp. 503–514. Junk, Dordrecht.

Grime J.P. (1987) Dominant and subordinate components of plant communities. implications for succession, stability and diversity. In: *Colonisation, Succession and Stability* (eds A.J. Gray, M.J. Crawley & P.J. Edwards), pp. 413–428. Blackwell Scientific Publications, Oxford.

Grime, J.P. (1993) Ecology sans frontières. *Oikos* **68**, 385–392.

Grime, J.P. (1997) Biodiversity and ecosystem function. the debate deepens. *Science* **277**, 1260–1261.

Grime, J.P. (1998) Benefits of plant diversity to ecosystems: immediate, filter and founder effects. *Journal of Ecology* **86**, 902–910.

Grime, J.P. & Hunt, R. (1975) Relative growth-rate: its range and adaptive significance in a local flora. *Journal of Ecology* **63**, 393–422.

Grime, J.P., Brown, V.K., Thompson, K. *et al.* (2000) The response of two contrasting limestone grasslands to simulated climate change. *Science* **289**, 762–765.

Grime, J.P., Hunt, R. & Krzanowski, W.J. (1987a) Evolutionary physiological ecology of plants. In: *Evolutionary Physiological Ecology* (ed. P. Calow), pp. 105–125. Cambridge University Press, Cambridge.

Grime, J.P., Mackey, J.M.L., Hillier, S.H. & Read, D.J. (1987b) Floristic diversity in a model system using experimental microcosms. *Nature* **328**, 420–422.

Grime, J.P., Shacklock, J.M.L. & Band, S.R. (1985) Nuclear DNA. contents, shoot phenology and species coexistence in a limestone grassland community. *New Phytologist* **100**, 435–444.

Grime, J.P., Thompson, K., Hunt, R. *et al.* (1997) Integrated screening validates primary axes of specialisation in plants. *Oikos* **79**, 259–281.

Grubb, P.J. (1985) Plant populations and vegetation in relation to habitat, disturbance and competition: Problems of generalisation. In: *The Population Structure of Vegetation* (ed. J. White), pp. 595–621. Junk, Dordrecht.

Hector, A., Schmid, B., Beierkuhnlein, C. *et al.* (1999) Plant diversity and productivity in European grasslands. *Science* **286**, 1123–1127.

Hendry, G.A.F. & Grime, J.P. (1993) *Methods in Comparative Plant Ecology. A Laboratory Manual.* Chapman & Hall, London.

Hillier, S.H., Walton, D.W.H. & Wells, D.A. (1990) *Calcareous Grasslands. Ecology and Management.* Bluntisham, Huntingdon.

Hodgson, J.G. (1991) The use of ecological theory and autecological databases in studies of endangered plant and animal species and communities. *Pirineos* **138**, 3–28.

Hodgson, J.G., Thompson, K., Wilson, P.J. & Bogaard, A. (1998) Does biodiversity determine ecosystem function? The Ecotron experiment revisited. *Functional Ecology* **12**, 843–848.

Holdridge, L.R. (1947) Determination of world plant formations from simple climatic data. *Science* **105**, 367–368.

Hooper, D.U. & Vitousek, P.M. (1997) The effects of

plant composition and diversity on ecosystem processes. *Science* **277**, 1302–1305.

Hunt, R., Hand, D.W., Hannah, M.A. & Neal, A.M. (1991) Response to CO_2 enrichment in 27 herbaceous species. *Functional Ecology* **5**, 410–421.

Huston, M.A. (1997) Hidden treatments in ecological experiments: Re-evaluating the ecosystem function of biodiversity. *Oecologia* **110**, 449–460.

Huston, M.A., Aarssen, L.W., Austin, M.P. *et al.* (2000) Technical Comment on 'Plant diversity and productivity experiments in European grasslands'. *Science* **280**, 1255–1256.

Janssens, F., Peeters, A., Tallowin, J.R.B. *et al.* (1998) Relationship between soil chemical factors and grassland diversity. *Plant and Soil* **202**, 69–78.

Jurado, E., Westoby, M. & Nelson, D. (1991) Diaspore weight, dispersal, growth form and perenniality of central Australian plants. *Journal of Ecology* **79**, 811–828.

Karieva, P. (1994) Diversity begets productivity. *Nature* **368**, 686–287.

Karieva, P. (1996) Diversity and sustainability on the prairie. *Nature* **379**, 673–674.

Keddy, P.A. (1992) A pragmatic approach to functional ecology. *Functional Ecology* **6**, 621–626.

Lawton, J.G., Naeem, S., Woofin, R.M. *et al.* (1993) The Ecotron: a controlled environmental facility for the investigation of populations and ecosystem processes. *Philosphical Transactions of the Royal Society of London, Series B* **341**, 181–194.

Leishman, M.R. & Westoby, M. (1992) Classifying plants into groups on the basis of associations of individual traits: evidence from Australian semi arid woodlands. *Journal of Ecology* **80**, 417–424.

Leps, J., Osbomova-Kosinova, J. & Rejmanek, K. (1982) Community stability, complexity and species life-history strategies. *Vegetatio* **511**, 53–63.

Levitt, J. (1975) *Responses of Plants to Environmental Stresses*. Academic Press, New York.

Liddle, M. (1975) A selective review of the ecological effects of human trampling on natural ecosystems. *Biological Conservation* **7**, 17–36.

MacArthur, R. (1955) Fluctuations of animal populations and a measure of community stability. *Ecology* **36**, 533–536.

MacArthur, R.H. & Wilson, E.D. (1967) *The Theory of Island Biogeography*. Princeton University Press, Princeton.

MacGillivray, C.W. & Grime, J.P. and the ISP team (1995) Testing predictions of resistance and resilience of vegetation subjected to extreme events. *Functional Ecology* **9**, 640–649.

Macleod, J. (1894) Over de bevruchting der bloemen in het Kempisch gedeelte van Vlaanderen. Deel. *Botanische Jaarboek* **6**, 119–511.

Miller, R.E. & Fowler, N.L. (1994) Life history variation and local adaptation within two populations of *Bouteloua rigidiseta* (Texas grama). *Journal of Ecology* **82**, 855–854.

Naeem, S., Thompson, L.J., Lawler, S.P., Lawton, J.H. & Woodfin, R.M. (1994) Declining biodiversity can alter the performance of ecosystems. *Nature* **368**, 734–737.

Noble, I.R. & Slatyer, R., (1979) The use of vital attributes to predict successional changes in plant communities subject to recurrent disturbances. *Vegetatio* **43**, 5–21.

Pastor, J. & Cohen, Y. (1997) Herbivores, the functional diversity of plant species and the cycling of nutrients in ecosystems. *Theoretical Population Biology* **51**, 1–15.

Prentice, H.C., Lonn, M., Lefkovitch, L.P. & Runyeon, H. (1995) Associations between allele frequencies in *Festuca ovina* and habitat variation in the alvar grasslands on the Baltic island of Oland. *Journal of Ecology* **83**, 391–402.

Rabotnov, T.A. (1985) Dynamics of plant coenotic populations. In: *the Population Structure of Vegetation* (ed. J. White), pp. 121–142. Junk, Dordrecht.

Ramenskii, L.G. (1938) *Introduction to the Geobotanical Study of Complex Vegetations*. Selkozgiz, Moscow.

Raunkiaer, C. (1934) *The Life Forms of Plants and Statistical Plant Geography, being the collected papers of C. Raunkiaer*. Clarendon Press. Oxford.

Sankaran, M. & McNaughton, S.M. (1999) Determinants of biodiversity regulate compositional stability of communities. *Nature* **401**, 691–693.

Singer, D.K., Jackson, S.T., Madsen, B.J. & Wilcox, D.A. (1996) Differentiating climatic and

Stone, E.C. & Vasey, R.B. (1968) Preservation of coastal redwoods on alluvial flats. *Science* **159**, 157–161.

Thompson, K. (1994) Predicting the fate of temperate species in response to human disturbance and global change. In: *Biodiversity, Temperate Ecosystems and Global Change* (eds T.J.B. Boyle & C.E.B. Boyle), pp. 61–76. Springer-Verlag, Berlin.

Thompson, K. & Jones, A. (1999) Human population density and prediction of local plant extinction in Britain. *Conservation Biology* **15**, 1–6.

Thorpe, P.C., MacGillivray, C.W. & Priestman, G.H. (1993) A portable device for the simulation of air frosts at remote field locations. *Functional Ecology* **7**, 503–505.

Tilman, D. (1996) Biodiversity: Population versus ecosystem stability. *Ecology* **77**, 97–106.

Tilman, D. (1999) Ecological consequences of biodiversity: A search for general principles. *Ecology* **80**, 1455–1474.

Tilman, D. & Downing, J.A. (1994) Biodiversity and stability in grasslands. *Nature* **367**, 363–365.

Tilman, D., Knops, J., Wedin, D., Reich, P., Ritchie, M. & Siemann, E. (1997) The influence of functional diversity and composition on ecosystem processes. *Science* **277**, 1300–1302.

Tilman, D., Wedin, D. & Knops, J. (1996) successional influence in long-term developments of a marsh. *Ecology* **77**, 1765–1778.

Productivity and sustainability influenced by biodiversity in grassland ecosystems. *Nature* **379**, 718–720.

van der Valk, A.G. (1981) Succession in wetlands: A Gleasonian approach. *Ecology* **62**, 688–696.

van der Valk, A.G. & Davis, C.B. (1978) The role of the seed bank in the vegetation dynamics of prairie glacial marshes. *Ecology* **59**, 322–335.

Wardle, D.A. (1999) Is 'sampling effect' a problem for experiments investigating biodiversity - ecosystem function relationships? *Oikos* **87**, 403–407.

Wardle, D.A., Huston, M.A., Grime, J.P. *et al.* (2000) Biodiversity and ecosystem function: an issue in ecology. *Bulletin of the Ecological Society of America* **81**, 235–239.

Wardle, D.A., Zackrisson, O., Hornberg, O. & Gallet, C. (1997) The influence of island area on ecosystem properties. *Science* **277**, 1296–1299.

Wedin, D.A. & Tilman, D. (1990) Species effects on nitrogen cycling: A test with perennial grasses. *Oecologia* **84**, 433–441.

Westman, W.E. (1978) Measuring the inertia and resilience of ecosystems. *Bioscience* **28**, 705–710.

Willis, A.J. (1989) Effects of the addition of mineral nutrients on the vegetation of the Avon Gorge, Bristol. *Proceedings of the Bristol Naturalists Society* **49**, 55–68.

Wilson, P.J. (1998) *The causes and consequences of recent vegetation change in Britain*. PhD Thesis, University of Sheffield, UK.

Chapter 9
Effects of diversity and composition on grassland stability and productivity

D. Tilman

Introduction

The factors and processes that control the composition, diversity, dynamics and structure of plant communities, and of ecosystems, have been the focus of much research during the past two decades. Fundamental advances have come from the work of many (see, for example, other chapters in this volume and papers cited in this chapter). Such work has shown that community and ecosystem processes depend on which species are present (Pastor *et al.* 1984; Vitousek *et al.* 1987; Wedin & Tilman 1990; Estes & Duggins 1995; Power 1995; Sterner 1995; Ewel & Bigelow 1996; Hobbie 1996), on the interactions among these species (May 1973; Harper 1977; Grime 1979; Chapter 8, this volume; Tilman 1982, 1988; Davis 1986; Grubb 1986; Doak *et al.* 1998; Ives *et al.* 1999; Lehman & Tilman 2000) and on interactions between species and the physical environment (e.g. disturbance regimes, climate, soil chemistry; Likens *et al.* 1977, 1998; Vitousek & Matson 1984, 1985). This expanding observational, experimental and theoretical knowledge of the workings of communities and ecosystems has been instrumental in addressing questions that resurfaced in the 1990s concerning the possible effects of species diversity and composition on ecosystem functioning (e.g. Wilson 1992; Schulze & Mooney 1993).

It has long been hypothesized that diversity influences ecosystem processes. As noted by McNaughton (1993), Darwin (1859) suggested that increased diversity led to increased primary productivity. Elton (1958) hypothesized that population and ecosystem stability, and the resistance of ecosystems to invasion by exotic species, were greater at greater diversity. Odum (1959), MacArthur (1955), Margalef (1968) and others also suggested reasons why ecosystem processes would depend on diversity. May (1973, 1974) provided a formal theoretical treatment of the effects of diversity on stability in competitive communities. He found that increased diversity was associated with decreased population temporal stability ('if we concentrate on any one particular species our impression will be one of flux and hazard . . .'; May 1974), but that increased diversity tended to stabilize total com-

Department of Ecology, Evolution and Behaviour, University of Minnesota, 1987 Upper Buford Circle, St Paul, MN 55108, USA

munity biomass ('... if we concentrate on total community properties such as biomass in a given trophic level our impression will be of pattern and steadiness'; May 1974). This work was generally interpreted as suggesting that there would be no consistent relationship between diversity and stability (e.g. Goodman 1975), but McNaughton (1978), Pimm (1979, 1984), King and Pimm (1983), Ehrlich and Ehrlich (1981) and others continued to explore the issue. Interest in the effects of diversity on ecosystem processes was rekindled in the early 1990s, especially by the papers in Schulze and Mooney (1993).

Here I synthesize recent work on the ecosystem effects of diversity and composition, focusing especially on the underlying theoretical concepts that link community and ecosystem processes to biodiversity and species composition, and on tests of these ideas in grassland field experiments, especially those at Cedar Creek in Minnesota. Because this area became a major focus of research only within the past decade, it is not surprising that it is contentious (e.g. Givnish 1994; Aarssen 1997; Grime 1997; Huston 1997; Wardle et al. 1997a; Huston et al. 2000; Wardle et al. 2000). Some of the disagreements come from differing definitions of terms, or other semantic issues, but others will only be resolved by a larger number of long-term experiments and by appropriately careful observational studies. I will address these controversies by considering four questions that have arisen about the effects of diversity on ecosystem functioning. First, what is diversity, how can it be quantified, and what components of diversity might impact ecosystem functioning? Second, by what mechanisms can diversity affect ecosystem functioning? Third, how many species may be required to assure a given level of ecosystem functioning? Fourth, must diversity always have the same qualitative effect on a given ecosystem process?

What is diversity?

In its simplest form, diversity has often been considered to be just a count of the number of species in a habitat (called species number or species richness), often for a particular trophic level, or an index weighted by the relative abundances of these species, such as the Shannon H' index. These are appropriate ways to measure diversity when comparing neighbouring natural communities that probably had been assembled from the same regional species pool, or when comparing experimental communities randomly drawn from the same species pool. As discussed later, a second important component of diversity is the range of traits encompassed in the species pool, that is, the differences among species.

The familiarity of species richness and H' as indices of diversity may be misleading when asking if this component of diversity might impact ecosystem functioning because these are easily confounded with differences in species composition or other factors that also might impact ecosystem functioning. For instance, a series of neighbouring sites might differ in their species number, but also differ in soil fertility, species composition, disturbance frequency or other aspects. If observed differences in ecosystem functioning were ascribed to observed species number without appropriate statistical control for correlated shifts in soil fertility, species composition, disturbance, etc., erroneous relations could result. Just such concerns have been expressed by Tilman *et al.* (1997c) about the work of Wardle *et al.* (1997a; reply by Wardle *et al.* 1997b) and by Givnish (1994) and Huston (1997) about the work of Tilman and Downing (1994; replies by Tilman *et al.* 1994b; Tilman 1996, 1999). It is also possible to confound diversity and composition in experimental studies, as pointed out in critiques of some recent biodiversity experiments. For instance, Huston (1997), Garnier *et al.* (1997) and Hodgson *et al.* (1998) criticized the ECOTRON experiment (Naeem *et al.* 1994; reply by Lawton *et al.* 1998), because species compositions were not randomly chosen. Huston *et al.* (2000) criticized the BIODEPTH experiment (Hector *et al.* 1999; reply by Hector *et al.* 2000) because some species were not grown in monoculture.

To avoid such problems, it is necessary that there be concomitant and appropriate experimental or statistical control for other confounding variables, such as composition, disturbance, etc., when experimentally or observationally exploring the potential effects of diversity on population, community or ecosystem processes. Diversity effects must thus be considered as being those attributable to diversity once there has been simultaneous control for other potentially confounding variables. One simple way to assure such control in a completely randomized experimental study would be to have many replicates at a given level of species number, each with a species composition determined by a separate random draw from the same pool of species. Differences among mean responses observed at different levels of species number could then be appropriately attributed to species number. Such a design could be expanded to have two or more replicates of each randomly chosen species combination if one were interested in testing for the relative impact of species composition vs. species number on ecosystem functioning. Indeed, a much larger number of designs are possible, with each design having higher power to test some diversity-related hypotheses and lower power to test others, as has been discussed by Allison (1999).

What might effects attributable to species number measure in a study of the effects of diversity on ecosystem functioning? In essence, species number would be a simple, convenient way to estimate the range of species traits that were present in a given ecosystem. Given that all species assemblages are drawn from the same species pool, greater local species number or H' should lead, on average, to a greater range of traits within a site, i.e. to greater coverage of the actual range of traits occurring in the full species pool. This, though, suggests that the effect of diversity is likely to depend both on the number of species present in a system, and on the range of traits present in the species pool. If all the species present in a species pool are similar to each other, even large changes in local species number might have little impact on ecosystem functioning because the range of species traits would be little impacted. This means that there is another component to diversity, which is the range of variation among species, i.e. the functional diversity of the regional species pool. Both regional functional diversity and local species number are thus expected to interactively control ecosystem functioning.

Diversity and stability

The initial work by Elton (1958) and others on the diversity–stability hypothesis, the contributions by May (1973; 1974), and the summary by Goodman (1975) were followed by contributions from McNaughton (1978), Pimm (1979, 1984), King and Pimm (1983) and others. Recent interest in this topic was sparked by the series of papers in Schulze and Mooney (1993). McNaughton's (1993) paper, in particular, presented many new lines of evidence suggestive of greater diversity leading to greater ecosystem stability. Although each data set was small, often containing only a few plots, the preponderance of this new evidence supported the diversity–stability hypothesis, as did a study of the impact of drought on Yellowstone grasslands (Frank & McNaughton 1991).

A larger data set was provided by the Park Grass Experiment (Dodd *et al.* 1994), for which plant biomass (hay yield) has been annually determined since 1856 in plots receiving various rates and combinations of mineral fertilizers. The year-to-year variability in total plant standing biomass (coefficient of variation, or CV, of hay yield) was determined for a plot over an 11-year period centred around a year when species richness had been counted. Species richness was determined in 42 of the years from 1856 through 1991, with as few as five plots sampled some years, and with a maximum of 34 plots sampled (in 1991) for species richness. Dodd *et al.* (1994) determined the relationship between plant species richness and biomass variability (CV) using multiple regressions of CV on both plant species richness and on mean standing crop. They found negative correlations between CV and species richness in 29 of the 42 periods tested. Three of these were significantly negative, with the strongest effect observed for the most thoroughly sampled period 1991 (for which the regression included 34 data triplets). They concluded that the Park Grass Experiment tended to support the hypothesis that greater diversity led to greater stability as measured by lower year-to-year variability in total biomass, but noted that the relationship was often weak. They warned that the data could not be used to determine if diversity caused stability, if stability caused diversity, or if both resulted from some third unmeasured variable.

Another line of evidence was provided by the response of 207 Minnesota successional and native grassland plots to a once-in-50-year drought, in 1987–88 (Tilman & Downing 1994). These plots, which differed in diversity because of different rates of nitrogen addition, different successional statuses, and location, responded markedly differently to drought. The most diverse plots, which contained from 16 to 26 vascular plant species per $0.3\,m^2$, had their total above-ground living plant biomass fall to about half of their predrought biomass, but the total biomass of the least diverse plots, which had 1 or 2 species per $0.3\,m^2$, fell to 1/10 or 1/12 of predrought levels (Tilman & Downing 1994). Thus, the most diverse plots were 5–6 times more resistant to a major perturbation than the least diverse plots. Because diversity was not independently manipulated, a series of multiple regressions were performed to control for potentially confounding variables, including successional status, species composition, and rate of nitrogen addition. Many of these variables, themselves, had significant effects on stability. However, in each such analysis there remained a highly significant positive effect of diversity on stability (drought resistance). Additional analyses, inspired by concerns raised by Givnish (1994), Huston (1997) and others, have all also provided statistically significant support for the hypothesis that greater diversity causes greater stability of total community biomass (e.g. Tilman & Downing 1994; Tilman 1996, 1999) and demonstrated diversity to be one of several variables, including species and functional group composition, that influence stability.

These long-term data also allowed examination, for non-drought years, of the relationship between the year-to-year variability in the total biomass of each plot (as measured by the coefficient of variation in total biomass, CV) and the diversity of each plot (Tilman 1996). Plots with greater plant diversity had significantly lower year-to-year variation in total biomass (Fig. 9.1a), even after statistically controlling for many potentially confounding variables, several of which also influenced stability. However, this was not so for individual species, which were slightly, but significantly, more variable at higher diversity (Fig. 9.1b). The latter results supported models of May (1973), and the former results were consistent with the conjecture made in May (1974) that greater diversity might stabilize traits of entire communities. However, the caveats offered by Dodd et al. (1994) also apply to this study.

Such support for the diversity–stability hypothesis had led to further theoretical explorations. An important contribution came from Doak et al. (1998), who showed that greater diversity could lead to greater community stability because of a statistical averaging effect. In essence, the sum (or average) of a series of random variables can be less variable than the average element of which it is composed, and this effect increases as more variables are considered (i.e. as diversity increases). This portfolio effect depends on the way that the temporal variance, σ^2, in the abundances of individual species scales with their mean abundance, m (Tilman et al. 1998). In particular, if $\sigma^2 = cm^z$, where c is a constant, then greater diversity would lead to greater community stability if $z > 1$ (assuming no covariance in species abundances). Ives et al. (1999) showed that the effect of species number on community stability depended on interspecific differences, with there being no effect of species number on stability if species were identical in their responses to perturbations. Their result thus reinforces the view that species number matters precisely because species differ in their traits. Tilman (1999) and Lehman and Tilman (2000) showed that, in three different models of multispecies competition, a greater number of species increased the temporal stability of the entire community (total community biomass) but decreased the temporal stability of individual species (species biomass). For instance, for a model of competition and coexistence in a habitat in which the disturbance is year-to-year variation in temperature, with each species having maximal competitive ability at a different temperature, the temporal stability of the entire community (measured as the mean, μ, divided by the standard deviation, σ, of its temporal variability) is predicted to be a linearly increasing function of diversity, whereas the temporal stability of the average species is a slightly declining function of diversity (Fig. 9.2). For this model, temporal stability is a

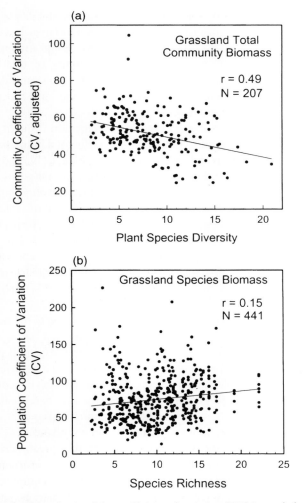

Figure 9.1 (a) The dependence of the coefficient of variation (CV) in total plant living biomass on plant species diversity for 207 grassland plots at Cedar Creek. Each point is the CV calculated for a plot, and represents the extent of year-to-year variation in total plant biomass for the plot. Points are adjusted for field differences in intercepts, based on general linear model regressions with field as a categorical variable. Note that plots with higher species number have significantly lower year-to-year variation in total biomass, i.e. are more stable. (Modified with permission from Tilman 1999.) (b) Similar analyses, using data for these 207 grassland plots, but determining CVs on a species-by-species basis for the common species of each plot for the non-drought years. These show a slight tendency for abundances of individual species to be less stable at higher diversity. (Modified with permission from Tilman 1996.)

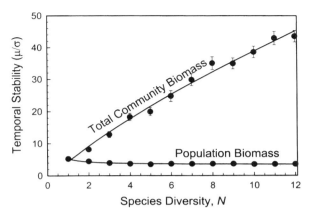

Figure 9.2 The dependence of temporal stability on diversity for a model of multispecies resource competition. The model assumes that all species compete for a single limiting resource, that species differ in the temperature at which they have optimal competitive ability, and that temperature varies through time. This variation in temperature allows multispecies coexistence. Temporal stability is defined as the ratio of mean abundance, μ, to its temporal standard deviation, σ, and is essentially the inverse of CV. Note that the temporal stability of total community biomass is predicted to be an almost linearly increasing function of species number, N. In contrast, the temporal stability of individual species is a slightly decreasing function of species number. The full model is presented in Tilman (1999), from which this figure was modified with permission. Related models are presented in Lehman and Tilman (2000).

measure of resistance to climatic (temperature) variation. In all three models, temporal stability is measured in response to a series of small stochastic perturbations, but should correlate with resistance to a single sustained perturbation, such as drought. The different effects of diversity on community vs. population stability agree both qualitatively and in relative magnitude with the responses observed during non-drought years in the 207 Minnesota grassland plots.

In total, analyses in May (1973), Doak et al. (1998), Tilman (1999) and Lehman and Tilman (2000) may explain why increased diversity destabilizes individual species but stabilizes the entire community. The destabilizing effect of increased diversity on individual species comes from the effects May (1973) discovered—in essence, greater diversity increases the number of feedback loops in a competitive community, causing a change in one species to impact many others. Total community biomass is stabilized by a negative covariance effect at lower diversity and by the portfolio effect at higher diversity. When species compete, a decrease in the abundance of one species leads to an increase in the abundance of another, causing the abundances of such species to negatively covary. This negative covariance, summed over all pairs of species, reaches a minimum as diversity increases from one to five or 10 species in the models of Tilman (1999) and Lehman and Tilman

(2000). For levels of diversity up to the diversity that leads to this minimum, increased diversity leads to lower temporal variation in total community biomass, because the CV of total community biomass depends on the sum of the variance and covariance terms for all species and pairs of species in the community (Tilman 1999). Although summed covariance becomes less stabilizing (approaches 0 from below) for increases in diversity beyond that of the covariance minimum, the portfolio effect strengthens more than this, causing community temporal stability to continue increasing as diversity increases further.

In total, the observational, experimental and theoretical work to date suggests a possible resolution to the diversity–stability debate. On average, greater diversity leads to greater temporal stability of an entire community trait, such as total community biomass, but simultaneously leads to lower stability of individual species. Thus, the work of both Elton (1958) and May (1973, 1974) is supported, with Elton's ideas applying to communities and ecosystems, and May's work applying to the dynamics of individual species in these communities and ecosystems.

Diversity and productivity

Let us consider two different ways that diversity might impact ecosystem primary productivity and the availability of a limiting resource: via a sampling effect and via niche differentiation. The sampling effect illustrates a fundamental way in which the number of species initially present in an ecosystem can influence its functioning (Aarssen 1997; Huston 1997; Tilman *et al.* 1997a; Loreau 1998a, 1998b). All else being equal, a given species has a greater chance of being present in a higher diversity community, i.e. it has a greater chance of having been sampled from the regional pool of species. If the traits of certain species lead them to be highly abundant and thus to have a disproportionate influence on ecosystem functioning, the sampling effect would assure that their traits would be increasingly likely at higher diversity, thus causing ecosystem functioning to move in that direction as diversity increased. The sampling effect emphasizes the role of a single dominant species, and its greater chance of being present (being sampled during community assembly) at higher diversity. Niche differentiation, in contrast, involves complementary effects that result from interactions among competing species that differ in the ways that they exploit a habitat.

Sampling effects

Two simple models lay bare how sampling effects can cause ecosystem functioning to depend on initial diversity. First, there is the case originally discussed by Huston (1997), Aarssen (1997) and Tilman *et al.* (1997a), which leads to increased primary productivity at higher diversity. Let species compete for a single limiting resource, R, and let all interactions rapidly go to equilibrium in a homogeneous habitat. In this case, the competitive winner would be the species with the lowest requirement (R^* of Tilman 1982) for the limiting resource. The species with the lowest R^* would thus acquire the most of the limiting resource. Assuming that all species were equally good at converting resource into biomass, the species with the lowest R^* would thus be the most productive species. Let the species pool consist of species with all possible R^* values ranging from a low of R^*_{min} to a high of R^*_{max}. Let compositions of communities be chosen by random draws of species from this pool. As is easily derived from Tilman et al. (1997a) and Lehman and Tilman (2000), the average biomass of a community, $B_{(N)}$, originally containing N randomly drawn species would be

$$B_{(N)} = c\left[S - \frac{2M-D}{2} - \frac{D}{N+1}\right] \quad (1)$$

Here c is a constant, S is the rate of supply of the limiting resource, M is the mean R^* value for the entire community, and D measures interspecific differences in R^* (here, with a uniform distribution of R^* values, $D = R^*_{max} - R^*_{min}$). This is defined for $N \geq 1$. If $N = 0$, $B = 0$.

This model predicts that average total community biomass at equilibrium increases both with the initial number of species, N, from which the community was assembled (Fig. 9.3a), and with the extent of interspecific differences, D. Both a greater number of species and greater differences among them (the two components of diversity) lead to greater productivity. If all species were identical (i.e. if $D = 0$), then Equation 1 shows that there would be no effect of the initial number of species on total community biomass. Note that D is a direct measure of the extent of interspecific differences in species traits for the entire species pool. A larger value of D leads to a greater effect of species number on ecosystem functioning. A characteristic of the predicted dependence of total community biomass on N for this sampling effect is that the upper bound of the observed variation is a flat line and the increased mean is caused solely by increases in the lower bound. The increase in average total community biomass that results from increased N is caused by the model assumption that the best competitors are the most productive species and by the greater chance that such species would be chosen ('sampled') at higher diversity. The upper bound is flat because it is possible for the best competitor to be present at any given level of diversity. The lower bound increases because each community comes from a random draw of species, and any such community is more likely to have a superior competitor present in it at higher diversity. This model can also be used to predict the variance in total community biomass at each level of diversity, and the average levels, and variance in, the unconsumed limiting resource (Tilman et al. 1997a).

There are many possible variants on the sampling effect model (Loreau 1998a,b). For instance, consider a community in which a single species again wins in competition at equilibrium but in which the outcome of a competitive interaction is determined by antagonistic interactions rather than by simple differences in the efficiency of resource exploitation. In such situations, it is plausible that species that are progressively better competitors would gain competitive ability by allocating their time and/or morphology to traits that maximized antagonistic efficiency at a direct cost to efficiency of resource capture and use. If so, progressively better

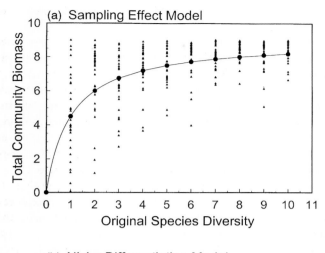

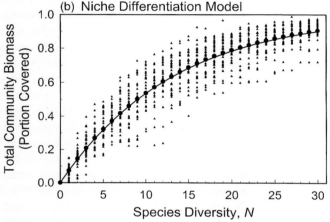

Figure 9.3 Predicted relations between productivity and diversity for a sampling effect model and a niche differentiation model. (a) This sampling effect model assumes that all species compete for a single resource and that interactions go to equilibrium, causing a single species, the best competitor present, to displace all others. Here it is assumed that better competitors use the limiting resource more effectively and produce greater biomass. The graph shows the effect of the number of randomly chosen species on the equilibrial biomass of the resulting communities, using the model of Tilman *et al.* (1997a). The analytical solution presented in that paper is the basis for the solid curve shown. Simulations were used to generate the points shown in the graph, which is derived from that of Tilman *et al.* (1997a). (b) This niche differentiation model assumes a spatially heterogeneous habitat and species that each perform optimally for a small region of the habitat. Random draws of various numbers of species gives the results shown, with results based on the proportion of the entire niche space that is 'covered' by the randomly chosen species. Tilman *et al.* (1997a) present an analytical solution to this model, which is the basis for the solid curve shown. Simulations (random draws and calculations of coverage) generated the data points shown. Note that the number of species required to saturate this relation depends on the portion of niche space covered by an average species.

competitors could be progressively less productive. Assuming a species pool with a suite of species ranging from those that were highly productive but poor competitors to those that were highly unproductive but strong competitors, then average total community biomass would decline as initial diversity increased. This would occur simply because any given species, including better competitors that are less productive, would be more likely to be sampled from the species pool at higher diversity (Loreau 1998a,b). The magnitude of the effect of initial diversity on total community biomass would depend directly on the extent of interspecific differences in the species pool, just as for the first sampling effect model.

In total, all else being equal, any given species is more likely to be present at higher diversity, i.e. to have been sampled by the random processes that are assumed to control the initial composition of communities for these idealized cases. If there are simple rules of interspecific interaction in these communities that cause species with particular traits to become dominant and to suppress (or perhaps even eliminate) species with other traits, then the traits of the competitive dominants would be increasingly expressed at higher diversity. If superior competitors produce more biomass, then such sampling effects would cause total community biomass to increase with diversity. If superior competitors drive limiting resources to lower levels, resource concentrations would decline, on average, as diversity increased. If superior competitors had lower biomass and left more resource unconsumed, then total community biomass would decrease and concentrations of resource would increase as diversity increased. These are all sampling effects—effects caused by the increased tendency for any given species to be present at higher diversity and by a correlation between species traits and dominance.

Huston (1997) and Huston et al. (2000) asserted that sampling effects were an experimental artefact, whereas Tilman et al. (1997a), Tilman (1999) and Naeem (2000) asserted that sampling effects were real results of diversity. This difference of opinion may result from disciplinary differences in perspective. In highly managed ecosystems, such as agro-ecosystems, it is common practice to determine the productivity of each crop in monoculture, and to consider a mixture of several species to be of agricultural importance only if it is more productive than the best monoculture. This is the perspective that may have led Huston and others to conclude that sampling effects were not 'true' effects of diversity. In contrast, because natural communities are not deliberately assembled by a manager, but rather assemble by various natural processes, Tilman et al. (1997a) considered the sampling effect to a real effect of diversity because, all else being equal, higher-diversity natural communities would be more likely to contain better-performing species, and thus to function differently than low diversity communities.

Niche differentiation effects
Sampling effect models assume that a single species is competitively superior, greatly suppressing or even excluding other species, and that its traits greatly influence community processes. In contrast to this assumption, there are many models that predict the long-term coexistence of many co-dominant competing species.

With few exceptions, such models of multispecies stable coexistence implicitly or explicitly assume some sort of interspecific differentiation. For instance, two species can persist indefinitely when competing for a fluctuating resource if one is better at exploiting the average availability of the resource and the other is better at exploiting resource pulses (e.g. Armstrong & McGehee 1980; Huisman & Weissing 1999). This can be extended to allow the indefinite persistence of a large number of competing species if each is better at exploiting some pattern or frequency of resource temporal variation (Armstrong & McGehee 1980; Chesson 1986; Huisman & Weissing 1999). Spatial heterogeneity in resource availability can also allow the coexistence of a large number of competing species. Models in which plants compete for two essential resources (such as light and soil water, or soil water and soil nitrogen) that are patchily distributed across a landscape predict that a large number of competing species can coexist if there are trade-offs in their abilities to compete for the two resources (e.g. Tilman 1982, 1988). Species may also be differentiated in their abilities to survive and compete in habitats that differ in physical characteristics, such as soil pH, or temperature. Spatial or temporal variation in such physical factors can allow coexistence of many species if they are differentiated with respect to these constraints. When the effects of diversity on total community biomass have been explored for such models of exploitative competition and multispecies coexistence in communities of species with interspecific trade-offs in their traits, a set of consistent results have emerged to date (e.g. Tilman et al. 1997a; Loreau 1998a,b; Tilman 1999; Lehman & Tilman 2000). In particular, average total community biomass is an increasing function of diversity, as are both the upper and lower bounds of composition-dependent variation in this (Fig. 9.3b). The increasing upper bound means that there are some higher-diversity combinations of species that are more productive than any possible combination of lower diversity. The average amounts of unconsumed limiting resources remaining in the habitat are decreasing functions of diversity, as are the upper and lower bounds of composition-dependent variation in these. Their decreasing lower bound means that there are some higher-diversity compositions that use resources more efficiently than any lower-diversity species combinations.

In both the niche and the sampling models, the variation in total community responses that occurs at a given level of diversity is caused by differences in community composition. As shown in Fig. 9.3, this variation can be large, indicating that composition can be a major determinant of ecosystem functioning. In the sampling effect model and the model of competition for two essential resources of Tilman et al. (1997a), the magnitude of the composition-dependent variance within a given level of diversity declined as diversity increased. However, for a third model (a generalized niche model) such variance was highest at intermediate diversity (Tilman et al. 1997a).

Experimental studies

The concepts summarized above were proposed as alternative potential explanations for patterns reported by studies of the effects of plant diversity on total plant community biomass or nutrient dynamics (Ewel *et al.* 1991; McNaughton 1993; Swift & Anderson 1993; Vitousek & Hooper 1993; Naeem *et al.* 1995, 1996; Tilman *et al.* 1996). Other, more recent studies have also shown significant effects of diversity on various community/ecosystem processes, including field experiments by Hooper and Vitousek (1997, 1998), Tilman *et al.* (1997b), Hector *et al.* (1999) and Knops *et al.* (1999). Additional insights have been provided by a variety of laboratory and greenhouse experiments (e.g. Naeem & Li 1997; McGrady-Steed *et al.* 1997; Symstad *et al.* 1998). As previously discussed, there has been controversy concerning such studies. Wardle *et al.* (2000) summarized critiques of six different diversity studies and concluded that no consensus could yet be reached on the effects of diversity on ecosystem processes. Naeem (2000) critiqued this summary and reached the opposite conclusion, as did an earlier report by Naeem *et al.* (1999a). A literature review (Schläpfer & Schmid 1999) and a survey of expert opinion (Schläpfer *et al.* 1999) both supported the hypothesis that diversity had significant and relatively consistent effects on various ecosystem processes. My synthesis of theory and experiments also supported this view (Tilman 1999).

Swift and Anderson (1993) and McNaughton (1993) provided early literature reviews that suggested greater diversity led, in some circumstances, to greater plant biomass/productivity. For agricultural studies of intercropping, such overyielding generally occurred mainly when a legume and a non-legume were grown together on a low-nitrogen soil, as would be expected. A greenhouse study (Naeem *et al.* 1995) in which plant community composition was determined by random draws from a pool of species provided a direct demonstration that total plant community biomass was an increasing function of diversity (Fig. 9.4a), a trend also suggested by the related ECOTRON study (Naeem *et al.* 1994; but see Huston 1997; Hodgson *et al.* 1998; Lawton *et al.* 1998). Two experiments in Minnesota grasslands similarly showed that plant total community abundance was greater, on average, at higher plant diversity and that the levels of soil nitrate, the limiting resource, were lower at higher diversity (Tilman *et al.* 1996, 1997b). In contrast, Hooper and Vitousek (1997, 1998) found that functional group composition, not functional group diversity, was the major controller of total plant community biomass. Tilman *et al.* (1997b) found that both species diversity and functional group composition were significant controllers of plant community abundance. In a field experiment that was replicated across eight different sites in Europe, Hector *et al.* (2000) found a highly significant pattern of greater total plant community biomass at higher plant diversity. In total, these studies of grassland communities have demonstrated that both plant diversity and functional group composition often have significant effects on total community biomass and on levels of unconsumed resources, but have not clearly resolved if the effects of diversity result from sampling effects or from niche differentiation.

The pattern of the dependence of biomass and resource levels on diversity can be used to determine the relative importance of the sampling effect vs. niche differentiation. Although both the sampling effect and niche differentiation models can predict qualitatively similar increases in productivity with increasing diversity, a

D. TILMAN

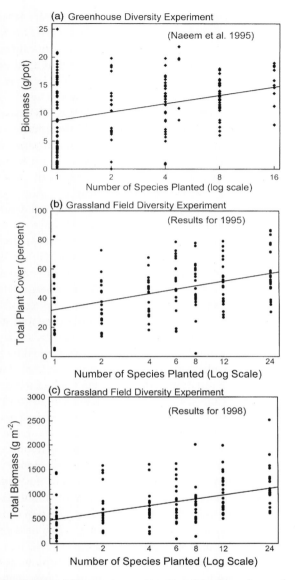

Figure 9.4 Results of some of the experimental studies of the dependence of ecosystem processes on diversity. (a) Results from the Naeem *et al.* (1995) greenhouse experiment in which 16 British grassland species were grown in various combinations for 10 weeks. Species combinations were chosen by separate random draws from the pool of 16 plant species. Note the flat, or even declining, upper bound, which supports sampling effects over niche effects. Data in graph provided by Shahid Naeem, whom we thank. (b) Early results (1995, the second field season) for the 'small' biodiversity experiment at Cedar Creek. In this experiment, 147 plots were planted to 1, 2, 4, 6, 8, 12 or 24 species, with about 20 replicates per diversity level (Tilman *et al.* 1996), and with the species composition for each

major difference between the sampling effect and niche differentation is the pattern of variation in this relationship. The flat upper bound of the composition-dependent variation in total biomass predicted by the sampling effect and the increasing upper bound predicted by niche models can be used to determine which is operational (given that an experiment is well replicated and has minimal measurement errors).

Inspection of available data suggests that the reported dependencies of total community biomass on plant diversity have resulted from both processes (Tilman et al. 2000). Interestingly, short-term studies, such as greenhouse studies or field studies that only lasted for 1–3 years, seem to have a signature more indicative of sampling effects than of niche differentiation effects. For instance, Naeem et al. (1995, 1996), Tilman et al. (1996), and Symstad et al. (1998) all reported increasing relations between plant total community biomass or cover, and these all had approximately flat upper bounds. For instance, in a 10-week long greenhouse experiment, Naeem et al. (1995, 1996) found that there were no eight- or 16-species plots (the two highest diversity treatments) with total biomass as high as that of more than 10 lower-diversity plots (Fig. 9.4a). Indeed, the upper bound of variation declined at higher diversity (Fig. 9.4a). In 1995, the second year of their field experiment, Tilman et al. (1996) found that only three higher-diversity plots had greater total plant community cover than the top monoculture, and these three plots were only a few percent cover greater (Fig. 9.4b). In contrast, after this same experiment had run for 5 years, the upper bound of the dependence of total community biomass on diversity was a clearly increasing function of diversity (Tilman et al. 2000; Fig. 9.4c). In 1998 there were 14 higher-diversity plots that had greater total community biomass than the monoculture with the highest biomass, and the most productive plot in the biodiversity experiment was in the highest-diversity treatment. It had 65% greater biomass than the highest monoculture (Tilman et al. 2000). The other Minnesota biodiversity field experiment (Tilman et al. 1997b) also had results after its first 3 years that were consistent with a sampling effect, but had results after 5 and 6 years that strongly supported niche differentiation effects (unpublished personal observations). Interestingly, the apparent support for sampling effects during the early stages of these experiments is not caused by dominance by a single species (which is an assumption made by the sampling effect model). Essentially all planted species coexisted. Rather, a more likely explanation, offered by Huston (1997) and explored mathematically by Pacala & Tilman (2000) is that the initial dynamics of these plots is determined by maximal growth rates.

---

plot being determined by a separate random draw. Total plant cover, visually estimated (to species), provided an index of total plant biomass. Note the apparent flat upper bound to these results, which supports the sampling effect model. (c) Later results (1998, fifth field season) for this same Cedar Creek biodiversity experiment. Note the clear increase in the diversity dependence of the upper bound of total plant biomass (which is the sum of above-ground biomass and below-ground plant biomass).

One or a few fast-growing species would obtain higher initial biomass, during the transient dynamics of establishment, than slow-growing species, and would probably obtain about the same biomass in monoculture as in high-diversity plots because of the same processes that led to Kira's law of constant yield (Pacala & Tilman 2000). Such species would thus set an initially flat upper bound for biomass. This would change as more slowly growing superior competitors, which were more efficient users of limiting resources, increased in abundance and interacted.

Because the predicted signature of niche differentiation models is an equilibrium result that depends on the competitive sorting of species and competition-dependent shifts in relative abundances, it may be that the initial effects of diversity on total community biomass are caused more by sampling effects and that the long-term effects come from various interspecific niche-like differences. If this were so, it would resolve much of the debate about the effects of diversity on productivity and resource levels. Huston (1997), Aarssen (1997), Huston et al. (2000) and Wardle et al. (2000) all suggested that sampling effects, rather than niche effects, could explain the results of Naeem et al. (1994, 1995, 1996) and of Tilman et al. (1996). Now that theory is well enough worked out to provide signatures that can differentiate these two models, sampling effects do seem more important than niche differentiation during the early (transient) stages of diversity experiments, whereas niche effects may take over later.

In addition to impacting plant community biomass and soil nutrient levels, greater plant diversity has been found to be associated with lower levels of foliar fungal pathogens and with a lower number and mean abundance of weedy plant species (Knops et al. 1999). Each of these apparent effects of diversity has a seemingly simple explanation. Analysis of the effects of plant diversity on the abundances and species diversity of invading weedy species suggested that these effects resulted from the lower levels of soil nitrate at higher diversity (Knops et al. 1999). Thus, it was mainly the effect of plant diversity on soil nutrient levels that caused weedy invader abundance and diversity to depend on plant diversity. The effect of diversity on foliar disease abundance seemed to be mainly ascribable to the lower density and greater distance among individuals of a given host species at higher diversity (Knops et al. 1999).

There have also been a few observational studies used to infer possible effects of diversity on ecosystem processes. For instance, Tilman et al. (1996) reported that greater plant diversity led to higher total plant community cover (an estimate of biomass) for 30 sites (with four plots each) spread across a single stand of native grassland, but did not measure potentially confounding variables. Wardle et al. (1997a) studied a suite of 50 Swedish islands that differed in size, fire frequency, diversity and community composition, and in a variety of ecosystem processes. They found strong effects of island size and fire frequency on vegetation composition, standing biomass, litter, plant diversity and nutrient dynamics. They noted that smaller, infrequently burned islands had lower standing biomass but higher plant diversity, but they did not present any analyses in which there had been statistical control for potentially confounding variables. Thus, observational studies have not given consistent results, perhaps because of the difficulty of controlling for potentially confounding variables that may covary with diversity. Another difficulty may be that two processes are operative in natural communities. As reported for many biome types (e.g. Dix & Smeins 1967; Holdridge *et al.* 1971; Whittaker 1975; Al-Mufti *et al.* 1977; Ashton 1977; Huston 1980, 1994; Tilman 1982; Bond 1983; Puerto *et al.* 1990; Tilman & Pacala 1993), plant diversity is often highest in habitats that have intermediate productivity. Long-term fertilization often leads to marked decreases in plant diversity (e.g. Thurston 1969; Tilman 1982; Tilman *et al.* 1994a). Thus, observations and experiments suggest that productivity influences diversity. Biodiversity experiments suggest that diversity also influences productivity. Both effects may operate simultaneously in natural communities. This hypothesis merits further consideration.

Implications and conclusions

Although the impact of diversity on ecosystem functioning is a relatively new area of inquiry, the conceptual, theoretical and experimental studies that have been performed to date suggest four possible generalizations that merit further study. First, there is no a priori reason to expect that diversity will have the same qualitative effect, from one ecosystem to another, on a given ecosystem process. For instance, as shown by alternative formulations of the sampling effect model, greater diversity could lead to greater or lesser total community biomass, or could have no effect on this or other processes. Rather, the effect of diversity on an ecosystem process depends on how the traits of species influence species abundances and on how these same traits, or other traits correlated with these, influence ecosystem processes. The consistent effects of grassland species diversity on grassland productivity suggest that there is a similar set of relationships among competitive ability, coexistence and efficiency of resource use for grassland species. Increases in productivity with increased diversity may occur, in general, in ecosystems in which species interactions are driven mainly by exploitative competition for a few limiting resources, as current theory suggests (Lehman & Tilman 2000). However, it is easy to visualize how antagonistic competitive interactions could cause higher-diversity ecosystems to have lower productivity. These two alternative possibilities suggest that we need to better understand the mechanisms allowing coexistence and controlling species abundances in natural communities if we are to better understand the effects of diversity on ecosystem processes.

Second, if there are effects of diversity on an ecosystem process, these effects are the result of differences in the traits of the species, with these differences being magnified by interspecific interactions that favour particular kinds or combinations of traits. Thus, the strength of the effect of diversity on an ecosystem process is expected to depend on the magnitude of interspecific differences and the extent to which such differences, or traits correlated with them, impact the relative abundances of species. If competition favours species that are efficient resource users and which thus produce more biomass, then, all else being equal, increased diversity would lead to increased total community biomass. This is, perhaps, the mainstream view of how terrestrial plant communities operate, but it would be worthwhile for it to be tested more deeply and broadly. Moreover, the magnitude of the increase in total community biomass from the average monoculture to the average high-diversity plot should depend on the extent of interspecific differences in relevant traits in the species pool. This lays bare the reason why diversity can impact ecosystem processes—diversity is a simple way to estimate the range of variation in the traits of species in an ecosystem. If two otherwise identical diversity experiments were started with species pools that differed in their ranges of species traits, the pool with the greater range of traits would be expected to have the greater effect of diversity. If species are very similar, there would likely be no detectable effect of diversity on ecosystem processes.

Does this second effect mean that species number, *per se*, is unimportant, as some have suggested (e.g. Ives *et al.* 1999)? To some extent, this is a semantic argument that hinges on the operational definition of diversity. However, there are also general principles (e.g. Equation 1) and empirical relationships that suggest the importance of diversity, *per se*. Clearly, it is possible to construct species pools that have almost any given level of interspecific differences. The relevant issue, though, is the extent of interspecific differences in pools of species that reasonably reflect the species of a given region. The effects of changes in species number observed in experiments that use such species pools should then reasonably reflect how the actual ecosystems containing these species would respond to changes in this component of diversity. Because field diversity experiments have used species pools containing the common species of their regions (Tilman *et al.* 1996, 1997b; Hooper & Vitousek 1997, 1998; Hector *et al.* 1999), as have many greenhouse or culture chamber experiments (e.g. Naeem *et al.* 1994, 1995, 1996; Symstad *et al.* 1998), these results, especially those from long-term studies, should indicate how the random loss of diversity would impact the ecosystems of these regions. The relatively large impacts observed in these experiments, especially the field experiments, thus probably indicate that the random loss of species should have strong effects on these ecosystems. To attribute this effect solely to the loss of traits, when the most obvious, easily measured event that has occurred is that the number of species has decreased, seems needlessly semantic.

Third, the community and ecosystem effects of a loss of species will depend on the number of species lost, on the identity of the species lost, and on the identity of the species (if any) that may replace them. The work performed to date has focused more on testing the possibility that species number may influence community and ecosystem processes than on determining the importance of diversity effects relative to effects of changes in composition, disturbance regime, etc. The study of randomly assembled experimental communities provides an unbiased look at the effects of species number and helps build knowledge of general principles that may govern diversity–functioning relations. However, the processes controlling both natural community assembly and human-influenced community disassembly are poorly known and are unlikely to be random (Grime 1997; Huston 1997; Wardle *et al.* 2000). Consider the range of non-random impacts that human actions can have on both community composition and diversity. First, human actions can lead to the selective loss of dominant species. For instance, accidental introductions of exotic pathogens, such as Chestnut blight and Dutch elm disease, reduced the tree diversity of US eastern deciduous forests via selective removals of chestnut and elm, which had been among the dominant species. Overhunting and massive destruction of US eastern deciduous forests during the 19th century led to the extinction of passenger pigeons, perhaps the most abundant bird species of those forests. Selective harvesting of particular fish species, often the original dominants of rivers, lakes or the oceans, has led to major shifts in the composition and diversity of fish communities. Selective harvesting of pines (especially *Pinus strobus* and *Pinus resinosa*) converted the vast pineries of the US Great Lakes states into depauperate *Populus*-, *Acer*- and/or *Quercus*-dominated hardwood forests, which still dominate the landscape 100 years later. However, human actions also impact rare species. For instance, rare species are more often listed as threatened or endangered by human action, and it is rare species, not common ones, that are most often missing from the remaining small fragments of North American prairie. Both theory and experiments indicate that loss of diversity and change in species or functional group composition should have approximately equal impacts on ecosystem processes (e.g. Hooper & Vitousek 1997; Tilman *et al.* 1997a,b), if species are lost at random. Depending on how it is biased, it seems plausible that the biased loss of species could accentuate, eliminate, or in some cases reverse the effects of diversity on ecosystem processes. Such possibilities, and the situations to which they may apply, merit additional study.

Fourth, the number of species needed to assure a given level of ecosystem functioning will depend on spatial scale. It might seem that this number would be provided in a straightforward manner by the existing diversity experiments, but the spatial scale used in such experiments, when compared to the spatial scale of habitats to which such information is likely to be applied, necessitates deeper analyses. The appropriate spatial scale for biodiversity experiments is a neighbourhood scale. This is so because the known mechanisms by which diversity might impact community or ecosystem processes are mechanisms of interactions among individual organisms (e.g. Naeem *et al.* 1999b). For organisms to interact, they must be in the proximity of each other. Both niche differentiation and sampling effects require interspecific competition. For instance, for niche differentiation, the soil nutrients not consumed by an individual plant of one species can only be consumed by an individual plant of another species if they are neighbouring individuals. Thus, the measure of diversity relevant to the mechanisms summarized in sampling and niche models is neighbourhood diversity—the number of different species that directly interact with each other in an average neighbourhood. For grassland communities, such neighbourhoods probably range in area from about

0.25 to 1 m², and contain approximately 10–100 individual plants. The physical area would be much larger for forested communities, but such areas would, again, represent the regions within which neighbouring trees interacted with each other, and probably would contain about 10–100 individual saplings and adults.

Although neighbourhood interactions occur on small spatial scales, and results of biodiversity experiments are thus appropriately measured on such scales, these are not the scales upon which the diversity of communities are likely to be measured. The successional grasslands of Cedar Creek Natural History Area, for instance, have estimated species richness that ranges from 31 to 87 vascular plant species, depending on field age (Inouye 1998). These jack-knife estimates use actual observations in 100 plots (each 0.5 m²) within an approximately 0.5 hectare area, and thus provide an estimate of total plant species richness per 0.5 hectare. As would be expected from the well-known species–area relationship, these data also show a dependence of diversity on scale. In particular, they show a strong interdependence of neighbourhood (0.5 m²) and regional (0.5 hectare) diversity (Tilman 1999) that can be reasonably described by the classical species–area relationship.

The species–area relationship (where S is species richness, A is area, and c and z are constants) is $S = cA^z$. Let the number of species within a neighbourhood within which local interactions occur be S_N, and let the number of species in some larger, regional area be S_R. Let the size (area) of the neighbourhood be A_N and the size of the region be A_R. How many species would then need to occur at the regional scale to allow there to be a given number of species at the local scale? Substitution into the species–area relationship predicts this to be

$$S_R = S_N (A_R/A_N)^z$$

Analysis of the relationship between neighbourhood (0.5 m²) species richness and species richness at the 0.5-hectare scale within successional grasslands at Cedar Creek gives $z = 0.21$. In general, a variety of other studies of terrestrial communities have generated z-values ranging from about 0.15 to 0.3 (e.g. MacArthur & Wilson 1967).

This information can be used to address the question of the number of species needed within a given region to assure a given level of ecosystem functioning. Although any such extrapolation is fraught with many untested assumptions, including the assumption that there is a causal link between regional and local diversity, it might help provide context for results and suggest future tests. For instance, the Cedar Creek biodiversity experiments show that about 12 species are needed locally (in about a 0.5 m² area) to assure maximal productivity and stability. The species–area relationship with $z = 0.21$ would then mean that about 83 plant species would occur within a 0.5-hectare area when the average 0.5 m² neighbourhood within it contains 12 plant species. Only one of the 20 successional grasslands studied at Cedar Creek is this species rich (Inouye 1998), and some fields have less than half this diversity. This suggests that local diversity may constrain ecosystem functioning in many successional grasslands at Cedar Creek. Comparable extrapolations using the species–area relationship suggest that a 1 km² area would contain about 340 species when average neighbourhood diversity was 12 plant species per 0.5 m². The species–area relationship could also be used (with admitted great trepidation) to extrapolate our small-scale diversity experiments up to the scale of the Great Plains of the US. The Great Plains of the US has an area of approximately 10^{12} m². Assuming z-values ranging from 0.15 to 0.2, the Great Plains would need to contain about 840–3500 vascular plant species to have average local (0.5 m²) diversity of 12 species. The described Great Plains flora of about 3000 vascular plant species (McGregor *et al.* 1977) falls near the upper end of this range. Thus, these estimates suggest that, on average, habitats within these regions that experience marked reductions of plant diversity could experience decreased stability and productivity. However, because ecosystem functioning depends on composition as much as on diversity, the above relationships provide but a rule of thumb. The actual impact of a given loss of diversity should depend both on which and how many species were lost.

Finally, let us consider, again, what diversity is and what it measures. As the theory and experiments reviewed here illustrate, diversity is a simple, convenient measure of the range of traits within an assemblage of species. All else being equal, the range of relevant traits within a given community should increase, probably toward an asymptote, as species number increased. Clearly, the functional impact of diversity should also depend on the range of species traits within the species pool from which species were drawn. The extent of variation within the species pool is a second component of diversity. The concept of a species pool applies strictly for well-designed diversity experiments, but also may provide a reasonable approximation for local, site-to-site variation within a given ecosystem for which there is a regional pool of species and for which local variation in both species number and composition result from random processes, such as random colonization events. Attempts to distinguish between the effects of species number vs. the effects of variation in species traits can become needless semanticism. The two are highly correlated. Moreover, much of the interest in the effects of diversity is rooted in the potential impacts of the loss of species on ecosystem functioning. Although the impact of the loss of species will depend on the extent of interspecific differences, the clear driving force is the loss of species. The observed impacts thus are reasonably attributed to this loss of diversity.

In total, work by a wide variety of researchers during the past decade has expanded our knowledge of the impacts of diversity on ecosystem processes and highlighted many new questions. In so doing, this work has led us to clarify what is meant by diversity, including the recognition that diversity has many components and that it can be confounded with many other factors in natural communities. Despite these advances, many issues remain unresolved. Chief among these is the relative importance of differences in diversity vs. differences in other correlated factors as controllers of ecosystem functioning in both natural systems and in systems impacted by human actions.

References

Aarssen, L.W. (1997) High productivity in grassland ecosystems: effected by species diversity or productive species? *Oikos* **80**, 183–184.

Allison, G.W. (1999) The implications of experimental design for biodiversity manipulations. *American Naturalist* **153**, 26–45.

Al-Mufti, M.M., Sydes, C.L., Furness, S.B., Grime, J.P. & Band, S.R. (1977) Aquantitative analysis of shoot phenology and dominance in herbaceous vegetation. *Journal of Ecology* **65**, 759–791.

Armstrong, R.A. & McGehee, R. (1980) Competitive exclusion. *American Naturalist* **115**, 151–170.

Ashton, P.S. (1977) A contribution of rainforest research to evolutionary theory. *Annals of the Missouri Botanical Garden* **64**, 694–705.

Bond, W. (1983) On alpha diversity and the richness of the Cape flora: A study in southern Cape fynbos. In: *Mediterranean-Type Ecosystems: the Role of Nutrients* (eds F.J. Kruger, D.T. Mitchell & J.U.M. Jarvis), pp. 337–356. Springer-Verlag, Berlin.

Chesson, P.L. (1986) Environmental variation and the coexistence of species. In: *Community Ecology* (eds J. Diamond & T. Case), pp. 240–256. Harper & Row, New York.

Darwin, C. (1859) *The Origin of Species by Means of Natural Selection*. Reprinted by The Modern Library, Random House, New York.

Davis, M.B. (1986) Climatic instability, time lags, and community disequilibrium. In: *Community Ecology* (eds J. Diamond & T. Case), pp. 269–284. Harper & Row, New York.

Dix, R. & Smeins, F. (1967) The prairie, meadow and marsh vegetation of Nelson County, North Dakota. *Canadian Journal of Botany* **45**, 21–58.

Doak, D.F., Bigger, D., Harding, E.K., Marvier, M.A., O'Malley, R.E. & Thomson, D. (1998) The statistical inevitability of stability-diversity relationships in community ecology. *The American Naturalist* **151**, 264–276.

Dodd, M.E., Silvertown, J., McConway, K., Potts, J. & Crawley, M. (1994) Stability in the plant communities of the Park Grass Experiment: the relationships between species richness, soil pH and biomass variability. *Philosophical Transactions of the Royal Society of London, Series B* **346**, 185–193.

Ehrlich, P.R. & Ehrlich, A.H. (1981) *Extinction. The Causes and Consequences of the Disappearance of Species*. Random House, New York.

Elton, C.S. (1958). *The Ecology of Invasions by Animals and Plants*. Methuen, London.

Estes, J.A. & Duggins, D.O. (1995) Sea otters and kelp forests in Alaska: generality and variation in a community ecological paradigm. *Ecological Monographs*. **65**, 75–100.

Ewel, J.J. & Bigelow, S.W. (1996) Plant life-forms and tropical ecosystem functioning. In: *Biodiversity and Ecosystem Processes in Tropical Forests* (eds G.H. Orians, R. Dirzo & J.H. Cushman), pp. 101–126. Springer-Verlag, New York.

Ewel, J.J., Mazzarino, M.J. & Berish, C.W. (1991) Tropical soil fertility changes under monocultures and successional communities of different structure. *Ecological Applications* **1**, 289–302.

Frank, D.A. & McNaughton, S.J. (1991) Stability increases with diversity in plant communities: empirical evidence from the 1988 Yellowstone drought. *Oikos* **62**, 360–362.

Garnier, E., Navas, M.-L., Austin, M.P., Lilley, J.M. & Gifford, R.M. (1997) A problem for biodiversity-productivity studies: how to compare the productivity of multispecific plant mixtures to that of monocultures? *Acta Œcologica* **18**, 657–670.

Givnish, T.J. (1994) Does diversity beget stability? *Nature* **371**, 113–114.

Goodman, D. (1975) The theory of diversity–stability relationships in ecology. *Quarterly Review of Biology* **50**, 237–266.

Grime, J.P. (1979). *Plant Strategies and Vegetation Processes*. John Wiley & Sons, Chichester.

Grime, J.P. (1997) Biodiversity and ecosystem function: the debate deepens. *Science* **277**, 1260–1261.

Grubb, P.J. (1986) Problems posed by sparse and patchily distributed species in species-rich plant communities. In: *Community Ecology* (eds J. Diamond & T. Case), pp. 207–225. Harper & Row Publishers, New York.

Harper, J.L. (1977) *Population Biology of Plants*. Academic Press, London.

Hector, A., Schmid, B., Beierkuhnlein, C. *et al.* (1999) Plant diversity and productivity experiments in European grasslands. *Science* **286**, 1123–1127.

Hector, A., Schmid, B., Beierkuhnlein, C. *et al.* (2000) Response. *Science* **289**, 1255a.

Hobbie, S.E. (1996) Temperature and plant species control over litter decomposition in Alaskan tundra. *Ecological Monographs* **66**, 503–522.

Hodgson, J.G., Thompson, K., Wilson, P.J. & Bogaard, A. (1998) Does biodiversity determine ecosystem function? The Ecotron experiment reconsidered. *Functional Ecology* **12**, 843–848.

Holdridge, L., Grenke, W., Hatheway, W., Liang, T. & Tosi Jr, J. (1971). *Forest Environments in Tropical Life Zones: a Pilot Study*. Pergamon Press, Oxford.

Hooper, D.U. & Vitousek, P.M. (1997) The effects of plant composition and diversity on ecosystem processes. *Science* **277**, 1302–1305.

Hooper, D.U. & Vitousek. P.M. (1998) Effects of plant composition and diversity on nutrient cycling. *Ecological Monographs* **68**, 121–149.

Huisman, J. & Weissing, F.J. (1999) Biodiversity of plankton by species oscillations and chaos. *Nature* **402**, 407–410.

Huston, M. (1980) Soil nutrients and tree species richness in Costa Rican forests. *Journal of Biogeography* **7**, 147–157.

Huston, M.A. (1994). *Biological Diversity: the Coexistence of Species on Changing Landscapes*. Cambridge University Press, Cambridge.

Huston, M.A. (1997) Hidden treatments in ecological experiments: re-evaluating the ecosystem function of biodiversity. *Oecologia* **110**, 449–460.

Huston, M.A., Aarssen, L.W., Austin, M.P. *et al.* (2000) No consistent effect of plant diversity on productivity. *Science* **289**, 1255a.

Inouye, R.S. (1998) Species–area curves and estimates of total species richness in an old-field chronosequence. *Plant Ecology* **137**, 31–40.

Ives, A.R., Gross, K. & Klug, J.L. (1999) Stability and variability in competitive communities. *Science* **286**, 542–544.

King, A.W. & Pimm, S.L. (1983) Complexity, diversity, and stability: a reconciliation of theoretical and empirical results. *American Naturalist* **122**, 229–239.

Knops, J.M.H., Tilman, D., Haddad, N.M. *et al.* (1999) Effects of plant species richness on invasions dynamics, disease outbreaks, insect abundances and diversity. *Ecology Letters* **2**, 286–293.

Lawton, J.H., Naeem, S., Thompson, L.J., Hector, A. & Crawley, M.J. (1998) Biodiversity and ecosystem function: Getting the Ecotron experiment in its correct context. *Functional Ecology* **12**, 848–852.

Lehman, C.L. & Tilman, D. (2000) Biodiversity, stability, and productivity in competitive communities. *American Naturalist* **156**, 534–552.

Likens, G.E., Bormann, F.H., Pierce, R.S., Eaton, J.S. & Johnson, A.M. (1977). *Bio-Geo-Chemistry of a Forested Ecosystem*. Springer-Verlag, New York.

Likens, G.E., Driscoll, C.T., Buso, D.C. *et al.* (1998) The biogeochemistry of calcium at Hubbard Brook. *Biogeochemistry* **41**, 89–173.

Loreau, M. (1998a) Biodiversity and ecosystem functioning: a mechanistic model. *Proceedings of the National Academy of Science* **95**, 5632–5636.

Loreau, M. (1998b) Separating sampling and other effects in biodiversity experiments. *Oikos* **82**, 600–602.

MacArthur, R.H. (1955) Fluctuations of animal populations and a measure of community stability. *Ecology* **36**, 533–536.

MacArthur, R.H. & Wilson, E.O. (1967). *The Theory of Island Biogeography*. Princeton University Press, Princeton, NJ.

Margalef, R. (1968) *Perspectives in Ecological Theory*. University of Chicago Press, Chicago.

May, R.M. (1973) *Stability and Complexity in Model Ecosystems*. Princeton University Press, Princeton, NJ.

May, R.M. (1974). *Stability and Complexity in Model Ecosystems*, 2nd edn. Princeton University Press, Princeton, NJ.

McGrady-Steed, J., Harris, P.M. & Morin, P.J. (1997) Biodiversity regulates ecosystem predictability. *Nature* **390**, 162–165.

McGregor, R.L., Barkley, T.M., Barker, W.T. *et al.* (1977). *Atlas of the Flora of the Great Plains*. The Iowa State University Press, Ames, Iowa.

McNaughton, S.J. (1978) Stability and diversity of ecological communities. *Nature* **274**, 251–253.

McNaughton, S.J. (1993) Biodiversity and function of grazing ecosystems. In: *Biodiversity and Ecosystem Function* (eds E.-D. Schulze & H.A. Mooney), pp. 361–383. Springer-Verlag, Berlin.

Naeem, S. (2000) Reply to Wardle *et al. Bulletin of the Ecological Society of America* **81**, 241–246.

Naeem, S. & S.Li. (1997) Biodiversity enhances ecosystem reliability. *Nature* **390**, 507–509.

Naeem, S., Chapin III, F.S., Costanza, R., et al. (1999a) Biodiversity and ecosystem functioning: Maintaining natural life support processes. *Issues in Ecology*, Number 4. Ecological Society of America.

Naeem, S., Håkenson, K., Lawton, J.H., Crawley, M.J. & Thompson, L.J. (1996) Biodiversity and plant productivity in a model assemblage of plant species. *Oikos* **76**, 259–264.

Naeem, S., Thompson, L.J., Lawler, S.P., Lawton, J.H. & Woodfin, R.M. (1994) Declining biodiversity can alter the performance of ecosystems. *Nature* **368**, 734–737.

Naeem, S., Thompson, L.J., Lawler, S.P., Lawton, J.H. & Woodfin, R.M. (1995) Empirical evidence that declining species diversity may alter the performance of terrestrial ecosystems. *Philosophical Transactions of the Royal Society of London, Series B* **347**, 249–262.

Naeem, S., Tjossem, S.F., Byers, D., Bristow, C. & Li, S. (1999b) Plant neighborhood diversity and production. *Ecoscience* **6**, 355–365.

Odum, E.P. (1959). *Fundamentals of Ecology*, 2nd edn. Saunders, Philadelphia.

Pacala, S. & Tilman, D. (2000) (in press) Chapter 7. In: *Functional Consequences of Biodiversity: Experimental Progress and Theoretical Extensions* (eds A. Kinzig, S. Pacala & D. Tilman). Princeton University Press, Princeton, NJ.

Pastor, J., Aber, J.D., McClaugherty, C.A. & Melillo, J.M. (1984) Aboveground production and N and P cycling along a nitrogen mineralization gradient on Blackhawk Island, Wisconsin. *Ecology* **65**, 256–268.

Pimm, S.L. (1979) Complexity and stability: another look at MacArthur's original hypothesis. *Oikos* **33**, 351–357.

Pimm, S.L. (1984) The complexity and stability of ecosystems. *Nature* **307**, 321–326.

Power, M.E. (1995) Floods, food chains, and ecosystem processes in rivers. In: *Linking Species and Ecosystems* (eds C.G. Jones & J.H. Lawton), pp. 52–60. Chapman & Hall, New York.

Puerto, A., Rico, M., Matias, M.D. & Garcia, J.A. (1990) Variation in structure and diversity in Mediterranean grasslands related to trophic status and grazing intensity. *Journal of Vegetation Science* **1**, 445–452.

Schläpfer, F. & Schmid, B. (1999) Ecosystem effects of biodiversity: a classification of hypotheses and exploration of empirical results. *Ecological Applications* **9**, 893–912.

Schläpfer, F., Schmid, B. & Seidl, I. (1999) Expert estimates about effects of biodiversity on ecosystem processes and services. *Oikos* **84**, 386.

Schulze, E.D. & Mooney, H.A. (1993). *Biodiversity and Ecosystem Function*. Springer-Verlag, Berlin.

Sterner, R.W. (1995) Elemental stoichiometry of species in ecosystems. In: *Linking Species and Ecosystems* (eds C.G. Jones & J.H. Lawton), pp. 240–252. Chapman & Hall, New York.

Swift, M.J. & Anderson, J.M. (1993) Biodiversity and ecosystem function in agricultural systems. In: *Biodiversity and Ecosystem Function* (eds E.-D. Schulze & H.A. Mooney), pp. 15–41. Springer-Verlag, Berlin.

Symstad, A.J., Tilman, D., Willson, J. & Knops, J.M.H. (1998) Species loss and ecosystem functioning: effects of species identity and community composition. *Oikos* **81**, 389–397.

Thurston, J. (1969) The effect of liming and fertilizers on the botanical composition of permanent grassland, and on the yield of hay. In: *Ecological Aspects of the Mineral Nutrition of Plants* (ed. I. Rorison), pp. 3–10. Blackwell Scientific Publications, Oxford.

Tilman, D. (1982) *Resource Competition and Community Structure. Monographs in Population Biology*. Princeton University Press. Princeton, NJ.

Tilman, D. (1988) *Plant Strategies and the Dynamics and Structure of Plant Communities*. Princeton University Press. Princeton, NJ.

Tilman, D. (1996) Biodiversity: Population versus ecosystem stability. *Ecology* **77**, 350–363.

Tilman, D. (1999) The ecological consequences of changes in biodiversity: a search for general principles. *Ecology* **80**, 1455–1474.

Tilman, D. & Downing, J.A. (1994) Biodiversity and stability in grasslands. *Nature* **367**, 363–365.

Tilman, D. & Lehman, C. (2000) Biodiversity, composition, and ecosystem processes: Theory and concepts. In: *Functional Consequences of Biodiversity: Experimental Progress and Theoretical Extensions* (eds A. Kinzig, S. Pacala & D. Tilman). Princeton University Press, Princeton, NJ.

Tilman, D. & Pacala, S. (1993) The maintenance of species richness in plant communities. In: *Species Diversity in Ecological Communities* (eds R.E. Ricklefs & D. Schluter), pp. 13–25. University of Chicago Press, Chicago.

Tilman, D., Dodd, M.E., Silvertown, J., Poulton, P.R., Johnston, A.E. & Crawley, M.J. (1994a) The Park Grass Experiment: Insights from the most long-term ecological study. In: *Long-Term Experiments in Agricultural and Ecological Sciences* (eds R.A. Leigh & A.E. Johnston), pp. 287–303. CAB International, Wallingford.

Tilman, D., Downing, J.A. & Wedin, D.A. (1994b) Reply. *Nature* **371**, 114.

Tilman, D., Wedin, D. & Knops, J. (1996) Productivity and sustainability influenced by biodiversity in grassland ecosystems. *Nature* **379**, 718–720.

Tilman, D., Lehman, C.L. & Thomson, K.T. (1997a) Plant diversity and ecosystem productivity: theoretical considerations. *Proceeding of the National Academy of Science of the USA* **94**, 1857–1861.

Tilman, D., Knops, J., Wedin, D., Reich, P., Ritchie, M. & Siemann, E. (1997b) The influence of functional diversity and composition on ecosystem processes. *Science* **277**, 1300–1302.

Tilman, D., Naeem, S., Knops, J. *et al.* (1997c) Biodiversity and ecosystem properties. *Science* **278**, 1866–1867.

Tilman, D., Lehman, C.L. & Bristow, C.E. (1998) Diversity–stability relationships: statistical inevitability or ecological consequence? *American Naturalist* **151**, 277–282.

Tilman, D., Knops, J., Wedin, D. & Reich, P. (2000) Experimental and observational studies of diversity, productivity and stability. In: *Functional Consequences of Biodiversity: Experimental Progress and Theoretical Extensions* (eds A. Kinzig, S. Pacala & D. Tilman). Princeton University Press, New Jersey.

Vitousek, P.M. & Hooper, D.U. (1993) Biological diversity and terrestrial ecosystem biogeochemistry. In: *Biodiversity and Ecosystem Function* (eds E.-D. Schulze & H.A. Mooney), pp. 3–14. Springer-Verlag, Berlin.

Vitousek, P.M. & Matson, P.A. (1984) Mechanisms of nitrogen retention in forest ecosystems: a field experiment. *Science* **225**, 51–52.

Vitousek, P.M. & Matson, P.A. (1985) Disturbance, nitrogen availability, and nitrogen losses in an intensively managed loblolly pine plantation. *Ecology* **66**, 1360–1376.

Vitousek, P.M., Walker, L.R., Whiteaker, L.D., Mueller-Dombois, D. & Matson, P.A. (1987) Biological invasion by *Myrica faya* alters ecosystem development in Hawaii. *Science* **238**, 802–804.

Wardle, D.A., Zackrisson, O., Hörnberg, G. & Gallet, C. (1997a) The influence of island area on ecosystem properties. *Science* **277**, 1296–1299.

Wardle, D.A., Zackrisson, O., Hörnberg, G. & Gallet, C. (1997b) Response. *Science* **278**, 1867–1869.

Wardle, D.A., Huston, M.A., Grime, J.P. *et al.* (2000) Biodiversity and ecosystem function: an issue in ecology. *Bulletin of the Ecological Society of America* **81**, 235–239.

Wedin, D. & Tilman, D. (1990) Species effects on nitrogen cycling: a test with perennial grasses. *Oecologia* **84**, 433–441.

Whittaker, R.H. (1975) *Communities and Ecosystems*, 2nd edn. Macmillan, New York.

Wilson, E.O. (1992) *The Diversity of Life*. Belknap Press of Harvard University Press, Cambridge, MA.

Part 3
Ecology of changing environments

Chapter 10
Climate change and steady state in temperate hardwood forests

M. B. Davis

Introduction

Temperate deciduous and mixed deciduous–coniferous forests in the northern United States are dynamic systems, a mosaic of stands in various stages of recovery from patchy disturbances (Bormann & Likens 1979; Runkle 1985). Although individual stands may be changing, forest composition averaged over the entire landscape is often considered a steady-state system. An exception might be a landscape recovering from a widespread, stand-initiating disturbance, but these are rare events in the upper Great Lakes region States, with return times greater than 1000 years (Lorimer 1983; Canham & Loucks 1984). Research in forest dynamics has focused on rapid changes in small gaps, where saplings and newly established seedlings compete for light (e.g. Runkle 1985; Webb 1988; Frelich & Lorimer 1991). Succession on larger spatial scales following catastrophic disturbance has also received attention (e.g. Heinselman 1973; Pickett & White 1985; Whitney 1986; Foster 1988; Foster et al. 1997).

A more complete understanding of forest dynamics comes from direct observations in permanent plots (Woods 2000a,b). Data recorded over several decades are extended to centuries through reconstructions of forest history from tree rings and fallen logs (Stearns 1949; Henry & Swan 1974; Oliver & Stephens 1977; Whitney 1986). Recent direct observations for 30 years at one site and 60 years at another show that slow but significant changes in composition are occurring (Woods 2000a,b). Past disturbance is often inferred from changes like these, and unusual climatic events are sometimes invoked to explain similar changes in many stands, i.e. the apparent synchroneity of hypothesized disturbances (Stearns 1949, 1991). By assuming that a disturbance occurred sometime in the past, changing composition can be reconciled with the idea of a steady-state landscape. Woods points out, however, that changes in composition are still occurring several centuries after the most recent possible disturbance, given the ages of trees in the forest. If recent changes in composition are a response to disturbance centuries after the event, recovery time approaches the same magnitude as time between disturbances, resulting in a system that seldom reaches stand-level equilibrium (Woods 2000a).

Department of Ecology, Evolution and Behavior, University of Minnesota, St. Paul, MN 55108, USA

Palaeo-ecologists who emphasize the effect of climate on vegetation consider forests on a regional scale and assume that dynamic changes in stand composition average out over the landscape as a whole. If responses to disturbance are considered, they are believed to occur on a shorter time scale than adjustment to climate, which occurs in decades to centuries. This 'dynamic equilibrium' view is similar to a steady state with the proviso that species composition averaged over the landscape drifts slowly in time in concert with the trajectory of climate change (Webb 1986). Assuming a steady state of any kind—stable or drifting—influences the way data are interpreted. Consequently the hypothesis that a steady state exists needs to be tested. However, the ponderous rates of change in forest composition recorded in long-term plots mean that a very long record at a number of sites is required. The record has to be several times longer than the lifespan of trees (Lertzman 1995), in other words much longer than feasible with permanent plots or reconstructions from tree rings.

Fossil pollen has potential to provide this record, now that palaeo-ecologists have developed techniques for recording forest composition at the stand scale. The new methods utilize fossil pollen in very tiny swales and bogs—'forest hollows'. Forest hollows, typically about 10 m across, sample forest with a resolution of 1–3 hectares (Sugita 1993, 1994; Calcote 1995). Interpretation depends on a suite of surface samples from hollows, accompanied by data on the composition of nearby and regional forest, for comparison with fossil pollen assemblages (Calcote 1998; Sugita 1994, 1998; Sugita *et al.* 1997; Parshall & Calcote 2001).

Pollen studies from small hollows at Sylvania, in northern Michigan, trace the stand-scale history of forests dominated by hemlock (*Tsuga canadensis*) and sugar maple (*Acer saccharum*)—similar to the forests in the Huron Mountains where permanent plots record changes over the past 30 years. The same species, plus beech (*Fagus grandifolia*) dominate the forests at Dukes where permanent plots record forest composition over the past 60 years (Woods 2000a,b) (Fig. 10.1). Fossil pollen assemblages corroborate data from permanent plots by showing that changes are occurring in the species composition of forest stands. The pollen record, however, extends farther back in time by several millennia, showing that the forest changes are directional, and consistent at many sites within the forest (Davis *et al.* 1998). Thus, the forest composition is changing from one century to the next and from one millennium to the next. In part the changes result from succession following disturbances, but synchronous changes at many sites that are directional over long periods are more likely to reflect climate forcing. Over such long spans of time, changes in regional climate are large enough to influence the outcome of competition and thus to affect forest composition at stand and landscape scales (Davis *et al.* 2000a). Furthermore, in northern Michigan disturbance regimes have changed, complicating forest response to climate change. Only a few thousand years ago, fires occurred more frequently, conferring advantage on species with particular life-history attributes (Loehle 2000), and altering landscape patterns of forest communities.

In this paper I contrast the view of northern hardwoods–hemlock forests that

EFFECTS OF CLIMATE CHANGE ON TEMPERATE FORESTS

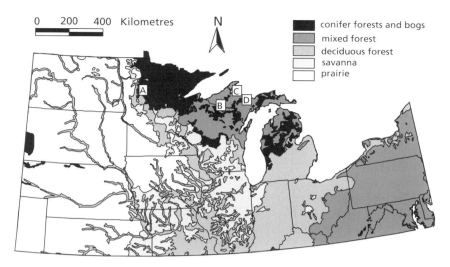

Figure 10.1 Map of upper Great Lakes Region, USA, including Michigan, Wisconsin and Minnesota, showing major potential vegetation types Locations mentioned in text are shown: A, Itasca Park; B, Sylvania Wilderness; C, Huron Mountains; D, Dukes Forest. (Modified with permission from Kuchler 1964.)

emerges from studies that use tree rings and long-term plots, with the insights gained from the longer record in fossil pollen. The combined sources of information lead to the conclusion that, in forests with long-lived trees, internal processes such as competition, succession and recovery from disturbance are influenced by the trajectory of climate change, including the changing disturbance regime (Lertzman 1995). A better understanding of the effects of climate on forest dynamics is essential now that climate is beginning a series of changes likely to accelerate in the coming century (Houghton et al. 1996).

Disturbance

Disturbance is one of the primary factors driving forest dynamics (Pickett & White 1985). Disturbance by wind is particularly important in hemlock–hardwood forests in the upper Great Lakes region: storms that remove 10% of the canopy have a return time of about 50 years, while storms that remove 30% or more have return times on a century time scale (Frelich & Lorimer 1991). These small disturbances result in faster canopy turnover than would occur from mortality due to ageing alone; further large multi-tree gaps are created by trees falling against one another, especially in hemlock stands where root grafts can pull down adjacent trees. Multi-tree gaps maintain diversity in the forest by providing opportunities for establishment for species of intermediate shade tolerance like yellow birch

(*Betula alleghaniensis*) and basswood (*Tilia americana*). Tree-ring studies in the mapped plots at Sylvania show that damage from windthrow is patchy, affecting different parts of the landscape in different years, but affecting maple stands and hemlock stands with similar probability (Frelich & Graumlich 1991; Parshall 1995).

Steady-state species composition within the forest depends on a relatively constant rate of turnover from disturbance (Runkle 1985; Loehle 2000). It is questionable, however, that disturbance rates remain constant because both disturbance and response to disturbance are associated with interannual climate variability, which varies on decadal scales (Skaggs *et al.* 1995). Drought damages root systems of trees, and as a result gap formation and mortality from insect damage are greatest in years following drought years (Secrest *et al.* 1941; Parshall 1995). Thus, the turnover rate of the forest is influenced by the frequency of drought years. Recent analysis of weather records shows a gradient in interannual drought variability across the Midwest. The Palmer Drought Severity Index (PDSI) calculates the moisture surplus or deficit relative to the long-term mean at that site; the interannual standard deviation of the PDSI captures the frequency of severe drought years when fires are likely. The standard deviation for 1931–92 is highest in the prairie regions of the Great Plains, intermediate in the mixed conifer forests with a fire-dominated disturbance regime, decreasing eastward to the hemlock–hardwood forest region (Cook *et al.* 1999; Davis *et al.* 2000b). This gradient changed in the past. The frequency and severity of interannual variations was as high in northern Michigan 3000 years ago as in western Wisconsin today (M. B. Davis *et al.*, unpublished). Thus, on long time scales, at least, turnover rates can and do change, with the result that changes will occur in species abundances in the forest.

In the more distant past, the disturbance regime in northern Michigan was dominated by fire (Davis *et al.* 1992, 1998), as in central Minnesota before fire suppression policies came into force around 1910 (Frissel 1973; Clark 1990). Just 6000 years ago precipitation was slightly lower, summer temperatures higher, and the standard deviation of the drought index higher. All of these variables suggest climate similar to the modern climate at Itasca Park, in central Minnesota (Davis *et al.* 2000a and unpublished). At Itasca, patchy fires occurred every few decades (Fig. 10.2a). Fire frequency differed in nearby sites, depending on the location of firebreaks (Grimm 1984); alternative plant communities grew in fire-prone and fire-protected sites. Time since last fire was also important, resulting in a mosaic of forest communities (Frissell 1973; Heinselman 1973). The cover-type map of Itasca Park (Fig. 10.2b) provides an analogue for the distribution of forest communities at Sylvania 6000 years ago; the similarity of pollen assemblages suggests that the overall frequencies of species were similar.

In addition to frequent minor wind disturbances, catastrophic, stand-initiating windstorms occur in the Great Lakes regions in association with massive 'supercell' thunderstorms. They are rare, however, with a return time longer than a millennium (Canham & Loucks 1984). The frequency of such events in the more distant

EFFECTS OF CLIMATE CHANGE ON TEMPERATE FORESTS

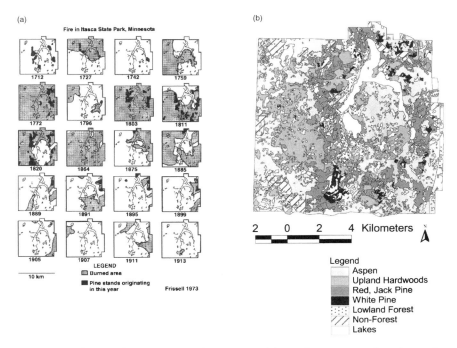

Figure 10.2 (a) Maps showing areas within Itasca Park burned in successive fires during the 18th and 19th centuries. (Reproduced with permission from Frissell 1973.) (b) Cover-type map of Itasca Park showing the mosaic of forest communities resulting from fire. (From Minnesota Department of Natural Resources, unpublished.)

past is unknown, although it seems likely the probabilities changed in the mid-Holocene when precipitation and summer and winter temperature changed.

Climate change involves not only slow changes in mean temperature and precipitation, but also changes in variability that change disturbance regimes and alter the relative importance of species with different life-history strategies (Loehle 2000). Changing variability makes responses to climate change complex and difficult to distinguish from disturbance dynamics.

Patch origin and persistence

Hemlock–maple forests are patchy on the landscape scale, but the stability of these patches had suggested that the relative abundance of maple and hemlock on the landscape as a whole remained in steady state for centuries or even millennia.

A mosaic of stands, dominated either by maple or by hemlock, was widespread in western upper Michigan before logging, as indicated by aerial photographs taken in the 1930s (Davis *et al.* 1996). The 10–30-hectare patch pattern is still evident in old-growth stands (Fig. 10.3). Hemlock and sugar maple are shade-

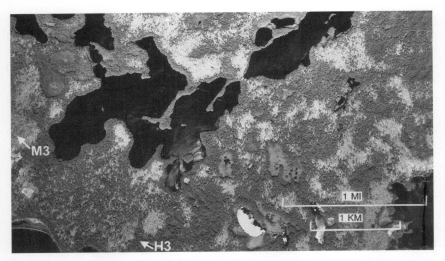

Figure 10.3 Leaf-off aerial photograph of old-growth forests in the Sylvania Wilderness, upper Michigan. Light patches are forests dominated by sugar maple; dark forest patches are hemlock stands. H3 and M3 are sites mentioned in the text.

tolerant species that can persist for many decades beneath the canopy; seedlings and subcanopy trees of each species are spatially associated with conspecific canopy trees. Consequently the probability of self-replacement is high for each species, favouring the persistence of patches dominated either by maple or by hemlock (Frelich et al. 1993). A shifting mosaic resulting from small wind disturbances occurs within the patches, maintaining basswood and ironwood (*Ostrya virginiana*) in sugar maple stands and yellow birch in both sugar maple and hemlock stands (Frelich & Graumlich 1991; Parshall 1995; Loehle 2000). Neighbourhood influences from litter affect nutrient availability, pH and other soil attributes that influence seedling success (Ferrari & Sugita 1996; Ferrari 1999), and seed dispersal is undoubtedly important as well (Ribbens et al. 1994; Woods 2000a). The MOSAIC model has been used to explore the power of positive neighbourhood feedbacks in producing spatial patterns in forests, demonstrating that very small differences in the physical environment that favour one species can be reinforced by neighbourhood feedbacks to produce very large patches dominated by that species (Frelich et al. 1999). Although we have not been able to measure extrinsic environmental differences between hemlock and maple patches at Sylvania (Pastor & Broschart 1990; Frelich et al. 1993), such differences may exist, explaining some of the very large patches (Fig. 10.3).

In contrast to the results from hemlock–maple forests in Sylvania that emphasize patch persistence, Woods' (2000a) observations in similar forests in the Huron Mountains suggest that hemlock has been gaining on sugar maple over the past 30 years, establishing colonies within maple patches and spreading from these local foci. Woods' observation fits with Pastor & Broschart's (1990) geographical information system (GIS) analysis of forest patterns at Sylvania: fractal geometry of hemlock patch edges suggested ongoing invasion of maple stands. Further, they observed that maple stands are generally surrounded by hemlock, and proposed that hemlock is a background vegetation on the landscape, upon which maple stands have been imposed. Pastor and Broschart's results seemed to suggest that maple stands are an ephemeral condition (perhaps on a very long time scale) that develops within hemlock stands. However, tree ages in mapped plot A at Sylvania gave no indication of hemlock invasion of the maple stand over the past century (Frelich & Graumlich 1991). To the contrary, in mapped plot C, tree ages suggest that maple replaced hemlock following wind disturbance of a mixed stand 60 years ago (Parshall 1995).

Fossil pollen can settle questions like this by recording patch dynamics on longer time scales. The ideal pollen records for this purpose would come from small hollows along the borders between hemlock and maple stands. Records from three border hollows are now being analysed (S. Sugita, unpublished data). Preliminary data from one site suggest that border stands are unstable on the millennial time scale relative to the centres of large patches (Davis *et al.* 1992). This result is compatible with Woods' observation of hemlock encroachment, but suggests as well that maple can on occasion replace hemlock. Certainly maple has an advantage over hemlock following logging (Davis *et al.* 1996), but the increase of deer density in recent decades and the negative effects of deer browse on hemlock seedlings makes observations in modern forests of limited use in explaining events in the past (Anderson & Loucks 1979; Frelich & Lorimer 1985; Alverson *et al.* 1988; Rooney *et al.* 2000).

Long-term histories of four hemlock stands and four hardwood stands appeared at first to contradict the idea that hemlock invades maple stands (Davis *et al.* 1998). Early results were interpreted as evidence that large patches are stable, remaining dominated by either hemlock or maple over the past 3400 years (Davis *et al.* 1992, 1994). We now have a larger data set of surface samples from hollows in a wider variety of forest communities (Parshall & Calcote 2001) and we are using them to reanalyse the fossil data from Sylvania. These new analyses lead to a modified view of patch stability.

Hemlock stands, as we had concluded earlier, originated as white pine stands that were invaded by hemlock 3400–2800 years ago. White pine stands were apparently distributed as patches throughout Sylvania, perhaps similar to Itasca today (see Fig. 10.2) (Davis *et al.* 2000a). Histories of three additional hemlock stands are similar to H3 (Davis *et al.* 1998). Figures 10.4(a),(b) show the trajectory of pollen assemblages at H3 between 6000 and 0 years ago, projected into multivariate space defined by canonical variates analysis (CVA) of surface pollen assemblages from 88 hollows in Michigan and Wisconsin. Clusters of surface pollen assemblages from similar forest types are outlined in the figure, with the dominant species (pine, oak, etc.) indicated. Six thousand years ago the pollen assemblages at H3 clustered in the same CVA space as surface samples from pine stands. Between 3500 and 3000 years

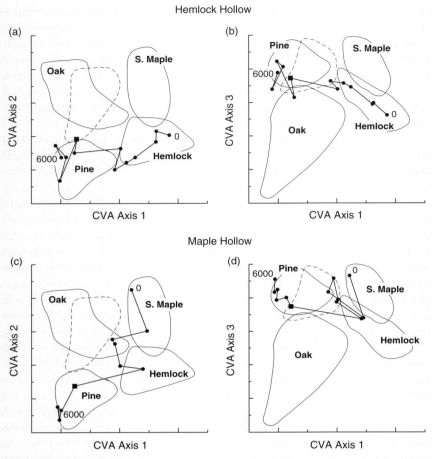

Figure 10.4 Canonical variates analysis (CVA) space defined by 88 pollen assemblages from different forest communities in Michigan and Wisconsin (Parshall & Calcote 2001); (a) and (c) show axes 1 and 2, while (b) and (d) show axes 1 and 3. Circles indicate groupings of assemblages from the same forest type; assemblages from maple forests are in two groups reflecting differences in background pollen in Michigan (group with solid outline) and Wisconsin (broken outline). (a, b) Fossil pollen assemblages dating at 500 year intervals from 6000 years ago to 0 years ago from a hemlock hollow (H3) are projected into the CVA space; they show a trajectory of change through time from white pine assemblages (on the left) to hemlock assemblages (on the right). The pollen assemblage dating from 3500 years ago, just before hemlock invasion, is indicated by a square symbol. (c, d) Fossil pollen assemblages from a maple hollow (M3) follow a similar trajectory from pine to hemlock between 6000 years ago and 3000 years ago (3500-year-old sample indicated by square symbol), and then change direction, moving toward the part of the CVA space defined by surface samples from maple forest.

ago they shifted to the part of CVA space defined by assemblages from mixed hemlock–pine stands, and later to the CVA space defined by surface samples from hemlock stands (Fig. 10.4a,b). The fossil assemblages just before hemlock invasion (3500 years before present, indicated by square symbol in the figure) resemble surface samples from pine forest, indicating that hemlock invaded a white-pine stand. The result was a mixed hemlock–white-pine stand, within which hemlock continued to increase until the stand became similar to modern hemlock-dominated forest (Fig. 10.4a,b).

Trajectories from fossil pollen in a sediment core from a hollow within maple stand M1 (Davis *et al.* 1998; not shown here) demonstrate that a forest stand with abundant oak and maple 3500 years ago was not invaded by hemlock, soon becoming dominated by sugar maple (Davis *et al.* 1998). But other maple stands had different histories. In our 1998 paper we described one maple stand (M4) that originated as a pine stand invaded by hemlock, which was later converted to a maple stand. Reanalysis has revealed a second example: Fig. 10.4(c),(d) show the trajectory in CVA space for pollen assemblages from M3, a hollow located in what is now a 20-hectare maple stand (see Fig. 10.3). The trajectory shows that a white-pine forest was invaded by hemlock between 3500 and 3000 years ago, resulting in a mixed hemlock–white-pine stand, just as in the case of H3. But 2500 years ago, the trajectory for M3 took a different course, indicating that the forest changed, for reasons not yet understood, from a hemlock–pine stand to a maple stand (Fig. 10.4c,d). These newer results suggest that hemlock and maple patches are much less stable than we thought previously. While the histories of many hemlock stands are similar to one another, the maple stands originated in several different ways—some from oak–maple stands that became dominated by sugar maple, and others from hemlock–pine or hemlock stands. While hemlock stands and maple stands can persist for centuries, on the millennium scale hemlock sometimes invades maple stands, resulting in mixed stands. Other, poorly understood processes can convert hemlock and mixed forest to maple-dominated stands. A hemlock–maple patch structure has been present continuously on the landscape for 3000 years, even though individual patches have occasionally switched from one community type to the other. However, species composition averaged over the landscape was not constant. Within hemlock patches, hemlock gradually displaced white pine, while within maple patches, sugar maple and basswood became more abundant at the expense of oak and red maple.

Invasion and competition between invading and resident species

Sixty-year records from permanent plots at the Dukes Forest, located at the western range limit for beech (*Fagus grandifolia*), show that beech is increasing in abundance (Woods 2000b). Local pollen records from humus profiles at Duke's show a corresponding increase in beech pollen during the 20th century (Davis 1989). Regional pollen records from nearby lakes indicate that beech extended its range to this part of Michigan within the past 500 years (Woods & Davis 1989). The increase in numbers of beech trees and saplings can be thought of as population expansion following an establishment phase after the first beech seeds were dispersed into the forest several hundred years ago. Alternatively, the range extension and population expansion can be viewed as a continuing response to an ongoing climate change that is making the environment ever more suitable for beech relative to other species in the forest.

The permanent plots show that beech is increasing most rapidly on fine-grained soil with drainage-impeding fragipan and a relatively high pH (Woods 2000b). This suggests that moisture is the limiting variable for beech, a finding of interest because climate reconstructions record increasing precipitation during the last 1000 years along the south shore of Lake Superior where Dukes Forest is located (Davis et al. 2000a). It seems possible that increasing precipitation is the ongoing change in climate that is allowing beech populations to expand in this part of northern Michigan, to colonize sites where survival was not possible previously, and to increase in population density in parts of the forest that were previously only marginally suitable. Woods' record in permanent plots of population increase within an undisturbed forest community is uniquely valuable as a record of what appears to be forest response to climate change.

The role of climate is considerably less clear in the record of hemlock expansion into Sylvania forests 3400 years ago, because hemlock populations may still have been increasing in density following the sudden decline that occurred throughout the range of hemlock 5500 years ago (Davis 1981; Webb 1982), apparently because of insect outbreaks (Filion & Quinty 1993, Bhiry & Filion 1996). Populations began to recover about 4500 years ago, reaching pre-decline levels in most of the hemlock range about 3500 years ago (Alison et al. 1986). At the time of the decline, the hemlock range limit was 50 km east of Sylvania (Davis et al. 1986). Although pollen analyses from lakes, which sample a larger area than hollows, record as much as 5% hemlock pollen 5000 years ago, suggesting hemlock presence at Sylvania, a recent, more detailed analysis shows that hemlock pollen is only about 1% of the total, supporting results from hollows that imply hemlock was not present locally (Brugam & Johnson 1997; Brugam et al. 1997; Davis et al. 1998; H. Ewing, unpublished data). Fossil stomates in lake and hollow sediment could give a more definitive answer regarding local presence (Parshall 1999).

Hemlock population expansion within Sylvania began about 3400 years ago. Hemlock populations continued to increase throughout the remainder of the Holocene in the forest stands where we have pollen records.

Climate changes at the time of hemlock range extension are reconstructed by Calcote (2000), who concludes that a combination of rising winter temperature and increasing moisture was important in relieving constraints on hemlock. These changes were occurring regionally during the late Holocene (Brugam & Johnson 1997; Davis et al. 2000a). As a tolerance threshold was exceeded, hemlock slowly extended its range westward and southward into Wisconsin, reaching its present range limit only a few centuries ago (Parshall 1999). At the same time the increasingly favourable climate within its range was allowing hemlock populations to expand. Within Sylvania, the result of the continuing climate change was gradual replacement of resident white pine by hemlock, a process that took 1200 years in one hemlock stand (H3) and is still continuing in another (H2) (Davis *et al.* 1998).

The initial colonization of Sylvania forests by hemlock was sensitive to resident vegetation: white-pine stands were invaded 3400–2800 years ago, whereas oak and maple stands were apparently not invaded at that time (Davis *et al.* 1998). In interpreting Sylvania data we make the assumption that the distribution of white-pine stands at the time of hemlock invasion was controlled by fire history, not substrate. The assumption seems justified because charcoal is abundant in sediment of that age, and pollen assemblages are similar to surface samples in regions of mixed pine forest where fire is important (see Fig. 10.2). Thus, current hemlock distribution appears to reflect fire history several thousands of years ago. In contrast, beech colonization of Dukes forests does not appear sensitive to forest composition, but can be explained better by substrate.

Conclusions

A better understanding of the effect of climate change on forests, including its effect on forest dynamics, is essential to predict the time course of changes in future forests responding to global change. The record of climate response in long-term plots, tree ring data and fossil pollen records is much more complete than generally appreciated. Retrospective studies document forest invasions by two forest dominants—beech and hemlock—that extended ranges in response to climate change. Further, they record the time course of displacement of a resident dominant white pine from forests invaded by hemlock. The beech invasion is recorded on the decadal scale in permanent plots that document population growth, and demonstrate that spatial variations in soils influence the pattern of beech invasion (Woods 2000b). Hemlock invasion, known only from fossil records, is documented in less detail, but over a larger landscape, with long stand histories that describe whether or not hemlock maintained dominance in the stands it invaded (Davis *et al.* 1998 and unpublished data). In both cases, only certain sites on the landscape were invaded. Fragmented landscapes may have posed problems for migrating species in the past as well as in the future, adding challenges to modellers attempting to predict rates of species advance across the landscape. It is interesting to speculate whether permanent plots now being set up across ecotones for the purpose of monitoring response to future climate change will be able to obtain as much useful information as these retrospective records.

Because the rate of change in forest systems composed of very long-lived species is so slow, we have no means other than retrospective studies or models to learn how climate change affects forest dynamics. Stand simulation models are an alternative to fossil materials, but to build useful models a profound understanding is needed of climate sensitivities. Although geographical ranges provide useful insights on climate constraints on establishment and survival, they seldom help in learning how establishment in gaps and the outcome of the ensuing competition will change as climate changes. Yet models must include responses at this level of detail to predict the time course of forest change as climate changes in future decades.

The very slow rates of change in hemlock–hardwood forest make it difficult to distinguish disturbance dynamics from responses to climate. Perhaps in some cases we have attributed the results of climate change to disturbance, but the converse may be true as well. Both drivers are affecting forests simultaneously, and they both affect forests at the stand scale. Models have been developed that simulate response to disturbance (e.g. Botkin *et al.* 1972; Shugart 1984; Loehle 2000), but to depict hemlock and maple forests, models need to incorporate the effects of neighbourhood feedbacks (Frelich & Reich 1999). The information I have reviewed here emphasizes that dynamic changes in Great Lakes forests occur on such long time scales that the trajectory of climate change can affect their rate and direction. Because mean temperature and precipitation affect species differently, changes in climate means can influence the course of forest succession following disturbance, as well as change the disturbance regime itself. Interannual climate variability can change from one decade to the next (Skaggs *et al.* 1995) or one millennium to the next (Woodhouse & Overpeck 1998), with both direct and indirect effects on vegetation. Changes in climate can shift return times of disturbances before the forest has completed its recovery from the last catastrophe.

In the Introduction I questioned attempts to explain forests as steady-state systems. I argue in this review that because climate changes continuously, and because forest systems respond to disturbance on similar time scales, the expectation of steady-state landscapes on a time scale similar to the lifetime of a forest tree is unrealistic, and can actually interfere with understanding forest dynamics as it is influenced by changing climate.

Acknowledgements

Particular thanks are given to Randy Calcote for the data analyses shown in Fig. 10.4. Christine Douglas criticized the text, and she, Sue Julson and Karen Walker prepared the figures. I appreciate the helpful comments on the paper provided by Kerry Woods, Lee Frelich, Kendra McLauchlan, Brian Huntley, Helene Muller-Landau, Phil Camill and Shinya Sugita. I am grateful to Simon Levin for his patience as editor. This work has been supported by the National Science Foundation and by the Mellon Foundation.

References

Alison, T.D., Moeller, R.E. & Davis, M.B. (1986) Pollen in laminated sediments provides evidence for a mid-Holocene forest pathogen outbreak. *Ecology* 67, 1101–1105.

Alverson, W.S., Waller, D.M. & Solheim, S.L. (1988) Forests too deer: edge effects in northern Wisconsin. *Conservation Biology* 2, 348–358.

Anderson, R.C. & Loucks, O.L. (1979) White-tailed

deer (*Odocoileus virginiana*) influence on the structure and composition of *Tsuga canadensis* forests. *Journal of Applied Ecology* **16**, 855–861.

Bhiry, N. & Filion, L. (1996) Mid-Holocene hemlock decline in eastern North America linked with phytophagous insect activity. *Quaternary Research* **45**, 312–320.

Bormann, F.H. & Likens, G.E. (1979) *Pattern and Process in a Forested Ecosystem*. Springer-Verlag, Berlin.

Botkin, D.B., Janak, J.F. & Wallis, J.R. (1972) Some ecological consequences of a computer model of forest growth. *Journal of Ecology* **60**, 849–872.

Brugam, R.G. & Johnson, S. McC. (1997) Holocene lake-level rise in the Upper Peninsula of Michigan USA as indicated by peat growth. *Holocene* **7**, 355–359.

Brugam, R.G., Giorgi, M., Sesvold, C., Johnson, S.McC. & Almus, R. (1997) Vegetation history of the Sylvania Wilderness Area of western Upper Peninsula of Michigan. *American Midland Naturalist* **137**, 62–71.

Calcote, R.R. (1995) Pollen source area and pollen productivity: evidence from forest hollows. *Journal of Ecology* **83**, 591–602.

Calcote, R. (1998) Identifying forest stand types using pollen from forest hollows. *Holocene* **8**, 423–432.

Calcote, R. (2000) *Modern analog interpretations of vegetation and climate from pollen in sediments*. PhD Dissertation, University of Minnesota, St Paul, MN.

Canham, C.D. & Loucks, O.L. (1984) Catastrophic windthrow in the presettlement forests of Wisconsin. *Ecology* **65**, 803–809.

Clark, J.S. (1990) Fire and climate change during the last 750 yrs in northwestern Minnesota. *Ecol. Monographs* **60**, 135–159.

Clark, J.S. (1994) Effects of long-term water balances on fire regime, northwestern Minnesota. *Journal of Ecology* **77**, 989–1004.

Cook, E.R., Meko, D.M., Stahle, D.W. & Cleaveland, M.K. (1999) Drought reconstructions for the continental United States. *Journal of Climate* **12**, 1145–1162.

Davis, M.B. (1981) *Outbreaks of forest pathogens in Quaternary history*. In: Proceedings IV International Conference of Palynology, Lucknow, Vol. 3 (ed. J.H.B. Birks), pp. 216–227.

Davis, M.B. (1989) Retrospective studies. In: *Long-Term Studies in Ecology* (ed. G.E. Likens), pp. 71–89. Springer-Verlag, New York.

Davis, M.B., Woods, K.D., Webb, S.L. & Futyma, R.P. (1986) Dispersal versus climate: Expansion of *Fagus* and *Tsuga* into the Upper Great Lakes region. *Vegetatio* **67**, 93–103.

Davis, M.B., Sugita, S., Calcote, R.R. & Frelich, L. (1992) Effects of invasion of *Tsuga canadensis* on a North American ecosystem. In:*Response of Forest Ecosystems to Environmental Changes* (eds A. Teller, P. Mathy & J.N.R. Jeffers), pp. 34–44. Elsevier Applied Science, London.

Davis, S., Sugita, R.R., Calcote, J.B., Ferrari & Frelich, L.E. (1994) Historical development of alternate communities in a hemlock-hardwood forest in northern Michigan, USA In: *Large-Scale Ecology and Conservation Biology* (eds P.J. Edwards, R. May & N.R. Webb), pp. 19–39. Blackwell Scientific Publications, Oxford.

Davis, M.B., Parshall, T.E. & Ferrari, J.B. (1996) Landscape heterogeneity of hemlock-hardwood forest in northern Michigan. In: *Eastern Old-Growth Forests* (ed. M. Davis), pp. 291–304. Island Press, Washington DC.

Davis, M.B., Calcote, R.R., Sugita, S. & Takahara, H. (1998) Patchy invasion and the origin of a hemlock-hardwoods forest mosaic. *Ecology* **79**, 2641–2659.

Davis, M.B., Douglas, C., Calcote, R., Cole, K., Winkler, M. & Flakne, R. (2000a) Holocene climate in the Western Great Lakes National Parks & Lakeshores: Implications for future climate change. *Conservation Biology* **14**, 968–983.

Davis, M.B., Douglas, C., Calcote, R., Skaggs, R.H. & Barnett, V. (2000b) Drought variability greater during the mid-Holocene. AMQUA 2000: Program and Abstracts of 16th Biennial Meeting, Fayetteville, AR, p. 58.

Ferrari, J.B. (1999) Fine-scale patterns of leaf litterfall and nitrogen cycling in an old-growth forest. *Canadian Journal of Forest Research* **29**, 291–302.

Ferrari, J.B. & Sugita, S. (1996) A spatially explicit model of leaf litterfall in hemlock-hardwood forests. *Canadian Journal of Forest Research* **26**, 1905–1913.

Filion, L. & Quinty, F. (1993) Macrofossil and tree-ring evidence for a long-term forest succession and mid-Holocene hemlock decline. *Quaternary Research* **40**, 89–97.

Foster, D.R. (1988) Species and stand response in central New England, USA. *Journal of Ecology* **76**, 135–151.

Foster, D.R., Aber, J.D., Melillo, J.M., Bowden, R.D. & Bazzaz, F.A. (1997) Forest response to disturbance and anthropogenic stress. *Bioscience* **47**, 437–445.

Frelich, L.E. & Graumlich, L.J. (1991) Age-class distribution and spatial patterns in an old-growth hemlock-hardwood forest. *Canadian Journal of Forest Research* **24**, 1939–1947.

Frelich, L.E. & Lorimer, C.G. (1985) Current and predicted long-term effects of deer browsing in hemlock forests in Michigan, USA. *Biological Conservation* **34**, 99–120.

Frelich, L.E. & Lorimer, C.G. (1991) Natural disturbance regimes in hemlock-hardwood forests of the upper Great Lakes region. *Ecological Monographs* **61**, 145–164.

Frelich, L.E. & Reich, P.B. (1996) Old growth in the Great Lakes region. In: *Eastern Old-Growth Forests* (ed. M.B. Davis), pp. 144–160. Island Press, Washington, DC.

Frelich, L.E. & Reich, P.B. (1999) Neighborhood effects, disturbance severity, and community stability in forests. *Ecosystems* **2**, 151–166.

Frelich, L.E., Calcote, R.R., Davis, M.B. & Pastor, J. (1993) Patch formation and maintenance in an old-growth hemlock-hardwood forest. *Ecology* **74**, 513–537.

Frelich, L.E., Sugita, S., Reich, P.B., Davis, M.B. & Friedman, S.K. (1999) Neighborhood effects in forests: implications for within-stand structure. *Journal of Ecology* **86**, 149–161.

Frissell, S.S. Jr (1973) The importance of fire as a natural ecological factor in Itasca State Park, Minnesota. *Quaternary Research* **3**, 397–407.

Grimm, E.C. (1984) Fire and other factors controlling the Big Woods vegetation of Minnesota in the mid-nineteenth century. *Ecological Monographs* **54**, 291–311.

Heinselman, M.L. (1973) Fire in the virgin forests of the Boundary Waters Canoe Area, Minnesota. *Quaternary Research* **3**, 329–382.

Henry, J.D. & Swan, J.M.A. (1974) Reconstructing forest history from live and dead plant material—an approach to the study of forest succession in southwest New Hampshire. *Ecology* **55**, 772–783.

Houghton, J.T., Meira Filho, L.G., Callander, B.A., Harris, N., Kattenberg, A. & Maskell, K. (1996) *Climate Change 1995: the Science of Climate Change.* Cambridge University Press, Cambridge.

Kuchler, A.W. (1964) Potential natural vegetation of the conterminous United States. *American Geographical Society Publications* **36**, 1–38.

Lertzman, K.P. (1995) Forest dynamics, differential mortality and variable recruitment probabilities. *Journal of Vegetation Science* **6**, 191–204.

Loehle, C. (2000) Strategy space and the disturbance spectrum: a life-history model for tree species coexistence. *American Naturalist* **156**, 14–33.

Lorimer, C.G. (1983) The presettlement forest and natural disturbance cycle of northeastern Maine. *Ecology* **58**, 139–148.

Oliver, C.D. & Stephens, E.P. (1977) Reconstruction of a mixed species forest in central New England. *Ecology* **58**, 562–572.

Parshall, T.E. (1995) Canopy mortality and stand-scale change in a northern hemlock-hardwood forest. *Canadian Journal of Forest Research* **25**, 1466–1478.

Parshall, T.E. (1999) Documenting forest stand invasion: fossil stomata and pollen in forest hollows. *Canadian Journal of Botany* **77**, 1529–1538.

Parshall, T. & Calcote, R. (2001) Effect of pollen from regional vegetation on stand-scale forest reconstruction. *The Holocene* **11**, 81–87.

Pastor, J. & Broschart, M. (1990) The spatial pattern of a northern conifer-hardwood landscape. *Landscape Ecology* **4**, 55–68.

Pickett, S.T.A. & White, P.S. (1985) *The Ecology and Natural Disturbance and Patch Dynamics.* Academic Press, Orlando, FL.

Ribbens, E., Silander, J.A. & Pacala, S.W. (1994) Seedling recruitment in forests: calibrating models to predict patterns of tree seedling dispersion. *Ecology* **75**, 1794–1806.

Rooney, T.P., Cormick, R.J.M.C., Solheim, S.L. & Waller, D.M. (2000) Regional variation in recruitment of hemlock seedlings and saplings in the Upper Great Lakes, USA. *Ecological Applications* **10**, 1119–1133.

Runkle, J.R. (1985) Disturbance regimes in temperate forests. In: *The Ecology of Natural*

Disturbance and Patch Dynamics (eds S.T.A. Pickett & P.S. White), pp. 17–33. Academic Press, Orlando, FL.

Shugart, H.H. (1984) *A theory of forest dynamics: The ecological implications of forest succession models.* Springer-Verlag, New York.

Secrest, H.C., MacAloney, H.J. & Lorenz, R.C. (1941) Causes of decadence of hemlock at the Menominee Indian Reservation, Wisconsin. *Journal of Forestry* **39**, 3–12.

Shugart, H.H. (1984) *A theory of forest dynamics: The ecological implications of forest succession models.* Springer-Verlag, New York.

Skaggs, R.H., Baker, D.B. & Ruschy, D.L. (1995) Interannual variability characteristics of the eastern Minnesota (USA) temperature record: implications for climate change studies. *Climate Research* **5**, 227.

Stearns, F.W. (1949) Ninety years' change in a northern hardwood forest in Wisconsin. *Ecology* **30**, 350–358.

Stearns, F.W. (1991) Forest history and management in the Northern Midwest. In: *Management of Dynamic Ecosystems* (ed. J.M. Sweeney), pp. 107–122. Wildlife Society, West Lafayette, Indiana.

Sugita, S. (1993) A model of pollen source area for an entire lake surface. *Quaternary Research* **39**, 239–244.

Sugita, S. (1994) Pollen representation of vegetation in Quaternary sediments: theory and method in patchy vegetation. *Journal of Ecology* **82**, 881–897.

Sugita, S. (1998) Modeling pollen representation of vegetation. In: *Paleoclimate Research*, Vol. 27, Special Issue: *ESF Project, European Paleoclimate and Man* 18 (ed. B. Frenzel), pp. 1–16.

European Science Foundation, Strasbourg, Austria.

Sugita, S., MacDonald, G.M. & Larsen, C.P.S. (1997) Reconstruction of fire disturbance and forest succession from fossil pollen in lake sediments: potential and limitations. In: *Sediment Records of Biomass Burning and Global Change* (eds J.S. Clark, H. Cachier, J.G. Goldammer, B. Stocks), pp. 387–412. Springer-Verlag, Berlin.

Webb, T. III (1982) Temporal resolution in Holocene pollen data. In: *Proceedings of the Third North American Paleontological Convention*, Vol. 2, pp. 569–572. Business and Economic Service, Toronto, Ontario, Canada.

Webb, T. III (1986) Is vegetation in equilibrium with climate? How to interpret late-Quaternary pollen data. *Vegetatio* **67** (2), 75–92.

Webb, S.L. (1988) Windstorm damage and microsite colonization in two Minnesota forests. *Canadian Journal of For Research* **18**, 1186–1195.

Whitney, G.G. (1986) Relation of Michigan's presettlement pine forests to substrate and disturbance history. *Ecology* **67**, 2548–1559.

Woodhouse, C.A. & Overpeck, J.T. (1998) 2000 years of drought variability in the cnetral United States. *Bulletin of the American Meteorological Society* **79**, 2693–2714.

Woods, K.D. (2000a) Dynamics in late-successional hemlock-hardwood forests over three decades. *Ecology* **81**, 110–126.

Woods, K.D. (2000b) Long-term change and spatial pattern in a late-successional hemlock-northern hardwood forest. *Journal of Ecology* **88**, 267–282.

Woods, K.D. & Davis, M.B. (1989) Paleoecology of range limits: beech in the upper peninsula of Michigan. *Ecology* **70**, 681–696.

Chapter 11

Experimental plant ecology: some lessons from global change research

Ch. Körner

Introduction

The planet is faced with a new diet in the sense that atmospheric CO_2 concentrations, now at 365 p.p.m., have never exceeded 290 p.p.m. during the past 0.42 million years (according to ice-core data by Petit *et al.* 1999). In addition, rates of human-derived soluble nitrogen deposition have reached unprecedented values, now exceeding natural nitrogen-fixation by the globe's biota (Vitousek *et al.* 1997).

Given that incorporation of carbon and nitrogen by plants are the two most significant components of biomass production, it seems highly likely that such fundamental changes in supplies will affect individual plants and whole ecosystems. A challenge for ecological research is the elucidation of the likely outcome of this global experiment by cleverly designed tests and through the use of mathematical models. In addition, researchers may record the behaviour of the globe's atmosphere using equally sophisticated analyses, which permit the understanding of biosphere responses as a whole. Here, I will reflect on contributions that have been made to this topic by experimental work with vegetation. The organizers of this 'millennial conference' invited presentations that critically reflected on past and future directions of ecological research, hence I take the liberty to express some personal views, largely at the conceptual level.

Much of the uncertainty with experimental data on global change questions rests on aspects of the conceptual framework of experimental ecology in general. I will depict three: (i) the significance and meaning of resource limitation; (ii) the response dynamics following any relief of what is assumed to be a resource limitation; and (iii) the handling of ecological data in the context of their precision and relevance (with regard to globally altered resource supply in particular). I will use plant responses to atmospheric CO_2 enrichment to underpin my arguments. In order to set the scene for discussing the likely biological outcome of atmospheric CO_2 experiments, and to illustrate the associated scaling dilemma, I will start with some simple background data on the global carbon balance, against which experi-

Institute of Botany, University of Basel, Schönbeinstrasse 6, CH-4056 Basel, Switzerland. E-mail: ch.koerner@unibas.ch

mental findings can be compared. For a recent synthetic account of the ecological facets of the CO_2 problem see Körner (2000).

Missing carbon

Atmospheric sciences reveal that not all the ≈ 7 Gt of carbon released by humans end up in the atmosphere. Part of it (≈ 2 Gt) becomes dissolved in the ocean surface waters and another less well quantified part (again ≈ 2 Gt) disappears on land (see Chapter 12). This 'carbon sequestration' by terrestrial biota is often referred to as 'missing carbon' (MC). Let us assume that current MC is exclusively attributable to CO_2 fertilization of the vegetation. Photosynthesis of C_3 plants (>90% of global biomass) is not CO_2 saturated at current concentrations. Thus, this assumption is not irrational, although perhaps somewhat overoptimistic. Distributed over the ≈ 100 million km^2 of densely vegetated land (disregarding 50 million km^2 of cold and hot deserts and semideserts), each square metre could then be assumed to fix an average of 20 g of carbon per year as a result of current CO_2 enrichment. This 20 g needs to be seen in light of the mean of ≈ 6 kg of biomass carbon m^{-2} and 24 kg of humus carbon per square metre. As small as this additional 20 g might appear to be, added annually over 100 years it could theoretically increase the biosphere's organic carbon stock of 3000 Gt (plants plus humus) by 200 Gt (equivalent to 28 years of carbon release at current rates) or add, on average, $\approx 7\%$ more carbon to every square metre of vegetated land. Given that carbon:nitrogen ratios are 40–60 in biomass and 10–15 in humus, four times more nitrogen would be required to bind this amount of carbon in soil compared with biomass. If more nitrogen were trapped in the soil by additional carbon input, plant nitrogen availability would diminish.

Since the beginning of the industrial revolution the atmospheric CO_2 concentration has increased by $\approx 30\%$. Extrapolating from a 30% to a future 100% increase, the assumed current CO_2-induced carbon binding may not triple as well because of the non-linearity of the photosynthetic CO_2 response and nutrient constraints, but perhaps a doubling would occur. 2 Gt $C a^{-1}$ due to CO_2 fertilization now, and ≈ 4 Gt $C a^{-1}$ in 100 years, average at ≈ 3 Gt $C a^{-1}$ for the whole 100-year period. This simplistic, back-of-the-envelope estimate is intended to illustrate the order of magnitude of carbon binding that ecological research has to account for. The resultant ≈ 300 Gt carbon binding that might result from this scenario is close to the output of a number of highly sophisticated models (≈ 250 Gt carbon, e.g. Cao & Woodward 1998).

Atmospheric vs. experimental signals

In reality, it is quite uncertain whether 'missing carbon' has anything to do with CO_2 or nitrogen fertilization. Forest regrowth in the temperate zone, arising from altered land use (mainly in the US and Europe), may explain much of the MC (Schimel 1995; Schimel et al. 2000). The effect of regional atmospheric nitrogen fertilization on carbon binding is now thought to be negligible (Nadelhoffer et al. 1999). If only half of current MC ($10\,g\,m^{-2}$) were due to CO_2 fertilization effects, the remaining MC would become quite marginal.

Compared with the above, optimistic estimate of an upper limit of the double CO_2 world mean biospheric net carbon binding of $40\,g\,m^{-2}\,a^{-1}$ (realistic estimates may be half as large), experimental CO_2 enrichment data suggest 5–10 times increases in the rates of carbon sequestration for a rise from 365 (not 280!) p.p.m. to 600 p.p.m. CO_2 (estimates of $200–400\,g\,m^{-2}\,a^{-1}$; Niklaus et al. 2000). Available carbon flux data for forests indicate a mean of $230\,g\,m^{-2}\,a^{-1}$ net carbon fixation at current CO_2 concentrations (Canadell et al. 2000; Chapter 12, this volume), which is ≈ 10 times the current MC. Even if this were representative of only half of the world's forests and woodlands (25 million km^2), atmospheric CO_2 concentration would be stabilized, and if applied to all forests, this would correspond to a current net drain of carbon from the atmosphere of $\approx 5\,Gt\,a^{-1}$ (only 7 of ≈ 12 Gt carbon balanced by carbon emissions, not accounting for ocean uptake/release). Any further stimulation of this process by ongoing atmospheric CO_2 enrichment (following the above rationale) could be expected to double these rates by the end of this century. Even the much more moderate current $70\,g\,m^{-2}\,a^{-1}$ net carbon fixation estimated by Phillips et al. (1998) via tropical forest inventories (50 permanent plots in South America) is still a very high rate and would mean that these neotropical forests alone would account for 40% (0.7 Gt) of all current MC. The discrepancy of one order of magnitude between locally measured and globally observed 'carbon disappearance' may result from one or more of the following four explanations:

1 the vegetation studied is not representative of the vegetated land area of the globe;
2 the vegetation is treated in a way that it is not consistent with a representative response;
3 there are very large unaccounted carbon sources in the biosphere (in addition to known deforestation);
4 there are measurement or calculation errors.

Only in case 3 (perhaps an unlikely option given the required flux rates and land area) can we trust the experimental data. The problem seems to be that experimentalists tend to select 'nice' sites and need to detect a very small net difference of otherwise large gas fluxes (signal to noise ratio problem, see Hungate et al. 1996; Niklaus et al. 2000). In addition, the predominant focus of CO_2-enrichment studies is the initial responses of rapidly growing plants. Furthermore, growth responses are often seen as representing carbon sequestration, which they are not. Why is confusion between growth rate and carbon sequestration still so common (e.g. DeLucia et al. 1999)? The following discussion will lead me to suggest that experimental ecology is, for very fundamental reasons, unable to contribute quantitative answers to the carbon-sequestration question. However, this field of research can offer very important and influential insights into mechanisms and other consequences of the changing diet of the globe. Yet, for the projections derived from these findings to become realistic, experimental approaches and data evaluation may require some reconsideration, the main topic of this paper.

'Limitation' in ecology

Ecology has been called the science of limiting resources. Light, water, nutrients, CO_2, oxygen, thermal energy and even space and time may be seen as resources that potentially constrain plants. Despite this, perhaps the most widely used term in ecology, 'limitation', is rarely defined. What is limited? It is commonly taken as self-evident that the meaning is the one used in agronomy, namely limitation of biomass production—but how relevant is this definition in ecology?

It is easy to demonstrate that whenever an assumed resource limitation is diminished and biomass production increases, most of the species that were present prior to the perturbation disappear. So, how can one claim that these plants were limited, if the removal of resource limitation terminates their existence in the treated area? From this it follows that no late successional plant community is limited by resources. The species present and their abundance reflect the given resource supply, although any single individual in this community is likely to be limited in growth, as compared with its potential when grown under optimal conditions in isolation. Hence, one or more resources almost always limit mass production, but the suite of taxa present in space over time is not resource limited. This distinction between the agronomic and the ecological dimension of limitation is crucial. With this in mind, no one would claim that the desert flora is water limited. If the desert was irrigated a swamp flora might develop in the long run. Similarly, fertilization of the 'nutrient-limited' tundra would result in tall herbfields replacing shrubs and mosses, and the creation of gaps in forests would lead to light-requiring pioneer species replacing shade specialists. Thus, any change in resource supply will change the players, as will CO_2 enrichment, if CO_2 turns out to be a limiting resource. The same applies to stress. Any relief of stress will commonly induce competitive exclusion by other species that cope well with these new conditions. So neither alpine nor arctic plants are stress limited. Warming their environment would, in fact, reduce their probability of persistence.

All this may sound trivial, but is perhaps not as widely appreciated in ecology as one might expect. We first have to prove that CO_2 is a limiting resource for plant growth, which requires that all other potential constraints operate while we test this. If we find out this is so, we then need to understand the dynamics that will be induced in plant communities by the mitigation of CO_2 'limitation'. This leads to the question of life-cycle responses, of which the initial response is only a small, although important component. The life-cycle response will determine the carbon pool size (Körner 2000). The latter cannot be determined experimentally for woody species because of time constraints and is still a difficult task even for perennial non-woody communities.

Growth rate vs. carbon pools

Several authors have recognized the inappropriateness of taking growth rate as a surrogate for carbon sequestration. What matters is the carbon pool size, and not the rate at which carbon cycles through this pool (Steffen et al. 1998). The so-called Kyoto process that aims to arrive at a carbon certificate trade for re- and afforestation (sequestration of fossil carbon emission by growing trees) partly reflects this misunderstanding. While there is no doubt that a growing forest binds carbon, periods of 100–300 years may be required to accumulate prelogging stocks of carbon, depending on the forest type and latitude. Thereafter, carbon recycles to the atmosphere via economic use or natural breakdown. The time lapse between initiation and use or breakdown may be seen as 'buying time' with respect to atmospheric CO_2 enrichment but carries with it a carbon release wave in the distant future. What is overlooked is the fact that the slow refilling of the 'carbon gap' created by previous logging is a minute counter-weight to the release of carbon by ongoing logging. If the 'Kyoto-process' were driven by attempts at carbon stocking efficiency, there is nothing more efficient than reducing old-growth forests logging (although such carbon certificate trading may be practically impossible). Reforestation of formerly logged areas is better viewed as a secondary, less efficient measure, and each dollar invested per hectare in this way is worth a small fraction, compared with a dollar's 'carbon-value' per unit of prevented forest clearing. Whenever the rate of deforestation exceeds regrowth, overall carbon stocks go down. Incentives of re- and afforestation on formerly cleared land are still very positive, because this may reduce the pressure on primary forests in the future (besides many other ecologically desirable consequences). If atmospheric CO_2 enrichment stimulates tree growth, this would allow faster forest rotation (provided the nutrient cycle could cope with this) and, thus, be beneficial to agroforestry. However, faster rotation would not affect long-term carbon stocking (as will be explained below). The expansion of forested land area does contribute to net carbon fixation in biomass in the long run, but only as long as the biomass pool has not reached a new steady state. In contrast, old forest preservation has an immediate and very large effect.

Whether carbon stocking of forests would be affected by increasing CO_2 concentration is a very different question that is not accessible by CO_2 enrichment research. As Fig. 11.1 illustrates, carbon stocking over large land areas depends on the relative length of various developmental phases of forests and the residence time of carbon in various soil carbon compartments. For instance, a reduction of the gap duration, with all other life stages unaffected, would enhance the carbon pool per land area. A similar stimulation at all life phases would only cause the developmental cycle to shorten. Forests with a short cycle tend to store less carbon than less vigorous, slowly cycling forests (Phillips 1996), hence the net carbon pool may even go down in case of a stimulated rotation—a signal that Phillips (1996) suggested is emerging from census data for tropical forests. With shorter forest rotation times, faster-growing species may become more dominant, lowering the mean residence time and overall pool size of carbon per unit of land area (e.g.

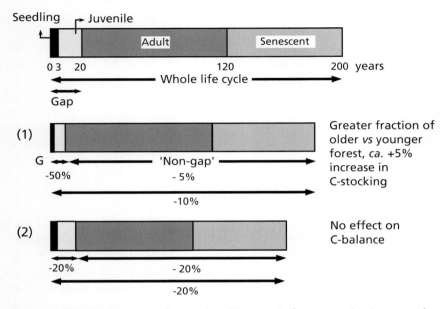

Figure 11.1 Whether biomass carbon stocks will increase in forests experiencing a growth stimulation by atmospheric CO_2 enrichment depends on the developmental dynamics and the mean residence time of a forest in a given life stage (here addressed as gap, adult, senescent). A similar stimulation at all life stages would only shorten the forest cycle, but would not enhance stocks (case 2). A stimulation (shortening) of the gap phase alone would increase the biomass carbon pool in the landscape (case 1).

poplar vs. oak). All available data on CO_2 enrichment effects are for rapidly growing young trees or even-aged plantations of one particular age class, and do not permit conclusions to be drawn with respect to forest carbon stocking in response to CO_2 with all age classes of trees included. Yet, these experiments yield many other important insights and are still greatly needed, but for different reasons, as will be discussed below.

Experimental CO_2 enrichment interacts with plant development. Unless shown otherwise, one needs to assume that: (i) different developmental phases have different carbon demands and hence will differ in growth sensitivity; (ii) a step increase in CO_2 induces a phase-shift in development compared with controls (whether or not the two transients merge at a later stage is debatable); and (iii) the CO_2 effect itself may exert a non-linear transient growth response, even if plant development is unaffected. Under all three conditions, the magnitude of the CO_2 response will depend on the time of sampling (Fig. 11.2), a serious problem in CO_2 research with long-lived plants (Loehle 1995). Although photosynthesis drives the

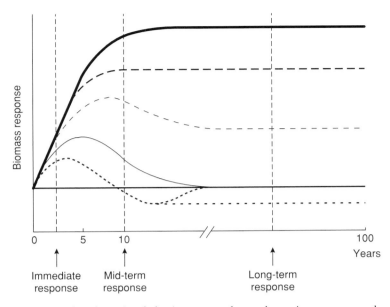

Figure 11.2 Examples of transient behaviour to any abrupt change in resource supply, including CO_2 concentration. Numbers indicate a few selected, potential response transients. It is clear that results depend on the sampling (e.g. harvest) time.

global biological carbon cycle, the formation of recalcitrant carbon pools is not immediately linked with the biochemistry of green leaves.

In light of the above, the ideas expounded by some gene engineers that greater photosynthetic efficiency of crop plants will contribute to carbon sequestration is completely unfounded (a recent press release related to the introduction of C_4-specific genes to rice, a C_3 plant, in Japan). The world's crops represent $\approx 1.6\%$ of the globe's biomass carbon pool and their yield is commonly consumed and hence recycled within a year. Irrespective of how fast a crop grows, consumption and carbon storage are mutually exclusive. Whether associated root turnover would add to the soil humus carbon pool is more a question of soil management. It should be added that breeders stopped selecting for high photosynthetic rates several decades ago, once it became obvious that processes other than leaf carbon assimilation determine harvestable yield (morphology, plant internal resource allocation, developmental processes such as leaf ageing, pathogen resistance, etc.). So neither rates of photosynthesis nor rates of growth should be confused with carbon storage at the landscape scale.

Drivers for scientific research

Available methodology has driven research in almost every field. For instance, the introduction of the infra-red gas analyser to plant sciences in Berlin in 1955 and the further refinement of this technique, led three to four generations of experimental ecologists prioritizing photosynthesis as the key to understanding plant growth. Plant development, plant architecture, root biology and interactions between organisms received less emphasis until very recently. This dependence of research direction on technology is still reflected in the contents of current textbooks. In water relations, for example, the invention of the pressure chamber led to the assumption that leaf water potential is a major driver of plant activity, stomata in particular. It took more than 25 years for a more moderate view to arrive, namely that plants (at least in humid regions) are little affected, unless a critical (rarely surpassed) threshold is reached, and that reductions in water potential under such conditions are commonly associated with high plant activity (transpiration).

In CO_2 enrichment research, recent decades have seen the development of different research tools, from small to larger growth chambers and greenhouses, from tiny to large open-top chambers and their variants, and to free-air CO_2 enrichment (FACE), the latest invention. Yet, the constraints of even the latter, the currently best available technology, are obvious. In order to remain financially viable, FACE rings require homogeneous, even aged and/or relatively low stature vegetation. In forest research, this commonly limits tests to young, rapidly expanding stands, i.e. to a particularly responsive phase of the life cycle (Loehle 1995). This is not meant as a criticism of pioneering forest FACE studies but projections from the results obtained should recognize that only part of the forest life cycle has been examined. It is not by principle, but by the sheer exploding costs of this technology when applied to the rough canopy of tall natural forests (>80% of the global biomass carbon pool) that these could not be considered for study to date. We are now experimenting with a new technique in our laboratory (web-FACE) where 0.5–1 km of laser-punched tubes are woven into each tree crown and which release pure CO_2 within the canopy. Control is much more unreliable than with conventional FACE, but this may be considered as balanced by the gain of so far inaccessible understanding and by the much more reasonable CO_2 consumption and costs ($\approx 2-3\,t\,d^{-1}$ CO_2 for 15 adult trees of 32 m height in a diverse natural forest).

The need for complex and large experiments

What we observe when plants receive enhanced CO_2 is a response reflecting the complex interaction of all drivers of plant life over the period of investigation. In the case of long-living plants such as trees the prehistory matters as well. The examples below illustrate that these interactions are not just modulating the CO_2 response, but that they may even reverse it. For this reason it is unfortunate that much research has been conducted at scales that do not bear up-scaling potential (Mooney et al. 1991; Körner 1995, 2000). The use of simplistic test system was driven by the deep 'last century' belief that the understanding of the properties of simple and 'clean' subsystems was the gateway to more complex ones (e.g. see the critique by Schindler 1998 and the discussion in Navas et al. 1999). I will present four such examples that challenge this belief.

CO_2 and light

Bazzaz and Miao (1993) were the first to show that CO_2 responses of tree seedlings depend on light availability. Subsequent studies in, for example, the understorey of humid tropical and temperate deciduous forests have supported this assertion (see e.g. Würth *et al.* 1998; Fig. 11.3). In the deciduous forest the interaction was so marked that the species ranking with respect to CO_2 sensitivity was reversed when seedlings of five common European forest trees experienced either 1 or 4% of full sunlight. This is much less light than is commonly supplied in growth chamber or greenhouse experiments, limiting the usefulness of such data (e.g. those analysed by Kerstiens 1998) for situations in which chances of success in a future forest gap are distributed among the reservoir of seedlings and young trees slowly establishing under a closed canopy. A general observation in this sort of experiment is that the relative effect of CO_2 enrichment is greater at very low light although absolute gains may be greater at high light.

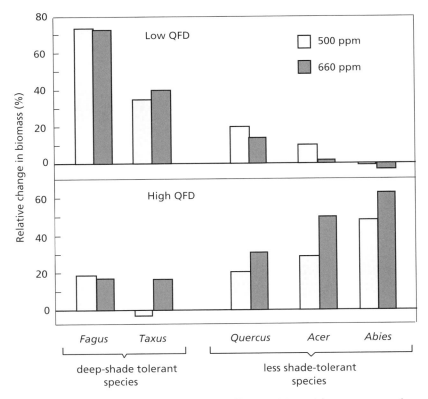

Figure 11.3 The *in situ* response of forest tree seedlings to CO_2 enrichment compared to 365 p.p.m CO_2 depends on microhabitat light conditions. Species were grown at 4 vs. 1% of above-forest photon flux (QFD) density for two seasons; note the ranking of species' responsiveness was reversed. (Reproduced with permission from Hättenschwiler & Körner 2000.)

CO_2 and soil

Substrate quality and nutrition of plants may reverse CO_2 effects. A most striking example is the CO_2 response of European beech grown on two types of natural substrate in assemblages with spruce and a natural understorey flora. When grown on a rich, calcareous alluvial soil, beech showed a significant positive growth response to CO_2 enrichment. However, when the same assemblage was grown on an acidic brown earth, the growth response of beech was negative (Egli *et al.* 1998; Fig. 11.4). It should be added that these substrates were taken from a field site where all the species examined co-occurred naturally. So, by selecting only one of two substrate types, opposing answers can be obtained with respect to likely future responses of beech. This example illustrates two things; first, the danger of using artificial substrates; and second, that repetition across sites (in addition to replication within sites) is essential.

As a second example of the interaction with substrate quality, I refer to the significance of phosphate availability for the CO_2 response of legumes. Without exception, legumes are reported in the literature to be at least as responsive to CO_2 enrichment as the most responsive non-legumes (e.g. Poorter 1993; Soussana & Hartwig 1996; Tissue *et al.* 1997; Lüscher *et al.* 1998). However, a study of five wild *Trifolium* species in seminatural calcareous grassland, revealed no such response

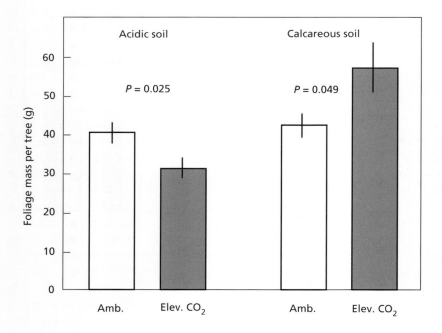

Figure 11.4 The growth response of young beech trees (*Fagus sylvatica*) to elevated CO_2 grown on calcareous or acidic soil. (Reproduced with permission from Egli *et al.* 1998.)

over several years of in situ CO_2 enrichment (Leadley et al. 1999). The reason for this unexpected non-responsiveness was revealed in a greenhouse test with model communities on natural substrate, part of which received very small amounts of phosphate fertilizer (Stöcklin & Körner 1999). Total community biomass (and with it, legume biomass in particular), increased by 45% when co-supplied with 600 p.p.m. CO_2, but by a small degree, if at all, with phosphorus addition alone or when legumes were replaced by non-legume dicots. These observations suggest that many previous studies were undertaken with legumes that grew under ample phosphorus supply, not a very common situation in nature.

CO_2 and soil moisture

CO_2 enrichment has been observed to reduce stomata aperture in plants and to cause transpiration to decline, compared with controls (see e.g. Woodward et al. 1991; Field et al. 1995; Wand et al. 1999). This phenomenon has been incorporated into global change models (Haxeltine & Prentice 1996; Jarvis et al. 1999). Some exceptions to this apparent general response have been noted (e.g. Hileman et al. 1992) and more recently it has emerged that stomatal response is largely restricted to certain groups of plants. These include young tree seedlings (Norby et al. 1999), grassland plants (Jackson et al. 1994; Knapp et al. 1996) and crops (Tuba et al. 1994); plants that can be readily manipulated using common experimental facilities. Response also seems to vary greatly with species and ambient humidity (Will & Teskey 1997; Heath 1998). Small and species-specific effects (on average −15% leaf diffusive conductance) are seen in adult broad-leaved trees, but are absent in tall conifers across a fair number of species and sites (Fig. 11.5). It seems as though the stomatal effects get smaller, the more closely leaves are coupled to the atmosphere (reduction of the aerodynamic boundary, cf. Aphalo & Jarvis 1993; Field et al. 1995). So, what has been seen as a common response in potted, isolated seedlings (with notable exceptions) and some grassland species, may not scale to forest canopies.

Evidence is accumulating that much, if not all, of the biomass response to CO_2 enrichment reported from field experiments in grasslands is due to small reductions in water loss, resulting from otherwise substantial stomatal responses. Small savings in evapotranspiration such as the ones recorded by Stocker et al. (1997, ≈6%), may accumulate over dry periods, leading to significantly less depletion of soil moisture (Niklaus et al. 1998; Owensby et al. 1999). This, in turn, may facilitate sustained microbial activity and nutrient uptake. Thus, the 'CO_2 effect' becomes almost fully explicable through impacts on water use (Volk et al. 2000), making the moisture regime critical in determining and predicting the effects of CO_2 on grasslands.

CO_2 and biodiversity

All the above examples reveal that CO_2 enrichment induces species-specific responses (e.g. Fig. 11.3). Moreover, since individualistic responses are co-determined by abiotic and biotic variables, it may be difficult to distinguish

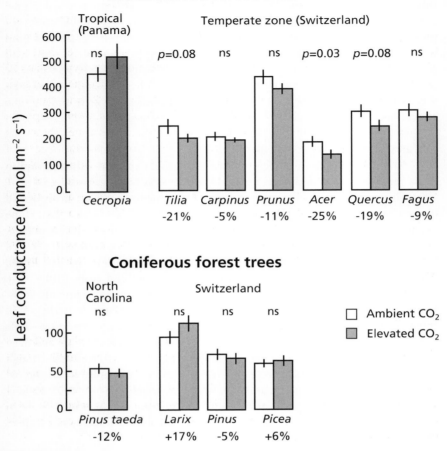

Figure 11.5 Examples of stomatal responses to a doubling of CO_2 concentration in adult forest trees (tropics, Körner & Würth 1996; *Pinus taeda*, Ellsworth 1999; temperate forest in Switzerland unpublished data by S. Guillod, S. Pepin & Ch. Körner).

between species-specific and treatment-specific effects. Even those few CO_2 effects that have been observed irrespective of growth conditions, namely carbohydrate accumulation and nitrogen depletion in leaves, show great variability among species. Given the high costs of field CO_2 enrichment operations, it seems like a wasteful approach to exclude biological variability (species, genotypes, functional types) from such trials. The presence of a specific neighbour, for example, is enough to impact on the results (e.g. Navas *et al.* 1999). Hence, as Dukes and Mooney (1999) stated, 'it is risky to predict which species will "win" or "lose" in high CO_2 conditions . . . in the absence of other species'. In fact, this is exactly the area where experimental ecology can greatly contribute to the understanding of the biological consequences of CO_2 enrichment. This includes all aspects of interactions of plants with other organisms, above and below ground.

One of the most surprising discoveries in recent years, relevant to CO_2 research, comes from mycorrhizal research: genotypes of endomycorrhizal fungi were found to switch on and off certain plant species within a community (van der Heijden et al. 1998; Chapter 5, this volume). The presence of one fungal genotype enhanced growth of one plant species, whose vigour reduced that of another. In view of such interactions, and given that CO_2 enrichment influences both ecto- and endomycorrhizas (e.g. Ineichen et al. 1995; Sanders et al. 1998; Rillig et al. 1999; Chapter 5, this volume), CO_2 enrichment experiments with artificial substrates should be interpreted with caution. Below-ground biotic interactions and processes may well determine whether terrestrial ecosystems accumulate or release carbon (Cardon 1996). Answers on plant and ecosystem responses to elevated CO_2 with a broader meaning are likely to be restricted to natural growth conditions, with a natural soil microflora (Curtis et al. 1994; Diaz 1996; Staddon & Fitter 1998). Across-site comparisons ('repetition') will reveal whether there is generality in certain responses or not, and under which conditions. The price to be paid for using such realistic experimental settings is what is commonly termed greater 'experimental noise', which raises the issue of precision.

What is precision in ecology?

Published research on biological effects of CO_2 enrichment in c. 3000 publications (Körner 2000) reflect a strong trend toward studies conducted in simple systems with low complexity. Ninety-five per cent of all publications are from experimental systems, where plants did not grow by their own choice, but were planted in artificial or highly disturbed substrates, often in isolation and in almost all cases, received fertilizer and were even aged (mostly young). There is a certain logic behind this, which leads me to my final, possibly controversial theme.

Complexity first or second?

There are two divergent ways to organize research in functional ecology. They share, however, the linking of processes at different hierarchical levels with the ultimate aim of understanding processes at higher degrees of complexity (population, vegetation, ecosystem or community). The first and commonly preferred approach, which is also academically considered to be more appealing, starts from first principles (or the lowest ones that one can handle, e.g. the response of a leaf). This bottom-up organization of research rests on the assumption that responses of parts are conserved across complexity levels. The alternative approach starts at the upper level and seeks explanations by down-scaling to whatever lower level provides answers of interest. The second avenue is unpopular, although much more likely to arrive at timely answers of the calibre necessary in research on environmental issues (Mooney et al. 1991). However, it is much less 'precise'.

Most scientist are very concerned that their experimental units bear some element of 'realism' with respect to 'natural' life conditions for their study organisms (see below). However, there is an asymmetric awareness and appreciation of the relative importance of environmental and biological variables in experimental designs. Once again, this may reflect the tools available for experimentation. Great care is commonly taken with air conditioning, and the whole development of CO_2 enrichment technology was driven by the aim of reducing climatic artefacts. Other, equally or even more significant facets of 'realism' are much more difficult to handle and are often neglected (Fig. 11.6). Why should soils, plant age and organismic interactions be less significant than, for instance, a 2-K variation in temperature?

In view of such determinants of plant responses to CO_2, it seems safer to study plants were they grow naturally, with the least possible anthropogenic interference. However, the scientific community has established certain conventions and certain precision criteria that discriminate against the top-down approach advocated here, favouring simple experimental systems. This is because units are easier to replicate at the lower end of the complexity cascade. For the two basic requirements of good science, precision and relevance, only the first has seen a scholarly development (statistics). Since precision only matters if the units tested are of relevance, it is surprising how little attention the latter has received (but see the excellent essay by Schindler 1998). The question of relevance is central to the progress of ecological research in general, and global change work in particular. Offending statistical rules (e.g. pseudo-replication) stigmatize, while offending the fundamentals of what I would call 'ecological relevance' does not. Below, I attempt to lay down some elements of a convention of ecological relevance, which might serve as a preliminary guideline as to whether the additional statistical confidence we seek addresses curiosity or relevant issues. I do realize that this is slippery terrain, possibly the major reason why there is so little debate in this direction.

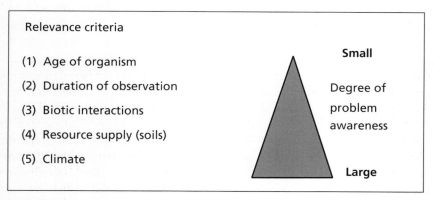

Figure 11.6 Elements of realism in experimental ecology and the common ranking of their importance by experimenters (as driven by the technical possibilities to handle them).

'Relevance' in experimental research in ecology

What is relevance? 'The' relevant test does not exist, but there are certain aspects that make studies more or less relevant to what might happen in the 'real world' (again, a term subject to interpretation). Further, we can ask the question, 'relevant' for what? The answer depends on aims and time scales. If one seeks answers as to how a certain species might respond, it will be the conditions under which the species is examined that will determine our understanding of its most likely response in its natural environment (a defined part of it). If the aim is the study of responses of a whole forest, a set of additional criteria will apply. Words like 'real' and 'natural' run the risk of being exclusive and idealistic. Here they subsume no more than 'being out and rooting in an unmanaged matrix'. Given the influence of humans on the globe's biota, there is little that could be called natural in the sense of being free from any human influence. Hence, 'natural' is meant to also include systems like rangelands and old fields with self-established plant communities, secondary forests and so on.

It seems counter-productive to criticize those who attempt to achieve more relevant test conditions for not having gone further, especially when the majority of tests are performed under conditions that bear little resemblance to the appropriate natural environment. In turn, it is equally dangerous when those who attempt more realistic test conditions overlook their remaining limitations and pretend to have captured 'the real thing'. In the ideal case one would perhaps like to see a combination of complex and simplified subsystems studied. However, the important point is that the subsystem is not defined from *a priori* reasoning about certain processes, but that is derived from, and nested in, the questions about, and conditions in, the larger system (top-down approach). Among the relevance criteria listed below, those relating to age, biological interactions and soils are most important. This is in clear contrast to what is commonly thought to be of foremost importance, namely (micro)climate (Fig. 11.6). Aware of the subjectivity of such a catalogue, I hope that this will develop into some future framework of conventions, which may help to channel funds and publications towards a more realistic representation of life processes in experimental plant ecological research (Körner 1995).

A preliminary checklist of relevance in experimental ecology

Although the list of conditions below is not exhaustive, fully accounting for all of the priority conditions would probably prohibit research. Hence, the purpose of this collection of criteria is more one of creating an awareness. In my experience, explanations for the abandonment of natural substrate, the growth of isolated plants, the light and nutrient regime imposed, and the duration of a measurement, for example, are missing in 90% of publications. Peers tend to be quite generous in this respect, but get tough on statistical matters even on material where statistics become an academic issue, given the irrelevance of experimental conditions. What we need is a more balanced account of relevance and statistical criteria in research and the inclusion, rather than the exclusion, of interactions of various environmental and biological factors; there is no single life process that depends on one driver alone.

Many of the criteria below refer to conditions that will be difficult to meet. However, deviations from what might be considered most realistic deserve some explanation in each case when results are reported. This alone would influence the way experiments were planned. It would be oversimplistic to come up with stereotype excuses like 'money' and 'time'. Very often there are feasible alternatives, which neither cost more, nor take more time than is available. However, a critical aspect would be the loss of statistical power as realism seeps in, which has to be weighted against the gain in relevance; both are required.

1 Overall experimental setting.
   (a) *In situ*, i.e. in a place were plants grow by their own 'choice'?
   (b) Planted outdoors on a substrate where plants would be found naturally?
   (c) Typical neighbours present?
   (d) Size (area) of experimental units relative to size of test plants and species richness?
   (e) Duration of test relative to the duration of the season and/or life cycle of test plants?

Neither test unit size nor experimental duration can be rated in absolute terms. A 1-year test in the humid tropics may yield as much relevant information as a 5-year test in the tundra with its 10-week season. A 1-m size experimental unit in short grassland or peat moss may be as adequate as a 50-m diameter unit in a forest. Hence, there is no 'best size'. In low vegetation, most FACE rings are much larger than they would have to be for the studied system, but technology requires such dimensions. The added space may be convenient (for additional experiments) but does not add to the confidence, for which a broader spatial replication might be more desirable.

2 Atmospheric environment.
   (a) The natural one?
   (b) Artificial, but dynamic, matching outside or long-term reference conditions?
   (c) Light conditions typical for the considered plant life phase (data)?
   (d) Temperature range and means reflecting the real ones in the field (data)?
   (e) Root temperature equal to shoot temperature?
   (f) Moisture supply constant or variable, matching typical field situations?

Most common deviations from natural atmospheric growth conditions are insufficient photon flux density, inappropriate light quality, evaporative conditions (wind) and precipitation/irrigation regimes. Air temperature is the easiest component to handle.

3 The below-ground environment.
   (a) Natural undisturbed soil (*in situ*)?
   (b) Monoliths excavated (e.g. in grassland, tundra)?

(c) Natural substrate, but disturbed (e.g. sieved)?
   (d) Rooting volume or depth unconstrained or only partially so?
   (e) Natural symbionts present?
   (f) Sterile substrate, inoculation by chance or with known inoculum (success)?
   (g) Fertilizer added, which, how much, when exactly applied and why then?

In controlled environment studies, taking soil from the field is superior to the use of standardized, artificial substrates, but excavation commonly activates substrates, brakes up root–soil interactions and alters soil structure. These artefacts are least pronounced in sandy substrates that have supported an intermittent weed crop. The use of containers is often problematic (unless plants are flooded with nutrients, which has other drawbacks), because any response of plants, which leads to indifferent size, affects the plant:substrate volume ratio. Equal-sized pots for unequally sized plants are all but equal in terms of growth conditions.

4 The plant.
   (a) Several vs. one species tested (why these ones)?
   (b) Full life cycle vs. only one life stage (only possible in herbaceous plants)?
   (c) Test plants varying genetically (provenances, ecotypes, cultivars)?
   (d) Test species belonging to several functional types?
   (e) Test species phylogenetically widely separated?

Given that species (cultivars) tend to respond differently, there is a danger in extrapolating from single-species studies. Phylogenetic relatedness among test species increases the risk of bias (Westoby 1999). It seems unlikely that *a priori* defined functional types will exhibit common responses (Leadley & Körner 1996), but it still adds to confidence if tests include contrasting morpho- or physiotypes of plants.

5 The plant community.
   (a) Natural community?
   (b) Late (vs. early) successional community?
   (c) Designed multispecies assemblage?
   (d) Diverse age structure of plants?
   (e) Trial with at least pairs or triplets of species?

Successional state has to do with the degree of plant–soil coupling. In late successional communities, soils can be expected to reflect the long-term presence, and interaction with, plants. In early successional or disturbed systems or in systems where plants do not grow by their 'own choice', the treatments interfere at an unknown transient state of the system, which limits the predictive value of results. Because responses are likely to be age specific, tests with equal-aged (herbaceous) communities may not capture response characteristics important in the long run.

The above list could be expanded, by including the presence of herbivores, pathogens, pollinators, earthworms, etc. or the occurrence of disturbances, variations in wind speed, or air pollutants. These components may not always be important, and could add more variability than could be handled.

Conclusion

Biological consequences of atmospheric CO_2 enrichment are used here to illustrate a number of conceptual aspects of experimental plant ecology that deserve reconsideration. The concept of limiting resources has its right only when mass production (irrespective of the 'players') is concerned and applies best to unstable, typically agronomic growth conditions or monocultures. It bears little relevance to diverse late successional plant communities whose genetic composition is the consequence of a given resource status. As experimentalists we can manipulate dynamic variables, rates of processes at a given time window, while long term effects on pool size, such as that of carbon, are more intractable, largely because of the non-linearity (transient nature) of responses and signal to noise problems (Niklaus *et al.* 2000). As a consequence of this, it would be more relevant and useful to understand the dynamics of the transient behaviour under observation, rather than the end product at a randomly selected (grant-driven?) termination of an experiment. Allowing complex interactions to occur during factorial treatments often ends with surprises, responses that are not predicted by any theory. This leads to what should become a central theme in ecology, introducing more realism to experiments (Schindler 1998). The likelihood that experimental findings have up-scaling potential should receive similar attention to statistical soundness. This should continue to reduce the abundance in the literature of tests that are well replicated but perhaps of limited ecological relevance.

Over the past 100 years experimental plant ecology has undergone impressive technological/analytical advances and accumulated an enormous body of evidence on plant responses to their environment. However, the sheer luxury by which modern technology now allows us to study, for instance, baselines of plant gas exchange in the field, seems to increasingly contrast with the added value obtained. In my opinion the advances to be achieved in the new century will come largely from conceptual advances and less so from technology. For instance, an advanced ability to handle, both experimentally and analytically, complex biotic and abiotic interactions, and by applying the available technology to test systems that bear a greater up-scaling potential. The latter will largely depend on whether funders will be willing to pay the extra costs of more realistic experiments, and whether the scientific community is willing to accept that very often (although not always) the price for more relevance is: (i) an enhanced experimental effort; (ii) work at higher hierarchical levels of organismic oragnization; and (iii) lower precision. The confidence that we seek and that society expects from global change research will emerge through the global replication of tests performed under the highest degree realism possible.

Although I could only briefly touch on these issues here, I suggest that in terms of themes the three most important fields of experimental plant ecology in the sphere of global change research will be: (i) the ecology of plant development; (ii) organismic interactions; and (iii) root and rhizosphere biology with its many facets. Syntheses, particularly the advanced form of 'meta-analysis' should become more of a priority for funding agencies than they were in the past. The analyses will then need to account to a much greater extent for experimental conditions, e.g. following the above relevance criteria for data ranking. Modelling, I believe, will only advance to the extent that experimentalists will be able to produce a more useful substrate for parametrization. Currently I feel modellers are 'left alone' with quite naive 'first-order principle' response functions, a fact that may explain much of the emergent discrepancies among current predictions.

References

Aphalo, P.J. & Jarvis, P.G. (1993) The boundary layer and the apparent responses of stomatal conductance to wind speed and to the mole fractions of CO_2 and water vapour in the air. *Plant, Cell and Environment* **16**, 771–783.

Bazzaz, F.A. & Miao, S.L. (1993) Successional status, seed size, and responses of tree seedlings to CO_2, light, and nutrients. *Ecology* **74**, 104–112.

Canadell, J., Mooney, H.A., Baldocchi, D.D. et al. (2000) Carbon metabolism of the terrestrial biosphere: a multitechnique approach for improved understanding. *Ecosystems* **3**, 115–130.

Cao, M. & Woodward, I. (1998) Net primary and ecosystem production and carbon stocks of terrestrial ecosystems and their responses to climate change. *Global Change Biology* **4**, 185–198.

Cardon, Z.G. (1996) Influence of rhizodeposition under elevated CO_2 on plant nutrition and soil organic matter. *Plant and Soil* **187**, 277–288.

Curtis, P.S., O'Neill, E.G., Teeri, J.A., Zak, D.R. & Pregitzer, K.S. (1994) Belowground responses to rising atmospheric CO_2: implications for plants, soil biota and ecosystem processes. *Plant and Soil* **165**, 1–6.

DeLucia, E.H., Hamilton, J.G., Naidu, S.L. et al. (1999) Net primary production of a forest ecosystem with experimental CO_2 enrichment. *Science* **284**, 1177–1179.

Diaz, S. (1996) Effects of elevated CO_2 at the community level mediated by root symbionts. *Plant and Soil* **187**, 309–320.

Dukes, J.S. & Mooney, H.A. (1999) Does global change increase the success of biological invaders? *Trends in Ecology and Evolution* **14**, 135–139.

Egli, P., Maurer, S., Günthardt-Goerg, M.S. & Körner, Ch. (1998) Effects of elevated CO_2 and soil quality on leaf gas exchange and aboveground growth in beech-spruce model ecosystems. *New Phytologist* **140**, 185–196.

Ellsworth, D.S. (1999) CO_2 enrichment in a maturing pine forest: are CO_2 exchange and water status in the canopy affected? *Plant, Cell and Environment* **22**, 461–472.

Field, C.B., Jackson, R.B. & Mooney, H.A. (1995) Stomatal responses to increased CO_2: implications from the plant to the global scale. *Plant, Cell and Environment* **18**, 1214–1225.

Hättenschwiler, S. & Körner, Ch. (2000) Tree seedling responses to *in situ* CO_2 enrichment differ among species and depend on understorey light availability. *Global Change Biology* **6**, 213–226.

Haxeltine, A. & Prentice, I.C. (1996) A general model for the light-use efficiency of primary production. *Functional Ecology* **10**, 551–561.

Heath, J. (1998) Stomata of trees growing in CO_2-enriched air show reduced sensitivity to vapour pressure deficit and drought. *Plant, Cell and Environment* **21**, 1077–1088.

Hileman, D.R., Bhattacharya, N.C., Gosh, P.P., Biswas, P.K., Lewin. K.F. & Hendrey, G.R. (1992) Responses of photosynthesis and stomatal conductance to elevated CO_2 in field-grown cotton. *Critical Reviews in Plant Sciences* **11**, 227–231.

Hungate, B.A., Jackson, R.B., Field, C.B. & Chapin, F.S. (1996) Detecting changes in soil carbon in CO_2 enrichment experiments. *Plant and Soil* **187**, 135–145.

Ineichen, K., Wiemken, V. & Wiemken, A. (1995) Shoots, roots and ectomycorrhiza formation of pine seedlings at elevated atmospheric carbon dioxide. *Plant, Cell and Environment* **18**, 703–707.

Jackson, R.B., Sala, O.E., Field, C.B. & Mooney, H.A. (1994) CO_2 alters water use, carbon gain, and yield for the dominant species in a natural grassland. *Oecologia* **98**, 257–262.

Jarvis, A.J., Mansfield, T.A. & Davies, W.J. (1999) Stomatal behaviour, photosynthesis and transpiration under rising CO_2. *Plant, Cell and Environment* **22**, 639–648.

Kerstiens, G. (1998) Shade-tolerance as a predictor of responses to elevated CO_2 in trees. *Physiologia Plantarum* **102**, 472–480.

Knapp, A.K., Hamerlynck, E.P., Ham, J.M. & Owensby, C.E. (1996) Responses in stomatal conductance to elevated CO_2 in 12 grassland species that differ in growth form. *Vegetatio* **125**, 31–41.

Körner, Ch. (1995) Towards a better experimental basis for upscaling plant responses to elevated CO_2 and climate warming. *Plant, Cell and Environment* **18**, 1101–1110.

Körner, Ch. (2000) Biosphere responses to CO_2 enrichment. *Ecological Applications* **10**, 1590–1619.

Körner, Ch. & Würth, M. (1996) A simple method for testing leaf responses of tall tropical forest trees to elevated CO_2. *Oecologia* **107**, 421–425.

Leadley, P.W. & Körner, Ch. (1996) Effects of elevated CO_2 on plant species dominance in a highly diverse calcareous grassland. In: *Carbon Dioxide, Populations and Communities* (eds Ch. Körner & F.A. Bazzaz), pp. 159–175, Academic Press, San Diego.

Leadley, P.W., Niklaus, P.A., Stocker, R. & Körner, Ch. (1999) A field study of the effects of elevated CO_2 on plant biomass and community structure in a calcareous grassland. *Oecologia* **118**, 39–49.

Loehle, C. (1995) Anomalous responses of plants to CO_2 enrichment. *Oikos* **73**, 181–187.

Lüscher, A., Hendrey, G.R. & Nösberger, J. (1998) Long-term responsiveness to free air CO_2 enrichment of functional types, species and genotypes of plants from fertile permanent grassland. *Oecologia* **113**, 37–45.

Mooney, H.A., Medina, E., Schindler, D.W., Schulze, E.D. & Walker, B.H. (1991). *Ecosystem Experiments*. John Wiley & Sons, Chichester.

Nadelhoffer, K.J., Emmett, B.A., Gundersen, P. *et al.* (1999) Nitrogen deposition makes a minor contribution to carbon sequestration in temperate forests. *Nature* **398**, 145–148.

Navas, M.L., Garnier, E., Austin, M.P. & Gifford, R.M. (1999) Effect of competition on the responses of grasses and legumes to elevated atmospheric CO_2 along a nitrogen gradient: differences between isolated plants, monocultures and multi-species mixtures. *New Phytologist* **143**, 323–331.

Niklaus, P.A., Spinnler, D. & Körner, Ch. (1998) Soil moisture dynamics of calcareous grassland under elevated CO_2. *Oecologia* **117**, 201–208.

Niklaus, P.A., Stocker, R., Körner, Ch. & Leadley, P.W. (2000) CO_2 flux estimates tend to overestimate ecosystem carbon sequestration at elevated CO_2. *Functional Ecology* **14**, 546–559.

Norby, R.J., Wullschleger, S.D., Gunderson, C.A., Johnson, D.W. & Ceulemans, R. (1999) Tree responses to rising CO_2 in field experiments: implications for the future forest. *Plant, Cell and Environment* **22**, 683–714.

Owensby, C.E., Ham, J.M., Knapp, A.K. & Auen, L.M. (1999) Biomass production and species composition change in a tallgrass prairie ecosystem after long-term exposure to elevated atmospheric CO_2. *Global Change Biology* **5**, 497–506.

Petit, J.R., Raynaud, D., Barkov, N.I. *et al.* (1999) Climate and atmospheric history of the past 420,000 years from the Vostok ice core, Antarctica. *Nature* **399**, 429–436.

Phillips, O.L. (1996) Long-term environmental change in tropical forests: increasing tree turnover. *Environmental Conservation* **23**, 235–248.

Phillips, O.L., Malhi, Y., Higuchi, N. *et al.* (1998) Changes in the carbon balance of tropical forests: evidence from long-term plots. *Science* **282**, 439–442.

Poorter, H. (1993) Interspecific variation in the growth response of plants to an elevated ambient CO_2 concentration. *Vegetatio* **104/105**, 77–97.

Rillig, M.C., Field, C.B. & Allen, M.F. (1999) Soil biota responses to long-term atmospheric CO_2 enrichment in two California annual grasslands. *Oecologia* **119**, 572–577.

Sanders, I.R., Streitwolf-Engel, R., Van der Heijden, M.G.A., Boller, T. & Wiemken, A. (1998) Increased allocation to external hyphae of arbuscular mycorrhizal fungi under CO_2 enrichment. *Oecologia* **117**, 496–503.

Schimel, D.S. (1995) Terrestrial ecosystems and the carbon cycle. *Global Change Biology* **1**, 77–91.

Schimel, D., Mellilo, J., Tian, H.Q. *et al.* (2000) Contribution of increasing CO_2 and climate to carbon storage by ecosystems in the United States. *Science* **287**, 2004–2006.

Schindler, D.W. (1998) Replication versus realism: The need for ecosystem-scale experiments. *Ecosystems* **1**, 323–334.

Soussana, J.F. & Hartwig, U.A. (1996) The effects of elevated CO_2 on symbiotic N_2 fixation: a link between the carbon and nitrogen cycles in grassland ecosystems. *Plant and Soil* **187**, 321–332.

Staddon, P.L. & Fitter, A.H. (1998) Does elevated atmospheric carbon dioxide affect arbuscular mycorrhizas? *Trends in Ecology and Evolution* **13**, 455–458.

Steffen, W., Noble, I., Canadell, J. *et al.* (1998) The terrestrial carbon cycle: Implications for the Kyoto Protocol. *Science* **280**, 1393–1394.

Stocker, R., Leadley, P.W. & Körner, Ch. (1997) Carbon and water fluxes in a calcareous grassland under elevated CO_2. *Functional Ecology* **11**, 222–230.

Stöcklin, J. & Körner, Ch. (1999) Interactive effects of CO_2, P availability and legume presence on calcareous grassland: results of a glasshouse experiment. *Functional Ecology* **13**, 200–209.

Tissue, D.T., Megonigal, J.P. & Thomas, R.B. (1997) Nitrogenase activity and N_2 fixation are stimulated by elevated CO_2 in a tropical N_2-fixing tree. *Oecologia* **109**, 28–33.

Tuba, Z., Szente, K. & Koch, J. (1994) Response of photosynthesis, stomatal conductance, water use efficiency and production to long-term elevated CO_2 in winter wheat. *Journal of Plant Physiology* **144**, 661–668.

Van der Heijden, M.G.A., Klironomos Jn, Ursic, M. *et al.* (1998) Mycorrhizal fungal diversity determines plant biodiversity, ecosystem variability and productivity. *Nature* **396**, 69–72.

Vitousek, P.M., Mooney, H.A., Lubchenco, J. & Melillo, J.M. (1997) Human domination of earth's ecosystems. *Science* **277**, 494–499.

Volk, M., Niklaus, P. & Körner, Ch. (2000) Soil moisture effects determine CO_2 responses of grassland species. *Oecologia* **125**, 380–388.

Wand, S.J.E., Midgley, G.F., Jones, M.H. & Curtis, P.S. (1999) Responses of wild C4 and C3 grass (Poaceae) species to elevated atmospheric CO_2 concentration: a meta-analytic test of current theories and perceptions. *Global Change Biology* **5**, 723–741.

Westoby, M. (1999) Generalization in functional plant ecology: The species sampling problem, plant ecology strategy schemes, and phylogeny. In: *Handbook of Functional Plant Ecology* (eds F.I. Pugnaire & F. Valladares), pp. 847–872. Marcel Dekker, New York.

Will, R.E. & Teskey, R.O. (1997) Effect of irradiance and vapour pressure deficit on stomatal response to CO_2 enrichment of four tree species. *Journal of Experimental Botany* **48**, 2095–2102.

Woodward, F.I., Thompson, G.B. & McKee, I.F. (1991) The effect of elevated concentrations of carbon dioxide on individual plants, populations, communities and ecosystems. *Annals of Botany* **67**, 23–38.

Würth, M., Winter, K. & Körner, Ch. (1998) In situ responses to elevated CO_2 in tropical forest understorey plants. *Functional Ecology* **12**, 886–895.

Chapter 12
Keeping track of carbon flows between biosphere and atmosphere

J. Grace, P. Meir and Y. Malhi

Introduction

The burning of fossil fuels and the clearing of forests has released CO_2 into the atmosphere, with a consequent increase in temperature. This link between CO_2 concentration and temperature was postulated over a century ago, and is now widely accepted (Arrhenius 1896; Lotka 1956; IPCC 1995). It is the basis for the international agreement to limit the production of gases which absorb infra-red radiation (radiatively active gases, or 'greenhouse gases'), to reduce the rate of global warming, according to the Kyoto Protocol of the UN Framework Convention on Climate Change (Grubb *et al.* 1999). Over the last three decades the importance of global warming has become widely realized, and the issue now ranks alongside the conservation of species as the most important environmental concern of our time (Tickell 1977). The total release of CO_2 from the burning of fossil fuel now exceeds 6 Gt of carbon annually (Marland *et al.* 1999; Grillot 2000), whilst another 1–2 Gt of carbon is released from forest clearance (Houghton 1999). Altogether, this total release of CO_2 to the atmosphere represents more than 1 tonne of carbon per year for each man, woman and child alive today.

The record of atmospheric CO_2 concentrations during the last millennium is revealed by analysis of air found in ice and snow taken from polar regions (Etheridge *et al.* 1996) coupled with the high-precision instrumental record from the 1950s to the present day (Keeling 1958; Tans *et al.* 1990; Keeling & Whorf 1999). The CO_2 concentration was more or less constant at 280 p.p.m. over hundreds of years, but increased sharply following the onset of the industrial revolution, reaching the current value of around 370 p.p.m. (Etheridge *et al.* 1996).

The total anthropogenic emissions of carbon from fossil fuel burning, cement manufacture and deforestation is about 8 Gt per year. This exceeds the average rate at which CO_2 is increasing in the atmosphere, which is 3 Gt per year (Schimel 1995). To account for the disparity, it is necessary to postulate the existence of carbon sinks distributed over the Earth's surface. Much of the work on the distribution and intensity of these sinks has taken place during the last decade, spurred

Institute of Ecology and Resource Management, University of Edinburgh, Darwin Building, King's Buildings, Edinburgh EH9 3JU

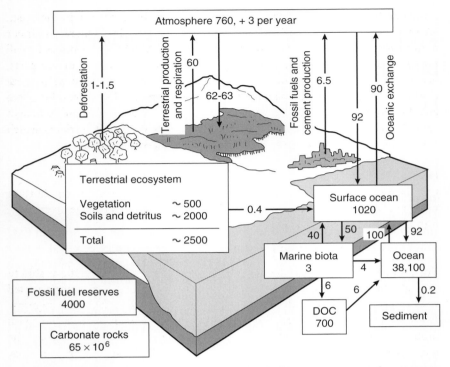

Figure 12.1 The global carbon cycle during the 1990s. The carbon stocks (shown in boxes) are in billions of tonnes of carbon (Gt C); the carbon fluxes (labelled arrows) are in Gt C yr^{-1}. There is additional organic carbon in the Earth's crust which is not in the estimated fossil fuel reserves, as it is too dilute to extract. Note that both the terrestrial ecosystem and the ocean are net absorbers of CO_2. The atmospheric stock is increasing by 3 Gt C yr^{-1}.

on by the need to understand the carbon cycle in the context of global warming. Recent representations of the carbon cycle usually show the global sink to be equally divided between the ocean and the terrestrial surface (Fig. 12.1). There are however, important and poorly understood variations between years in the extent to which the CO_2 emissions are absorbed: the proportion of emitted carbon remaining airborne varies widely (Rayner *et al.* 1999; Bender *et al.* 2000). The advancement of our understanding of this variation in the strength of the carbon sink, as well as the location of the sink, are important scientific challenges. Without this understanding, any attempt to manage either the ocean or the terrestrial biosphere, as has often been suggested, seems quite premature.

Nature of the carbon sinks

Arrhenius noted that the air over the ocean contained lower concentrations of CO_2 than air over the land, and inferred that the ocean must be a major sink for CO_2. He estimated that five-sixths of the CO_2 emissions must be dissolving in the ocean (Arrhenius 1896). Today the distribution of sinks is inferred by a somewhat similar argument, but based on far more numerous and accurate measurements, as we will see below. It is sometimes convenient to consider the carbon cycle in two parts, ocean and land, whilst remembering that there is, in reality, *only one carbon cycle* consisting of fluctuating exchanges of carbon between the biosphere, lithosphere and the two geochemical fluids *water* and *air* (Fig. 12.1). Moreover, these exchanges are influenced not only by climate, but also by fluxes of water and nutrients, which are themselves part of biogeochemical cycles (see Schlesinger 1997).

Oceans

CO_2 is readily soluble in water. In aqueous solution, there is an equilibrium mixture of carbonic acid, bicarbonate and carbonate ions, which make up the fraction 'dissolved inorganic carbon':

$$CO_2 + H_2O \Leftrightarrow H_2CO_3 \Leftrightarrow H^+ + HCO_3^- \Leftrightarrow 2H^+ + CO_3$$

The proportions of each species depends on pH. At high pH the reactions shift to the right. At a pH of around 8, in the ocean, the water is saturated or supersaturated with respect to $CaCO_3$ and carbonate is freely precipitated from the solution. However, most of the world's limestone since Cambrian times has a biological origin, consisting of shells and exoskeletons of marine life. In present-day seas the main groups of organisms responsible for carbonate formation are molluscs, corals, echinoderms, foraminifera and calcareous algae.

The magnitude of the oceanic sink has been estimated using knowledge of the CO_2 partial pressure in the surface waters. The rate of uptake or loss of CO_2 is proportional to the small difference in the partial pressure of CO_2 between the water and the air, and to the exchange coefficient which depends on wind speed. Tens of thousands of such measurements have been made annually. In high latitudes the surface waters are less than saturated with CO_2, so CO_2 dissolves from the atmosphere. In equatorial regions many waters are saturated with CO_2 and so there is an efflux to the atmosphere. The ocean is currently absorbing CO_2 at an average rate of about $2\,\text{Gt}\,\text{C}\,\text{yr}^{-1}$, with strong sinks in the north Atlantic and Pacific, and a source area in the mid-Pacific corresponding to an efflux of $0.5-1\,\text{Gt}\,\text{C}\,\text{yr}^{-1}$ (Peng *et al.* 1998). Continental shelf regions require further study, and may turn out to be more important than realized (Tsunogai *et al.* 1999). Significant interannual variability in the sink strength of the ocean may occur as a result of variations in currents which influence sea-surface temperatures, and this may influence the extent to which CO_2-rich water is brought to the surface. For example, the efflux from the equatorial waters is increased during El Niño years, when the surface temperature of the Pacific increases. Over geological time scales the atmosphere–oceanic sink is likely to have fluctuated enormously, being a strong sink in glacial periods, and a source during the glacial–interglacial transition periods (Raven & Falkowski 1999).

Marine organisms are thought by many to play a significant role in the carbon cycle through the so-called 'biological pump', acting as follows. Organisms occupy the well-mixed surface layers of the ocean, and photosynthesize and grow at a rate that varies according to the nutritional state of the ocean. The phytoplankton consists of short-lived photosynthetic organisms which are consumed by the zooplankton. Dead biota and faeces fall through the water column, thus removing carbon from the surface layers and reducing the partial pressure of CO_2 there. This enables uptake of fresh CO_2 from the atmosphere, whilst the decomposition of the detritus releases carbon and nutrients at depth. Thus, the ocean's sink strength is increased by biological activity (Louanchi et al. 1999). However, the relative importance of biological and physical processes in determining the sink strength of the oceans is still disputed (e.g. see Sarmiento et al. 1995).

Terrestrial ecosystems

There are profound differences between terrestrial and marine systems in the magnitude of the stocks and fluxes. Many terrestrial systems have a much larger biomass per area, and a much greater capacity to take up CO_2 by net photosynthesis than marine systems (see Whittaker & Likens 1975; Atjay et al. 1979; Field et al. 1998). Terrestrial systems can be thought of as biological pumps transferring carbon from the atmosphere to the soil through net primary productivity P_n where

$$P_n = P_g - R_a$$

Net primary productivity is the difference between P_g, the gross productivity (photosynthesis), and R_a, the autotrophic respiration.

P_n is notoriously difficult to measure using classical techniques (see Long et al. 1989), but values for different biomes have nevertheless been tabulated several times (Whittaker & Likens 1975; Atjay et al. 1979). A much simplified summary of P_n on a world scale is given in Table 12.1. The net production is partly appropriated by humans as food, fibre or timber (perhaps 40% of the total according to Vitousek et al. 1986), but a large fraction directly enters the soil as detritus where it is subject to comminution by the soil fauna and decomposition by microbial organisms. Some fractions of this soil organic matter are resistant to decay and have a very long residence time before the carbon is released as CO_2 or CH_4. Over periods of decades and centuries carbon may accumulate in the soil (Koutika et al. 1997). In cold climates substantial stores occur as accumulated peat, and even soils of hot climates such as rain forests and savannas have large stores of carbon (Table 12.1). When decomposition occurs, minerals are released and assimilated by roots and mycorrhizas. The rate of photosynthesis of terrestrial ecosystems is typically limited by the supply of nitrogen, but phosphorus is often in short supply also, and may frequently be limiting.

The net carbon balance of an ecosystem, termed the net ecosystem productivity P_e, is what remains of net primary production after heterotrophic respiration R_h:

$$P_e = P_n - R_h$$

Net ecosystem productivity P_e is much more relevant to discussions of carbon sinks than net primary productivity, and, as we shall see, somewhat more easily

Table 12.1 Current estimates of global carbon stocks in vegetation and soils, and the range of net primary productivity, NPP (WBGU, 1998; IPCC 2000). The real NPP figures are likely to be higher than these, even though they have been widely accepted for many years.

Biome	Area (10^6 km^2)	Global Carbon Stocks (Gt C) Vegetation	Soils	Total	NPP (t C ha^{-1} yr^{-1})
Tropical forests	17.6	212	216	428	11.0 (5.0–17.5)
Temperate forests	10.4	59	100	159	6.3 (2.0–12.5)
Boreal forests	13.7	88	471	559	4.0 (1.0–7.5)
Tropical savannas	22.5	66	264	330	4.5 (1.0–10.0)
Temperate grasslands	12.5	9	295	304	3.0 (1.0–7.5)
Deserts & semideserts	30.0	8	191	199	0.05 (0.0–0.1)
Tundra	9.5	6	121	127	0.1 (0.0–0.4)
Wetlands	3.5	15	225	240	0.9 (0.1–3.9)
Croplands	16.0	3	128	131	1.6 (0.2–3.9)
Total	**135.6**	**466**	**2011**	**2477**	

measured. It is reasonable to expect that P_e would be zero for the case of undisturbed ecosystems under a steady climate (Woodwell & Whittaker 1968). In the last few years it has been possible to measure P_e using the micrometeorological technique of eddy covariance (Moncrieff *et al.* 1997; Aubinet *et al.* 2000). In 1996 the International Geosphere–Biosphere Programme launched FLUXNET, to promote cooperation for studies of CO_2 uptake by the terrestrial vegetation (Fig. 12.2). Currently over 70 science teams over the world are measuring the net uptake of carbon using sensors mounted above vegetation on towers. Conclusions so far are outlined here.

1 Most of the boreal and temperate forests in the study are sinks for carbon, in the range 0.5–8 t C ha^{-1} yr^{-1} (Goulden *et al.* 1997; Jarvis *et al.* 1997; Valentini *et al.* 2000).

2 The sink strength shows notable interannual variability, depending on the weather (e.g. Lindroth *et al.* 1998). This provides insights into climatic control of carbon fluxes, and such data are increasingly being used for the development of models (Williams *et al.* 1998).

3 Old and undisturbed forests in Amazonia are sinks of 0.5–6 t C ha^{-1} yr^{-1} (Grace *et al.* 1995; Malhi *et al.* 1998). These data from eddy covariance measurements have been corroborated using quite independent data from permanent forest plots (Phillips *et al.* 1998). The discovery suggests that the remaining rain forests are a strong sink for carbon, even being sufficient to compensate for the loss of carbon brought about by tropical deforestation (see Malhi & Grace 2000).

4 In all forests, the net carbon gain is a small difference between two large fluxes, the incoming photosynthesis and the outgoing respiration. Of the respiratory

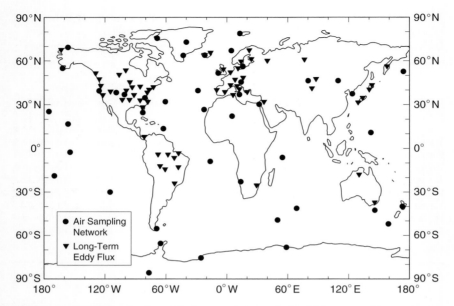

Figure 12.2 Global distribution of flask network for analysis of CO_2 composition and location of eddy-covariance sites for measuring CO_2 fluxes. (Reproduced with permission from Canadell et al. 2000.)

fluxes, the flux from the soil far exceeds the flux from the above-ground plant material.

With some important assumptions and additional data, it is possible to produce summary diagrams for the flows of carbon associated with forest ecosystems (Fig. 12.3). Although these are only examples, and neglect the relatively small carbon flows associated with volatile organic compounds such as isoprene and methane (Lerdau & Throop 1999), they provide a first glimpse of the quantitative behaviour of forests as sinks. It is perhaps unsurprising that forests that are actively managed for timber production are strong sinks. They contain predominantly young trees. But it is much more surprising to see a strong sink strength for undisturbed rain forests. It is widely thought that there is an important contribution from CO_2 fertilization, augmented by nitrogen deposition. Indeed, models based on physiological principles show this effect rather clearly (Lloyd & Farquhar 1996; Lloyd 1999).

It seems unlikely that terrestrial ecosystems are ever at a truly steady state with respect to their carbon content. The 'biological pump' transfers carbon from the atmosphere to the soil, but the rate at which the organic matter in the soil decomposes is unlikely to be exactly the same as the rate of gain of carbon. Climate acts in quite different ways on photosynthesis vs. decomposition. Some ecosystems such as bogs are capable of accumulating carbon over thousands of years. Data from 5000-year chronosequences of unmanaged forests in northern Finland show that

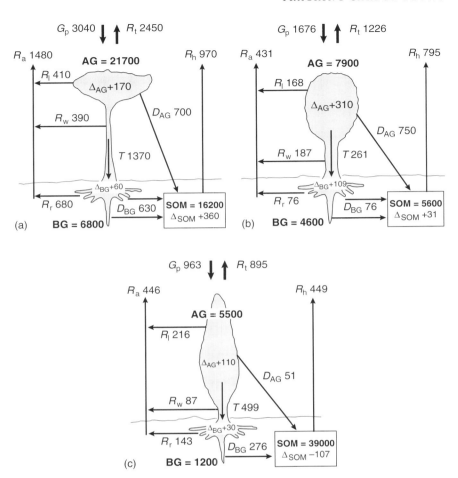

Figure 12.3 Carbon fluxes in three forest ecosystems: (a) humid tropical; (b) temperate; (c) boreal. G_p, gross primary productivity; R_t, total respiration; R_a, autotrophic respiration; R_h, heterotrophic respiration; R_l, leaf respiration; R_w, above-ground wood respiration; R_r, root respiration; D_{AG}, above-ground detritus; D_{BG}, below-ground detritus; T, below-ground carbon translocation. Stocks of carbon and their annual increments are shown in bold: AG, carbon in above-ground biomass and Δ_{AG} is the annual increment in AG; BG, carbon in below-ground biomass and Δ_{BG} is the annual increment; SOM, carbon in soil organic matter and Δ_{SOM} is its annual increment. Units are $g\,C\,m^{-2}\,yr^{-1}$ for fluxes and $g\,C\,m^{-2}$ for stocks. (Reproduced with permission from Malhi et al. 1999.)

carbon in the soil stabilizes after 2000 years (Liski et al. 1998), and on an agricultural scale, soils that have been fertilized with organic manure for over a century show an increase in carbon content (Smith et al. 1998a,b). Thus, the potential for soils to store carbon is considerable, and the 'steady state' may be the exception rather than the rule.

Controls on the sinks

In a world of rising CO_2 concentration and temperature it is expected that both photosynthesis and respiration will increase. Photosynthesis *per se* might be expected to increase as a result of the increased CO_2 diffusion gradient between atmosphere and leaf, although a doubling of the CO_2 concentration is unlikely to double the rate, as the number of stomata per area decreases with rising CO_2 in many species (Woodward 1987). There may additionally be down-regulation of photosynthetic capacity in C_3 plants (Medlyn *et al.* 1999), and C_4 plants are somewhat less responsive to elevated CO_2 because of their internal CO_2 concentrating mechanism. Photosynthesis is relatively insensitive to temperature but the annual duration of photosynthesis in mid and high latitudes is likely to increase as spring arrives earlier over large forested areas, e.g. in the boreal region (Myneni *et al.* 1997). Moreover, as warming proceeds there may be a shift in dominance from species using the C_3 mechanism of photosynthesis to those using C_4. This will also produce a stimulation in global photosynthesis as temperatures rise. However, as CO_2 rises, photosynthesis is ultimately likely to saturate at some unknown CO_2 concentration, depending partly on the supply of nutrients to the vegetation. Of these nutrients, nitrogen and phosphorus are usually singled out for discussion in temperate and tropical regions, respectively (Tanner *et al.* 1998; Lloyd 1999), although other nutrients may become limiting, and many soils have deficiencies in micronutrients. The supply of nitrogen and phosphorus is dependent on the rate at which mineralization of the soil organic matter occurs, and this is an area of major uncertainty (Grayston *et al.* 1998; Giardina & Ryan 2000).

Overall, the net primary productivity is likely to increase in a high CO_2 world. Experiments on the effect of elevated CO_2 on plants are sometimes misleading. Several hundred experiments have been conducted but many are now regarded as flawed, as conditions have been unnatural, and the plants have been juvenile. The most reliable data are those from plants grown in natural soil, preferably for many years. Medlyn *et al.* (1999) present a meta-analysis of recent data of this kind for temperate trees grown over periods of about 3 years. The analysis shows a rather small extent of down-regulation of photosynthesis, and provides the basis for parameterization of models of the effect of elevated CO_2 upon plants. Generally, the productivity of trees is increased from 30 to 60% when the CO_2 concentration is doubled. The most compelling data on the effect of elevated CO_2 on plant growth comes from naturally enriched sites, such as the famous CO_2 springs of Tuscany (Hättenschwiler *et al.* 1997), where mature trees exposed to elevated CO_2 concentration over their entire lifetime show a substantially smaller growth enhancement than that obtained in short-term experiments (Idso 1999; Fig. 12.4). Although the data set on the impact of elevated CO_2 upon plants is now huge, there are still important deficiencies. For example, there are relatively few data on the effect of elevated CO_2 on tropical trees, despite their importance in the global carbon cycle (Körner & Arnone 1992; Lin *et al.* 1999; Carswell *et al.* 2000), but it is unlikely that tropical trees will behave any differently from temperate trees. Tropical grasses, on the other hand, were considered to be less

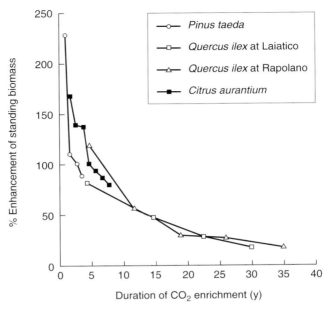

Figure 12.4 Percentage enhancement of standing tree biomass produced by a sustained 300 p.p.m. increase in atmospheric CO_2 concentration for three tree species growing out of doors and rooted in natural soil, plotted against the duration of the study. (Reproduced with permission from Idso 1999.)

responsive to elevated CO_2 than temperate grasses, as they have CO_2-concentrating mechanisms. However, they have turned out to be less different than supposed.

The gain in uptake of carbon by photosynthesis at elevated CO_2 may be offset by an enhanced respiration, especially microbial respiration associated with decomposition. It is usually assumed that efflux of CO_2 from the soil will increase with temperature. Indeed, analysis of eddy covariance data, or data from soil chambers show an apparent Q_{10} in excess of 2 (Lloyd & Taylor 1994). Such data may be used to parameterize simple models to predict the effect of elevated CO_2 and temperature on photosynthesis and respiration. The indication from this exercise is that ecosystems that are currently sinks will become sources as a result of the effect of temperature on soil (heterotrophic and autotrophic) respiration. Measurements of the isotopic composition of the CO_2 emanating from forest soil indicate that some very old carbon is now being broken down, suggesting that the carbon stored in the soil over hundreds or even thousands of years may be vulnerable to climatic warming (Oechel et al. 1993; Valentini et al. 2000).

Similar uncertainty is associated with the future of the ocean sink. Global warming is likely to cause a decrease in the high-latitude sink because of the effect of temperature on solubility, and because of the reduction in the rate of downwelling of cold dense water. On the other hand, the higher temperatures may be expected to stimulate the photosynthetic production and hence increase the biological pump.

The ocean sink is believed to be controlled by the supply of nutrients (Falkowski et al. 1998). The most well-known evidence to support this is from the IRONEX experiment, in which the phytoplankton in mid-ocean was shown to increase its rate of photosynthesis after the addition of iron in the form of iron citrate (Watson et al. 1994). The realization that anthropogenic nitrogen, from motor engines and agriculture, is transported world-wide and deposited everywhere is relevant to this discussion, as it may lead to an increase in the oceanic as well as the terrestrial carbon sink (Townsend et al. 1996). There may also be a response to elevated CO_2 in a manner similar to that of terrestrial systems, although the process is less well-studied than in the case of terrestrial systems. As the atmospheric CO_2 concentration increases, the pH of surface waters is expected to decrease because the bicarbonate–carbonate equilibrium is displaced to the right-hand side (see above). In fact, the pH of surface waters has decreased by 0.1 since 1800, and so the representation of carbonic acid, bicarbonate and carbonate is gradually changing and will continue to do so. However, whatever happens, carbon supply is likely to be high in relation to the supply of nutrients, and so (the argument goes) carbon is not limiting. Many marine phytoplankton have carbon-concentrating mechanisms, which is another reason why phytoplankton has traditionally been thought to be unresponsive to elevated CO_2 (Raven & Falkowski 1999). However, recent experiments in microcosms show otherwise: a direct stimulatory effect of CO_2 concentration on growth and shell formation which is at odds with theoretical considerations (Wolf-Gladrow et al. 1999).

Location of the sinks

Evidence for the distribution of sinks comes from analysis of the temporal and spatial concentrations of CO_2, $\delta^{13}C$ and O_2/N_2 in the atmosphere, obtained from 'flask networks' distributed at remote sites over the ocean and land. From a knowledge of the sources of CO_2, and using an atmospheric transport model, it is possible to infer where the sinks must have been in order to give the observed CO_2 concentrations. Information derived from $\delta^{13}C$ allows resolution between ocean sinks and terrestrial sinks (photosynthesis absorbs $\delta^{12}C$ more than $\delta^{13}C$, whereas dissolution is much less discriminating); and this resolution is enhanced by the measurement of O_2/N_2 (terrestrial photosynthesis emits O_2 but has no effect on N_2 but O_2 from marine photosynthesis is absorbed by bacterial respiration). Most inversions produce strong evidence for a northern hemisphere terrestrial sink, whilst the tropical region is a slight source (Fig. 12.5). Rayner et al. (1999) inferred a total ocean sink of 2.1 Gt C yr^{-1} and a net land sink of 0.7 Gt C yr^{-1}. Currently such calculations are poorly constrained, as the number of flask stations is small, and there are likely to be calibration errors between networks. Therefore, the uncertain-

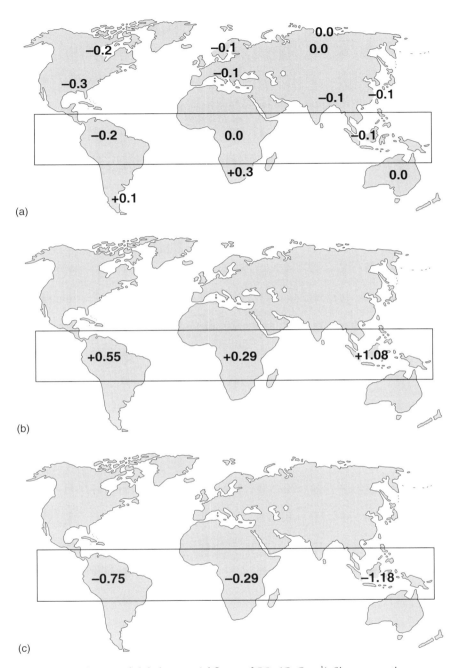

Figure 12.5 Estimates of global terrestrial fluxes of CO_2 ($Gt\,C\,yr^{-1}$). Sign convention: negative means uptake from the atmosphere to the vegetation. (a) Fluxes inferred from a flask network. (Rayner *et al.* 1999.) (b) Tropical deforestation fluxes. (Houghton 1999.) (c) Inferred fluxes to the remaining tropical vegetation (from a and b).

ties are considerable, and contrasting results have been reported between research groups.

Such calculations are especially uncertain in geographical regions where there are few sampling stations, or in areas where there is a significant rate of land-use change. Consequently, there is large uncertainty in the tropical region, which is poorly represented by the air sampling network (see Fig. 12.2). It is important to realize that the deforestation associated with land-use change is estimated as 1–2 Gt C yr^{-1} (Houghton 1999), and so the terrestrial biotic sink for the world as a whole must be 1.7–2.7 Gt C yr^{-1}, consistent with Fig. 12.1. In the inversion calculations, the tropical land is estimated as sometimes a small net sink, and sometimes a small net source, but as almost all of the world's deforestation is in the tropics, there must be a strong biotic sink there (Fig. 12.5). This is indeed what has been found in eddy covariance studies (reviewed by Malhi & Grace 2000).

All the inversions indicate an increase in the strength of sinks since 1991 (immediately following 1991 the rate of CO_2 increase was very low, indicating an exceptionally strong sink). The increased strength of the sink after 1991 has been linked to the major eruption of Pinatubo which caused slight cooling, presumably reducing the rate of respiration, and deposited minerals over both the ocean and land, which may have had a fertilizing effect on photosynthesis.

A recent development has been the establishment of regional networks, where concentration measurements can be linked to mesoscale models and to terrestrial eddy-covariance flux towers (Canadell et al. 2000). Data from flux towers may give an indication of geographical patterns in the sink strength, provided that the sites are well chosen as 'representative'. For example, the project EUROFLUX involved flux measurements at 15 sites over Europe (Valentini et al. 2000). The result showed a remarkable pattern with respect to latitude. The sink strength tended to decline at northerly latitudes, because the soil respiration rate was higher there (Fig. 12.6). It is a surprising result, as one would expect respiration to be least where the temperature is lowest (Grace & Rayment 2000). However, a substantial store of soil carbon has accumulated in northern soils, and this may now be decomposing at a faster rate than hitherto. In contrast, in the southern European soils the decomposition rate may be limited by the low soil moisture content during the summer.

There is now an effort to establish regional scale studies of carbon sinks. Examples include LBA (Large Scale Biosphere Atmosphere Experiment in Amazonia), a Brazilian initiative designed to link atmospheric and terrestrial measurements over the entire 4 million km^2 of the Amazon; and more recently CarboEurope (a Europe-wide project from the European Union). Web addresses for these projects are included in Appendix 1.

Future of the terrestrial carbon sink

It is proving remarkably difficult to predict the future of the carbon sink. The main reason for this is the uncertainty in the decomposition rate of the soil pool of

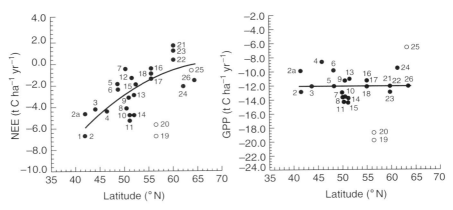

Figure 12.6 Evidence for a latitudinal gradient in sink strength over European forests. The net ecosystem exchange (NEE) is measured from eddy-covariance towers. The gross primary productivity (GPP, obtained by subtracting total respiration from NEE) does not vary with latitude. Location of sites: 1, 2, 4, Italy; 3, 5, 6, France; 7, 9, 10, 11, 13, 14, Germany; 8, Belgium; 15, Netherlands; 16, 17, 18, Denmark; 19, 20, Scotland; 21, 22, 23, 26, Sweden; 24, Finland; 25, Iceland. (Reproduced with permission from Valentini et al. 2000.)

carbon. Decomposition rates of organic matter are known to be sensitive to temperature. In short-term experiments on plant tissues or soils, maintained in strict laboratory conditions, a 10°C rise in temperature normally leads to a doubling in the rate of respiration (we say the Q_{10} is 2.0, see Lloyd & Taylor 1994). However, when eddy-covariance data are collected over an entire season, an apparent Q_{10} of much more than this has been observed (Boone et al. 1998). The range of Q_{10} in such experiments is 2–8. These Q_{10} values are 'apparent' in the sense that they are not equivalent to a physiological Q_{10} where the amount of biological material is held constant. Over an entire season, the quantity of substrate in the form of leaf litter and the quantity of microbial biomass are likely to vary considerably. Nevertheless, such observations suggest that the soil pool of carbon may be highly vulnerable to climate warming (Oechel et al. 1993; DETR 1999; Trumbore 2000). On the other hand, there are long-term experiments that suggest the contrary. Liski et al. (1998) showed that the organic soils of Finland contain much more carbon at southern sites than expected from a simple model, and he also showed that the rate at which pine needles are decomposed in litter bags is not an exponential function of temperature but a linear one. Very recent work (Giardina & Ryan 2000; Valentini et al. 2000) lends support to this view. It seems that when soil is warmed, the readily decomposed fractions of soil organic matter decompose rapidly, but the more recalcitrant fractions behave quite differently. In climate warming, we are likely to see a changing relationship between temperature and decomposition rate, as the easily decomposed fraction progressively decomposes leaving the resistant fraction. It seems that the likely conclusion is that the future of the soil pool of carbon is more secure than most people have imagined, and that re-parameterization of some of the global vegetation models is now required, to take this finding into account (Grace & Rayment 2000).

A second uncertainty in the future of the sink relates to the response of photosynthesis to the rate of nutrient supply. Photosynthesis under many natural conditions is limited by the availability of nitrogen (Wong 1979; Field & Mooney 1986), and the rise in deposition of nitrogen from anthropogenic sources continues (Pitcairn *et al.* 1995; Galloway *et al.* 1995; Townsend *et al.* 1996). To what extent is this global fertilization of the terrestrial and marine ecosystems causing an enhancement in the sink strength, at least in the northern hemisphere where the deposition of nitrogen is most intense? The relationship between the carbon cycle and the nitrogen cycle is still not well understood, but model calculations suggest that the process is at least significant (Lloyd & Farquhar 1996; Lloyd 1999). In general, it is likely that increased photosynthesis brought about by elevated CO_2 leads to an increase in carbon flux to the rhizosphere, which in turn stimulates the nitrogen-fixing micro-organisms, providing an enhanced supply of nitrogen. Moreover, the additional carbon flux to the rhizosphere probably enhances several processes that influence the availability of phosphorus. These include the extent of mycorrhizal development (Lovelock *et al.* 1996), the production of organic acids which increases mineralization of phosphorus, and a general enhancement of root growth relative to shoot growth. Another process that may be stimulated is the production of phosphatases by root and mycorrhizal surfaces, which mineralize organic phosphates in leaf litter, making the phosphorus available for uptake once more.

Keeping track of carbon on a variety of scales

The Kyoto Protocol of 1997 calls upon governments to put in place mechanisms to reduce the emissions of greenhouse gases, in order to reduce climate warming (Grubb *et al.* 1999). The mechanisms that are currently allowed include the use of young forests to absorb CO_2. The prospect of growing trees to sequester carbon is appealing for many reasons. First, the technology is established and 'safe'; second, the 'Kyoto' forests thus created may have significant other benefits, such as environmental enhancement, increase in biodiversity, and ultimate economic reward when timber is harvested. A simple calculation will, however, show that the forests need to be on a large scale to be globally useful. If we wanted to absorb the total annual emissions from fossil fuels in excess of 6 Gt, and if we assume fast-growing plantations ($5 \, t \, C \, ha^{-1} \, yr^{-1}$) we would need over 10^9 hectares of land, which is 40% more than the area of the entire Amazon basin. Calculations of the cost of acquiring such land suggest that the global project would be expensive, and there might not be sufficient foresters in the world to carry out such a project! There is also the question of what to do with the timber when the forests are mature. Many forest products, such as paper, have a short lifetime, but others such as structural timber are long-lived. Despite these difficulties, there is no doubt that plantations could make at least a contribution to the absorption of CO_2 over the next 50–100 years. Moreover, the Protocol will evolve, and may allow additional measures such as the conservation of forests which would otherwise be converted to agriculture, and the restoration of degraded forests. Under the terms of the Treaty, carbon absorbed would be rewarded by 'forest credits', and such credits could be traded, in much the same way as stocks and shares are traded. For example, a Brazilian landowner might in future be rewarded for the mass of carbon in the forest compared to what would have been there if the standard Amazonian deforestation rate had been applied. Such proposals are not without difficulty, as has often been pointed out (Singer 2000). For example, the landowner's brother on a nearby farm might decide to increase his deforestation rate to compensate the family for lost production on the farm where the project is running!

One of the most controversial aspects of the Kyoto Convention relates to the problem of verification. In any carbon-sequestering mechanism it will be necessary to put in place accounting and auditing mechanisms to monitor the stocks or fluxes of carbon from the atmosphere to the ecosystems, on a variety of scales (Table 12.2). At the local scale (hectares) a carbon farmer would need to be able to measure the changing stocks of carbon in the biomass and soil, so that income can be claimed per tonne of carbon sequestered. At a national scale, governments already keep national inventories of greenhouse gas emissions, but these are currently not very accurate. In the case of forests, for example, not enough is known about how the stock of carbon in the soil changes over the lifetime of the forest. Meanwhile, atmospheric measurements, possibly using sensors on board civil aircraft or other 'ships of opportunity' should enable carbon accounting at scales corresponding to nations or regions of the world. Finally, at the global scale, it will be necessary to monitor the distribution of sources and sinks, not only to ensure compliance but also to improve our overall understanding of the behaviour of the carbon cycle in response to the changing conditions.

The value of a global surveillance system might extend far beyond the need for verification. Several monitoring systems could be integrated, feeding data into a global model, in fact a General Circulation Model in which the land and sea surface are represented by a process-based model of the ecosystem (Cox *et al.* 1999). The model would progressively 'learn' based on a comparison of predictions with observations of CO_2 concentration made at local, regional and global scales. Such data assimilation is comparable to procedures used in weather forecasting, and indeed carbon fluxes and weather forecasting are likely to become one and the same operation. The 'refined' model would then be run off-line, to estimate the future sink strengths under various scenarios of emissions.

It is as well to remember that uncertainty in the future CO_2 emissions far outweigh the uncertainty in our capacity to predict the sink's behaviour.

Table 12.2 Some of the methods available or being developed to track flows of carbon between biosphere and atmosphere.

Scale	Technique	Notes	Example
Local 0.01–10 km^2	Inventory of biomass and soil carbon	Inherent spatial variation in soil carbon requires large sampling effort to establish reliable mean	Kurtz & Apps (1994), Phillips et al. (1998)
	Eddy covariance	Expensive, needs trained operators at present	Lindroth et al. (1998)
	Process models	To be parameterized with eddy covariance data. Soil submodel may be the difficult part	Williams et al. (1998), Lloyd (1999)
Regional 10 km^2–10^7 km^2	Remote sensing of biomass	Technical developments required to deal with vegetation with leaf area index >2	Hoekman & Quinones (2000)
	Remote sensing of fluxes	GPP and NPP can be estimated from reflectance, but NEE cannot unless soil respiration model is used	Prince & Goward (1995), Nichol et al. (2000)
	Planetary boundary layer measurements from aircraft	Promising technique which could be developed to an operational network	Lloyd et al. (1996)
	Remote sensing of CO_2 concentration using laser	Expect developments by 2010	Taczak & Killinger (1998)
	Air sampling and inversion	Needs more stations, improved precision	Rayner et al. (1999)
Global	Air sampling and inversion	See above	
	Modelling (GVM)	Critical assumption have to be made about vegetation transitions	Cao & Woodward (1998)
	Integration of scales	This is beginning	Raupach et al. (1992), Running et al. (1999)
	GCMs linked to process models	Expect developments by 2005	Cox et al. (1999)
	GCMs linked to process models with operational data assimilation	Expect developments by 2010	

GCM global circulation model; GPP gross primary productivity; GVM global vegetation model; NEE net ecosystem exchange; NPP net primary productivity.

Appendix 1: websites

The **CarboEurope** cluster of projects (http://www.bgc-jena.mpg.de/public/carboeur/) is designed to better understand, quantify and predict under current and future scenarios the carbon balance of Europe, from local ecosystem to regional and continental scale. This work will support policy-makers responsible for negotiating the implementation of the Kyoto Protocol. It includes the investigation of age-related dynamics of net ecosystem exchange throughout the life cycle of a forest ecosystem and the verification of the stock change approach vs. flux measurements. It also addresses possible incentives of the Kyoto Protocol that may conflict with those of the Biodiversity convention, such as the creation of plantations with high production levels but with attendant biodiversity loss.

The **Large Scale Biosphere–Atmosphere Experiment in Amazonia** (LBA) is an international research initiative led by Brazil. LBA is designed to create the new knowledge needed to understand the climatological, ecological, biogeochemical and hydrological functioning of Amazonia, the impact of land use change on these functions, and the interactions between Amazonia and the Earth system (http://www3.cptec.inpe.br/lba/index.html).

Data about CO_2 emissions may be found at http://www.eia.doe.gov/emeu/international/environm.html and http://cdiac.esd.ornl.gov/trends/emis/em_cont.htm.

Acknowledgements

We would like to acknowledge continuing support from the Natural Environment Research Council (NERC) and the European Commission (DG XII).

References

Arrhenius, S. (1896) On the influence of carbonic acid in the air upon the temperature of the ground. *London, Edinburgh & Dublin Philosophical Magazine and Journal of Science Series* **5**, 237–276.

Atjay, G.L., Ketner, P. & Duvigneaud J. (1979) Terrestrial primary production and phytomass. *The Global Carbon Cycle, SCOPE 13* (eds B. Bolin, E. Degens, S. Kempe & P. Ketner), pp. 129–182. John Wiley & Sons, Chichester.

Aubinet, M., Grelle, A., Ibrom, A. *et al.* (2000) Estimates of the annual net carbon and water exchange of forests: The EUROFLUX methodology. *Advances in Ecological Research* **30**, 113–1175.

Bender, M., Bender, M.L., Tans, P.P. *et al.* (2000) Global carbon sinks and their variability inferred from atmospheric O_2 and delta C^{13}. *Science* **287**, 2467–2470.

Boone, R.D., Nadelhoffer, K.J., Canary, J.D. & Kaye, J.P. (1998) Roots exert a strong influence on the temperature sensitivity of soil respiration. *Nature* **396**, 570–572.

Canadell, J.G., Mooney, H.A., Baldocchi, D.D. *et al.* (2000) Carbon metabolism of the terrestrial biosphere: a multitechnique approach for improved understanding. *Ecosystems* **2**, 1–16.

Cao, M.K. & Woodward, F.I. (1998) Dynamic responses of terrestrial ecosystem carbon cycling to global climate change. *Nature* **393**, 249–252.

Carswell, F.E., Meir, P., Wandelli, E.V. *et al.* (2000) Photosynthetic capacity in a central Amazonian rain forest. *Tree Physiology* **20**, 179–186.

Cox, P.M., Betts, R.A., Bunton, C.B., Essery, R.L.H.,

Rowntree, P.R. & Smith, J. (1999) The impact of new land surface physics on the GCM simulation of climate and climate sensitivity. *Climate Dynamics* **15**, 183–203.

DETR (1999). *Climate Change and its Impacts*. The Meteorological Office, Bracknell, London.

Etheridge, D.M., Steele, L.P., Langenfelds, R.L., Francey, R.J., Barnola, J.M. & Morgan, V.I. (1996) Natural and anthropogenic changes in atmospheric CO_2 over the last 1000 years from air in Antarctic ice and firn. *Journal of Geophysical Research—Atmospheres* **101**, 4115–4128.

Falkowski, P.G., Barber, R.T. & Smetacek, V. (1998) Biogeochemical controls and feedbacks on ocean primary production. *Science* **281**, 200–206.

Field, C., Behrenfeld, M., Randerson, J. & Falkowski, P. (1998) Primary production of the biosphere: integrating terrestrial and oceanic components. *Science* **281**, 237–240.

Field, C. & Mooney, H.A. (1986) The photosynthesis-nitrogen relationship in wild plants. *On the Economy of Plant Form and Function* (ed. T.J. Givnish), pp. 25–55. Cambridge University Press, Cambridge.

Galloway, J.N., Schlesinger, W.H., Levy, H., Michaels, A. & Schnoor, J.L. (1995) Nitrogen fixation: anthropogenic enhancement—environmental response. *Global Biogeochemical Cycles* **9**, 235–252.

Giardina, C.P. & Ryan, M.G. (2000) Evidence that decomposition rates of organic carbon in mineral soil do not vary with temperature. *Nature* **404**, 858–861.

Goulden, M.L., Daube, B.C., Fan, S.M. et al. (1997) Physiological responses of black spruce forests to weather. *Journal of Geophysical Research—Atmospheres* **102**, 28987–28996.

Grace, J. & Rayment, M. (2000) Respiration in the Balance. *Nature* **404**, 819–820.

Grace, J., Lloyd, J., McIntyre, J. et al. (1995) Carbon dioxide uptake by an undisturbed tropical rain forest in South-West Amazonia 1992–93. *Science* **270**, 778–780.

Grayston, S.J., Campbell, C.D., Lutze, J.L. & Gifford, R.M. (1998) Impact of elevated CO_2 on the metabolic diversity of microbial communities in N-limited grass swards. *Plant and Soil* **203**, 289–300.

Grillot, M. (2000) *Carbon Emissions*. National Energy Information Centre, Washington, USA. (http://www.eia.doe.gov/emeu/international/carbon.html)

Grubb, M., Vrolijk, C. & Brack, D. (1999) *The Kyoto Protocol*. Royal Institute of International Affairs, London.

Hättenschwiler, S., Miglietta, F., Raschi, A. & Korner, Ch. (1997) Thirty years of *in situ* tree growth under elevated CO_2: a model for future forest responses? *Global Change Biology* **3**, 463–471.

Hoekman, D.H. & Quinones, M.J. (2000) Land cover type and biomass classification using AirSAR data for evaluation of monitoring scenarios in the Colombian Amazon. *IEEE Transactions on Geoscience and Remote Sensing* **38**, 685–696.

Houghton, R.A. (1999) The annual net flux of carbon to the atmosphere from changes in land use 1850–1990. *Tellus* **50B**, 298–313.

Idso, S.B. (1999) The long-term response of trees to atmospheric CO_2 enrichment. *Global Change Biology* **5**, 493–495.

Intergovernmental Panel on Climate Change (1995). *Climate Change 1995: the Science of Climate Change*. Cambridge University Press, Cambridge.

IPCC (2000) *Land use, land use change, and forestry*. Cambridge University Press, Cambridge.

Jarvis, P.G., Massheder, J.M., Hale, S.E., Moncrieff, J.B., Rayment, M. & Scott, S.L. (1997) Seasonal variation of carbon dioxide, water vapour and energy exchanges of a boreal black spruce forest. *Journal of Geophysical Research* **102**, 28953–28966.

Keeling, C.D. (1958) The concentrations and isotopic abundances of atmospheric carbon dioxide in rural and marine air. *Geochimica Cosmochimica Actai* **13**, 332–334.

Keeling, C.D. & Whorf, T.P. (1999) Atmospheric CO_2 records from sites in the SIO air sapling network. *Trends: a Compendium of Data on Global Change*. Carbon Dioxide Information Analysis Centre, Oak Ridge National Laboratory, US Department of Energy, Oak Ridge, Tennessee.

Körner, C. & Arnone, J.A. (1992) Responses to elevated carbon dioxide in artificial tropical ecosystems. *Science* **257**, 1672–1673.

Koutika, L.S., Bartoli, F., Andreux, F. et al. (1997) Organic matter dynamics and aggregation in soils under rain forest and pastures of increasing age in the eastern Amazon Basin. *Geoderma* **76**, 87–112.

Kurtz, W.A. & Apps, M.J. (1994) The carbon budget of Canadian forests: a sensitivity analysis of changes in disturbance regimes, growth rates, and decomposition rates. *Environmental Pollution* **83**, 55–61.

Lerdau, M.T. & Throop, H.L. (1999) Isoprene emissions and photosynthesis in a tropical forest. *Ecological Applications* **9**, 1109–1117.

Lin, G.H., Adams, J., Farnsworth, B., Wei, Y.D., Marino, B.D.V. & Berry, J.A. (1999) Ecosystem carbon exchange in two terrestrial ecosystems under changing atmospheric CO_2 concentration. *Oecologia* **119**, 97–108.

Lindroth, A., Grelle, A. & Moren, A.S. (1998) Long-term measurements of boreal forest carbon balance reveal large temperature sensitivity. *Global Change Biology* **4**, 443–450.

Liski, J., Ilvesniemi, H., Makela, A. & Starr, M. (1998) Model analysis of the effect of soil age, fires and harvesting on the carbon storage of boreal forest soils. *European Journal of Soil Science* **49**, 407–416.

Lloyd, J. (1999) The CO_2 dependence of photosynthesis and plant growth in response to elevated atmospheric CO_2 concentration and their interrelationship with soil nutrient status 2. Temperate and boreal forest productivity and the combined effects of increasing CO_2 concentrations and increased nitrogen deposition at a global scale. *Functional Ecology* **13**, 439–459.

Lloyd, J. & Farquhar, G.D. (1996) The CO_2 dependence of photosynthesis and plant growth in response to elevated atmospheric CO_2 concentration and their interrelationship with soil nutrient status 1. General principles. *Functional Ecology* **10**, 4–32.

Lloyd, J. & Taylor, J.A. (1994) On the temperature dependence of soil respiration. *Functional Ecology* **8**, 315–323.

Lloyd, J., Kruijt, B., Hollinger, D.Y. *et al.* (1996) Vegetational effects on the isotopic composition of atmospheric CO_2 at local and regional scales- theoretical aspects and a comparison between rain-forest in Amazonia and a boreal forest in Siberia. *Australian Journal of Plant Physiology* **23**, 371–399.

Long, S.P., Garcia Moya, E., Imbamba, S.K. *et al.* (1989) Primary productivity of natural grass ecosystems of the tropics: a reappraisal. *Plant and Soil* **115**, 155–166.

Lotka, A.J. (1956). *Elements in Mathematical Biology* (first published in 1923). Dover, New York.

Louanchi, F., RuizPino, D.P. & Poisson A. (1999) Temporal variations of mixed-layer oceanic CO_2 at JGOFS-KERFIX time-series station. Physical versus biogeochemical processes. *Journal of Marine Research* **57**, 165–187.

Lovelock, C.E., Kyllo, D. & Winter, K. (1996) Growth responses to vesicular-arbuscular mycorrhizae and elevated CO_2 in seedlings of a tropical tree *Beilschmieda pendula*. *Functional Ecology* **10**, 662–667.

Malhi, Y. & Grace, J. (2000) Tropical forests and atmospheric carbon dioxide. *Trends in Ecology and Evolution* **15**, 332–337.

Malhi, Y., Baldocchi, D.D. & Jarvis, P.G. (1999) The carbon balance of tropical, temperate and boreal forests. *Plant, Cell and Environment* **22**, 715–740.

Malhi, Y., Nobre, A.D., Grace, J. *et al.* (1998) Carbon dioxide transfer over a central Amazonian rain forest. *Journal of Geophysical Research* **103**, 593–631.

Marland, G., Boden, T.A., Andres, R.J., Brenkert, A.L. & Johnston, C. (1999) Global, Regional, and National CO_2 Emissions. *Trends: a Compendium of Data on Global Change.* Carbon Dioxide Information Analysis Center, Oak Ridge National Laboratory, US Department of Energy, Oak Ridge, Tennessee.

Medlyn, B.E. & 20 others (1999) Effects of elevated CO_2 on photosynthesis in European forest species: a meta-analysis of model parameters. *Plant Cell and Environment* **22**, 1475–1495.

Moncrieff, J.B., Massheder, J.M., de Bruin, H. *et al.* (1997) A system to measure surface fluxes of momentum, sensible heat, water vapour and carbon dioxide. *Journal of Hydrology* **189**, 589–611.

Myneni, R.B., Keeling, C.D., Tucker, C.J., Asrar, G. & Nemani, R.R. (1997) Increased plant growth in the northern high latitudes from 1981 to 1991. *Nature* **386**, 698–702.

Nichol, C.J., Huemmrich, K.F., Black, T.A. *et al.* (2000) Remote sensing of photosynthetic-light-use efficiency of boreal forest. *Agricultural and Forest Meteorology* **101**, 131–142.

Oechel, W.C., Hastings, S.J., Vourlitis, G., Jenkins, M., Riechers, G. & Grulke, N. (1993) Recent change of arctic tundra ecosystems from a net carbon dioxide sink to a source. *Nature* **361**, 520–523.

Peng, T.H., Wanninkhof, R., Bullister, J.L., Feely, R.A. & Takahashi, T. (1998) Quantification of decadal anthropogenic CO_2 uptake in the ocean based on dissolved inorganic carbon measurements. *Nature* **396**, 560–563.

Phillips, O.L., Malhi, Y., Higuchi, N. *et al.* (1998) Changes in the carbon balance of tropical forests: evidence from long-term plots. *Science* **282**, 439–442.

Pitcairn, C.E.R., Fowler, D. & Grace, J. (1995) Deposition of fixed atmospheric nitrogen and foliar nitrogen content of bryophytes and *Calluna vulgaris* (L) Hull. *Environmental Pollution* **88**, 193–205.

Prince, S.D. & Goward, S.G. (1995) Global primary production: a remote sensing approach. *Journal of Biogeography* **22**, 815–835.

Raupach, M.R., Denmead, O.T. & Dunin, F.X. (1992) Challenges in linking atmospheric CO_2 concentration to fluxes at local and regional scales. *Australian Journal of Botany* **40**, 697–716.

Raven, J.A. & Falkowski, P.G. (1999) Oceanic sinks for atmospheric CO_2. *Plant Cell* &. *Environment* **22**, 741–755.

Rayner, P.J., Enting, I.G., Francey, R.J. & Langenfields, R. (1999) Reconstructing the recent carbon cycle from atmospheric CO_2, $\delta^{13}C$ and O_2/N_2 observations. *Tellus* **51B**, 213–232.

Running, S.W., Baldocchi, D.D., Turner, D.P., Gower, S.T., Bakwin, P.S. & Hibbard, K.A. (1999) A global terrestrial monitoring network integrating tower fluxes, flask sampling, ecosystem modelling and EOS satellite data. *Remote Sensing of the Environment* **70**, 108–127.

Sarmiento, J.L., Murnane, R. & Lequere, C. (1995) Air–sea CO_2 transfer and the carbon budget of the North-Atlantic. *Philosophical Transactions of the Royal Society of London* **348**, 211–219.

Schimel, D.S. (1995) Terrestrial ecosystems and the global carbon cycle. *Global Change Biology* **1**, 77–91.

Schlesinger, W.H. (1997). *Biogeochemistry*. Academic Press, San Diego.

Singer, S. (2000) *Sinks in the CDM? Implications and Loopholes*. World Wide Fund for Nature, European Policy Office, Brussels.

Smith, P., Andren, O., Brussaard, L. *et al.* (1998a) Soil biota and global change at the ecosystem level: describing soil biota in mathematical models. *Global Change Biology* **4**, 773–784.

Smith, P., Powlson, D.S., Glendining, M.J. & Smith, J.U. (1998b) Preliminary estimates of the potential for carbon mitigation in European soils through no-till. *Global Change Biology* **4**, 679–685.

Taczak, T.M. & Killinger, D.K. (1998) Development of a tunable, narrow-linewidth, cw 2.066-mu m Ho:YLF laser for remote sensing of atmospheric CO_2 and H_2O. *Applied Optics* **37**, 8460–8476.

Tanner, E.V.J., Vitousek, P.M. & Cuevas, E. (1998) Experimental investigation of nutrient limitation of forest. *Ecology* **79**, 10–22.

Tans, P.P., Fung, I.Y. & Takahashi, T. (1990) Observational constraints on the global atmospheric CO_2 budget. *Science* **247**, 1431–1438.

Tickell, C. (1977). *Climate Change and World Affairs*. Pergamon, London.

Townsend, A.R., Braswell, B.H., Holand, E.A. & Penner, J.E. (1996) Spatial and temporal patterns in terrestrial carbon storage due to deposition of fossil fuel nitrogen. *Ecological Applications* **6**, 806–814.

Trumbore, S. (2000) Age of soil organic matter and soil respiration: Radiocarbon constraints on belowground C dynamics. *Ecological Applications* **10**, 399–411.

Tsunogai, S., Watanabe, S. & Tetsuro, S. (1999) Is there a continental shelf pump for the absorption of atmospheric CO_2. *Tellus* **51B**, 701–712.

Valentini, R., Matteucci, G., Dolman, A.J. *et al.* (2000) Respiration as the main determinant of carbon balance in European forests. *Nature* **404**, 861–865.

Vitousek, P.M., Ehrlich, P.R., Erhlich, A.H. & Matson, P.A. (1986) Human appropriation of the products of photosynthesis. *Bioscience* **36**, 368–373.

Watson, A.J., Law, C.S., Vanscoy, K.A. *et al.* (1994) Minimal effects of iron fertilization on sea surface carbon dioxide concentrations. *Nature* **371**, 143–145.

WBGU (Wissenschaftlicher Beirat der Bundesregierung Globale Umweltveränderungen) (1998) *Die Anrechnung biologischer Quellen und Senken im Kyoto-Protokoll: Fortschritt oder Rückschlag für den globalen Umweltschutz. Sondergutachten 1998.* Bremerhaven.

Whittaker, R.H. & Likens, G.E., eds (1975) The biosphere and man. *Primary Productivity of the Biosphere*, pp. 305–328. Springer-Verlag, Berlin.

Williams, M., Malhi, Y., Nobre, A.D., Rastetter, E.B., Grace, J. & Pereira, M.G.P. (1998) Seasonal variation in net carbon dioxide exchange and evapotranspiration in a Brazilian rain forest: a modelling analysis. *Plant, Cell and Environment* **21**, 953–968.

Wolf-Gladrow, D.A., Riebesell, U., Burkhart, S. & Bijma, J. (1999) Direct effects of CO_2 concentration on growth and isotopic composition of marine phytoplankton. *Tellus* **51B**, 461–476.

Wong, S.-C. (1979) Elevated atmospheric partial pressure of carbon dioxide and plant growth 1. Interactions of nitrogen nutrition and photosynthetic capacity in C_3 and C_4 plants. *Oecologia* **44**, 68–74.

Woodward, F.I. (1987) Stomatal numbers are sensitive to increases in CO_2 from preindustrial levels. *Nature* **327**, 617–618.

Woodwell, G.M. & Whittaker, R.H. (1968) Primary production in terrestrial ecosystems. *American Zoologist* **8**, 19–30.

Chapter 13

Climate and plants: past, present and future interactions

F. I. Woodward

Introduction

This symposium celebrates the achievements and highlights the future challenges for ecology in the 21st century. Individual speakers present their views from particular areas of expertise and this chapter will provide a broad, rather than in-depth, view of interactions between plants and climate, with particular emphasis on the terrestrial carbon cycle. In contrast to the chapter on the contemporary carbon cycle, by Grace et al. (Chapter 12), here the time frame covers the last 450 million years, over which time plants have first colonized and then evolved to the very different contemporary forms and physiologies. The challenge, illustrated here, has been to understand past environments and plants based on current understanding. The achievement has been to demonstrate the validity of the approach, although, as it will be seen, it is often not possible to provide real validation of the approaches currently used.

Over the Phanerozoic period of the last 450 million years, the evidence from fossils indicates that about 90% of species on land have become extinct (Niklas 1997), making way for different, often more complex forms. Analysis of fossil plants indicates considerable increases in structural complexity and individual height over the earliest period of land colonization extending between about 470 and 350 million years ago (Edwards 1998). Contemporary vegetation differing in height exerts considerable impacts on climate (Betts et al. 1997; Hayden 1998), through variations in exchanges of energy (Charney et al. 1977) and mass, through evapotranspiration (Shukla & Mintz 1982) and CO_2 exchange (Heimann 1997). It is most likely therefore that similar feedbacks between climate and vegetation occurred in the past. If the feedbacks were large then plant activity would influence evolution by imposing changes on climate and on the activity of the carbon cycle. The fossil record indicates that the average life of fossil flowering plants is about 3.5 million years, with a new species appearing about every 0.4 million years (Niklas 1997). Therefore it is quite clear that climatic and evolutionary antecedents have strong contemporary relevance and so the understanding of contemporary

Department of Animal and Plant Sciences, University of Sheffield, Sheffield, S10 2TN, UK. E-mail: F.I.Woodward@Sheffield.ac.uk

ecology, as a result, must also consider the inheritance from the past. This paper will provide a number of examples to illustrate this premise, indicating important achievements while identifying contemporary challenges which still require to be answered.

Vegetation feedbacks on climate and the carbon cycle

The Phanerozoic

The climate and the carbon cycle over the Phanerozoic period have been influenced by a number of forcing agents. Over this period solar luminosity has been increasing and the tectonic forces of volcanism and continental drift have all influenced the Earth's climate and the fluxes of CO_2 into the atmosphere (Frakes *et al.* 1992). On a million-year time scale the dynamics of the carbon cycle are dominated by the weathering and deposition of organic carbon stored in rocks (Berner 1998). Surprisingly, given such large- and slow-scale processes, it has been found necessary to include the activity of plants in vegetation in order for the cycle to be balanced (Berner 1994, 1998). The particular vegetation activity is that of the weathering of silicate bed rock, which, when very active, leads to the removal of atmospheric CO_2 (Berner 1994). This process has been defined theoretically and also observed experimentally (Berner 1998). Increasing root penetration into soil, the incorporation of organic matter into the soil and the recycling of evapotranspired water as precipitation, all lead to increased weathering and a reduction in atmospheric CO_2 (Berner 1997). This reduction in atmospheric CO_2 reduces greenhouse warming and probably precipitation, and so also reduces the rate of weathering. These reduced rates of vegetation weathering, even with no changes in tectonic activity, then allow atmospheric CO_2 and global temperatures again to increase. This in turn leads to increased rates of vegetation activity and weathering and a decrease in atmospheric CO_2 (Berner 1997). Vegetation therefore appears to play a major long-term role in controlling atmospheric CO_2 concentration.

Over these geological time scales there is therefore an expectation of fluctuations in the carbon cycle resulting from changes in solar activity, tectonics and weathering which is affected by vegetation. Figure 13.1 shows a time series of such atmospheric CO_2 simulated by Berner (1997), and indicating the best estimate, and likely upper and lower bounds of the simulations. It should be noted that Berner indicates quite wide error limits to these simulations. The production of this time series of atmospheric CO_2 is a major achievement in its own right but it needs to be tested against independent data. Such a test is broadly provided by analysis of the stable isotopic composition of organic carbon and carbonate in rocks (Mora *et al.* 1996).

The trend in atmospheric CO_2 shows a remarkable drop from about 375 million years ago, during the Devonian period and into the Carboniferous period. Berner's model indicates that this is primarily due to enhanced vegetation weathering. Independent fossil data (Retallack 1997) indicate that Devonian forests spread considerably during this period. This expansion was allied with a deeper and thicker

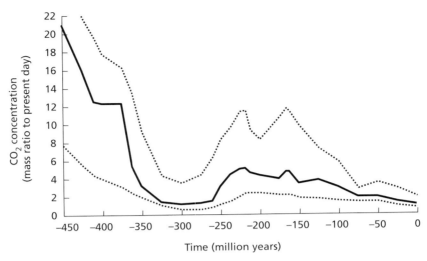

Figure 13.1 Best guess (——) and upper and lower error estimates (. . .) of atmospheric CO_2 through the Phanerozoic simulated with the GEOCARB II model. (Data from Berner 1997.)

rooting habit (Retallack 1997), and therefore considerable increases in weathering, and the accumulation of dead organic matter and decomposition-resistant lignin (Robinson 1990; Retallack 1997). All of these responses indicate a strong negative feedback of vegetation on atmospheric CO_2 concentration. Much of the CO_2 that was locked away as dead organic matter led to the large coal reserves of the Carboniferous and early Permian periods (330–260 million years ago).

Further evidence in support of Berner's simulations is provided on Fig. 13.2, which shows the timing of major cold or glacial events (based on Frakes et al. 1992). The most severe cold event was during the Carboniferous (from 340 million years ago) when atmospheric CO_2 concentrations were perhaps as low as the present day, or at least in the range of 300–600 p.p.m. (Berner 1997). According to the constructions of Frakes et al. (1992) it can be seen that cold events are long-lived and correspond with periods of low or declining atmospheric CO_2 concentrations, at the Carboniferous and early Permian (340–250 million years ago), the Jurassic and Cretaceous (175–105 million years ago) and the Cretaceous to the present (last 80 million years), although there is some controversy about the extent of the Jurassic and Cretaceous and most recent cold periods (Frakes et al. 1992).

Contemporary and future climates

One feature to emerge from the research on vegetation and climate over the Phanerozoic era is that the evidence points towards significant impacts of vegetation on climate. In general, increased vegetation activity appears to lead to the long-term draw-down of atmospheric CO_2, with a consequent reduction in the

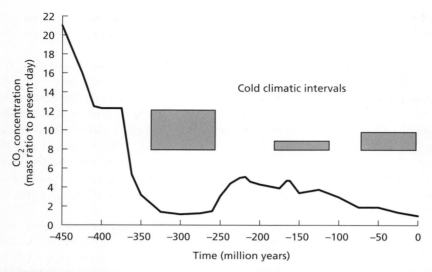

Figure 13.2 Major cold climatic periods, the larger the box the colder the climate. (Data from Frakes *et al.* 1992.)

greenhouse effect and global cooling. Global cooling leads to a reduction in vegetation activity, an increase in atmospheric CO_2, and subsequent warming. The evidence in support of these conclusions remains tantalizingly indirect and a certain requirement for future achievements in this area of research is more direct evidence, probably resulting from new techniques of measurement.

The time scale that is resolved in geological approaches to vegetation–climate feedbacks is in the order of a million years. The small pools of carbon in both the terrestrial and oceanic biospheres have minimal long-term consequences on the carbon cycle. The turnover time of carbon in the biosphere is in the order of decades (Heimann 1997) and so plays a critical part in the carbon cycle and climatic interactions over the contemporary time scale and over the next century. This is the era of discernible anthropogenic impacts on the carbon cycle and probably climate, and so it remains a major challenge to understand how the biosphere may, for example, be delaying the increase in the atmospheric CO_2 concentration and its subsequent feedback on climate (Heimann 1997).

The large size of the carbon pools of the oceans indicates turnover times and influences on the carbon cycle in the order of 1000–100 000 years (Watson & Liss 1998). By contrast, the surface ocean and associated biosphere will have turnover times of about a century, or less. Therefore changes in oceanic photosynthesis and the solution of CO_2 in sea water will play some part in the global carbon cycle responses over the next 100 years (Watson & Liss 1998). However, modelling studies (Sarmiento & Lequere 1996) indicate that changes in the oceanic thermohaline circulation and rates of upwelling of deep water might exert significant effects on the global carbon cycle. These aspects of the oceanic impact on the carbon cycle are of major interest and remain as major challenges to determine their impacts over the next century.

The decadal turnover times of the carbon pools in vegetation match the time frames of concern about the contemporary and future (the next century) carbon cycle and its human perturbations. The emphasis on the future is critical for influencing strategies which aim to mitigate against the continual increase in atmospheric CO_2 concentrations. This also means a reliance on model results, clearly a problem for experimentalists to swallow but the current state of dynamic vegetation models (e.g. Cramer et al. 2001) is surely a major achievement for this area of ecology. There is still much to do and perhaps foremost is the inclusion of the wide spectrum of human impacts on land cover, still an area under development (Ramankutty & Foley 1999).

Cao and Woodward (1998) and Cramer et al. (2001) have investigated the response of the terrestrial biosphere to one particular General Circulation Model (GCM) simulation of climate from 1860 to 2100, the UK Meteorological Office HadCM2 transient model (Mitchell et al. 1995; with additional comments in Cramer et al. 2001). The climatic simulation is one of many possible simulations for a future climate. The HadCM2 includes sulphate aerosol feedback and depends on the IS92a IPCC scenario (Houghton et al. 1992) of atmospheric CO_2. These simulations have only investigated likely responses of vegetation to changes in climate and atmospheric CO_2 and have not been fully coupled with the GCM to determine feedbacks on climate—a current challenge in this field of endeavour. The combination of global warming ($\approx 4°C$ from 1990 to 2100) and CO_2 increase (800 p.p.m. in 2100) lead to significant increases in net carbon exchange by terrestrial vegetation (Fig. 13.3). Net carbon exchange is simulated by a dynamic vegetation model (Cramer et al. 2001), which accounts for the dynamics of gaseous and nutrient exchange, phenology and succession, in addition to modelling the occurrence of fire. In this context net carbon exchange (NCE) is gross photosynthetic production, less plant and heterotrophic respiration and carbon losses by fire.

A critical feature of the NCE simulation (Fig. 13.3) is that the combined influences of changing climate and CO_2 increase the sink capacity of the terrestrial biosphere for atmospheric CO_2 (increases in atmospheric CO_2 with a constant climate increase the sink capacity even further). Vegetation therefore acts as a negative feedback on climate change. However, if only changing climate and a constant CO_2 is considered (Fig. 13.3) then vegetation releases CO_2 to the atmosphere, a positive feedback on climate change. A further important feature is that although vegetation acts as a negative feedback, by absorbing atmospheric CO_2 and sequestering it in soils and woody tissue, its capacity to increase carbon sequestration saturates, by about 2050 in this particular simulation. Other dynamic vegetation models (Cramer et al. 2001) show similar saturating responses. This appears therefore to be a robust and important result and will be a continued challenge to determine its reality in the future, particularly as such information is critical for defining future CO_2 mitigation strategies.

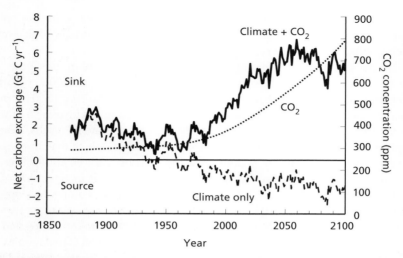

Figure 13.3 Simulated trends in net carbon exchange for the terrestrial biosphere, with (——) and without (– – –) the influence of increasing atmospheric CO_2 concentration (· · ·).

These contemporary simulations of the terrestrial component of the carbon cycle broadly support the inferences from the geological time scale. More detail can, of course, be investigated for the contemporary situation. One consideration is to investigate the geographical distribution of major areas where the terrestrial biosphere is simulated to be a negative (carbon sink) or positive (carbon source) feedback on atmospheric CO_2, over the 20th century. There is often considered to be a missing terrestrial sink for carbon, as this is an unaccounted component of the carbon cycle (Houghton et al. 1998). However, this simulation indicates (Fig. 13.4) that the major terrestrial sink is between the tropics, even when fires are taken into account (Cramer et al. 2001).

The simulations of the terrestrial carbon cycle for the 20th century were achieved with a new global field of observed and interpolated climate (New et al. 2000). These data are not exactly comparable with climate simulations from a GCM for the same period. However, this area, of simulating past and current climate, has improved remarkably over the last decade and, as an area of very active development, it is expected that simulations will continue to improve. Information about the climate of the next 100 years is only provided by GCMs and so it is critical that the simulation accuracy continues to improve. The map of future areas of negative and positive feedbacks on atmospheric CO_2 (Fig. 13.5) bear some similarities, but also some notable differences from the 20th century. A particular feature is an increase in the activity of South America as a positive feedback on atmospheric CO_2. This occurs because the GCM (UK Meteorological Office, Cramer et al. 2001) reduces precipitation significantly, leading to an increased drought

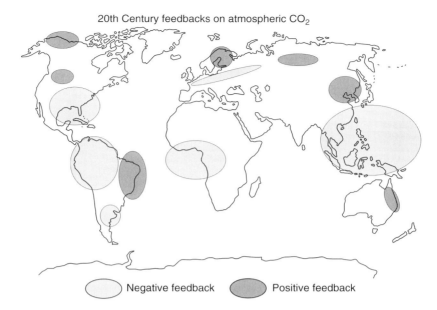

Figure 13.4 Simulations of the 20th-century feedbacks of vegetation on atmospheric CO_2 concentration. Simulations carried out with the Sheffield Dynamic Vegetation Model (Cramer *et al.* 2001) and using a global climate database (New *et al.* 2000). The areas of negative and positive feedbacks are those in which the annual flux exceeds $\pm 50\,\text{g}\,\text{C}\,\text{m}^{-2}\,\text{yr}^{-1}$, averaged over the whole of the 20th century.

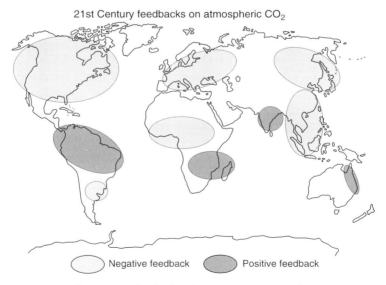

Figure 13.5 Twenty-first century feedbacks of vegetation on atmospheric CO_2 concentration. Details as for Fig. 13.4 except that the climate fields were derived from the UKMO HadCM2 transient GCM.

frequency and associated fall in net primary productivity, plus an increase in fire frequency, both of which will reduce net carbon exchange. The northern mid to high latitude sinks increase under this scenario, particularly because of increases in temperature.

Betts et al. (1997) and Woodward et al. (1998) investigated the feedbacks of vegetation on climate in a coupled simulation between a GCM and a vegetation model. The simulation investigated how vegetation would respond in terms of energy and mass exchange (radiation and water vapour) to the climatic changes associated with a doubling of atmospheric CO_2 concentration. The simulation accounted for feedbacks of vegetation on climate. At high latitudes (Fig. 13.6) the boreal forest spread into the dwarf tundra and led to warming, because of reduced solar reflectivity, particularly during the winter coverage of snow. This is a response which broadly parallels reconstructions from the warmer Tertiary periods (Graham 1999). Other positive feedbacks were modelled for India and South America, where reduced precipitation led to a reduction in forest and an increase in grassland, and associated reductions in evapotranspiration. In the northern mid latitudes vegetation change led to an opposing negative feedback on climate, through increases in vegetation leafiness and evapotranspiration. An important feature of the CO_2 feedbacks (see Figs 13.4 and 13.5) is that, although the areas of significant changes in activity can be defined by region, the impacts are global. This is the case because the atmosphere is very well mixed and so these regional changes are soon integrated into a global response. By contrast the regional feedbacks on climate (Fig. 13.6) are realized at the regional scale.

Plants as climatic indicators during extinction events

The occurrence of major extinction events in the geological past has continually attracted considerable attention (Erwin 1998). This is because of the complete losses of major life forms and because of considerable curiosity about the causes of these major events. An interesting feature of plant and animal extinctions (Fig. 13.7) is that the two rarely overlap (Niklas 1997). This curiosity opens a new potential, which is that if plants survived through whole extinction events then they may well carry signals that indicate the cause of the events; all that is required is the occurrence of plant fossils bearing interpretable environmental signals. McElwain et al. (1999) have capitalized on this approach to investigate the environmental changes during a major animal extinction event (Fig. 13.7), at the boundary of the Triassic and Jurassic eras, 205.7 million years ago.

The Triassic–Jurassic (T–J) extinction event was the third largest in the Phanerozoic with major extinctions of marine and terrestrial animals, but also with considerable turnover, but not necessarily extinction, of terrestrial plants (McElwain et al. 1999). Investigations of fossil leaf material indicated very significant reductions in leaf size (either whole lamina, or individual lobes of dissected leaves) of the chang-

CLIMATE AND PLANTS: PAST, PRESENT AND FUTURE

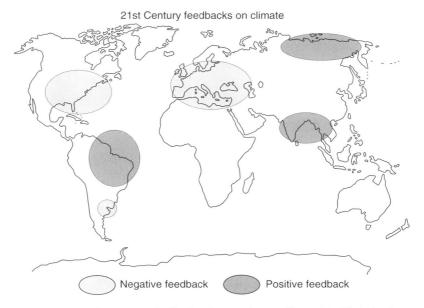

Figure 13.6 Twenty-first century feedbacks of vegetation on climate. Model-derived output from a coupled GCM (Hadley Centre) and vegetation model (Betts *et al*. 1997). Areas of identified feedback are where the impact on temperature is equal to or greater than ±0.5°C.

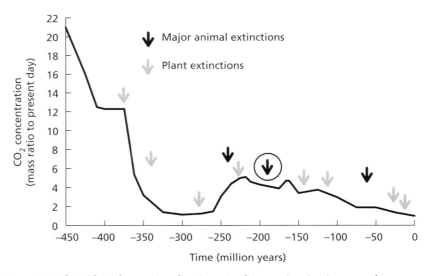

Figure 13.7 Plant (dotted arrows) and major animal (arrows) extinctions over the Phanerozoic. Circle indicates the Triassic–Jurassic extinction event. (From data compiled by Niklas 1997.)

279

ing and dominant forest species through this period of extinction (Fig. 13.8). Such changes could have been the result of a response to arid conditions; however, this is not the case because the fossil material originated from wet habitats. A further alternative is that plants in this tropical climate had been subjected to increasing temperatures through the period of extinction. The initially large-leaved dominant species would have therefore been subjected to an increasing heat load, perhaps reaching the threshold for leaf damage. By contrast smaller leaves could have avoided this damage because of their greater capacity for convective heat loss (Gauslaa 1984).

The lethal threshold for heat damage in extant non-succulent plant species is highly conserved, between about 45°C and 50°C (Gauslaa 1984). If it is assumed that plants 200 million years ago had not improved on this limit then it is possible to model the likely temperature excesses of leaves with different dimensions, in this tropical climate. Model results indicate that the large-leaved species from the Triassic would have reached lethal temperatures if the regional climate warmed by about 3–4°C, while the narrow-leaved species could have endured considerably higher ambient temperatures. Thus, a global temperature increase of about 3–4°C is what might have caused the loss of the broad-leaved species at the T–J. Can the fossil leaves provide any further information about the likely causes of such a temperature increase? Over the period of the Phanerozoic, fossil leaf stomatal densities have followed a pattern that is inversely related to the estimated atmospheric trend of CO_2 (Beerling & Woodward 1997; Fig. 13.9). Research has shown that, on average across many species, stomatal density is inversely proportional to atmos-

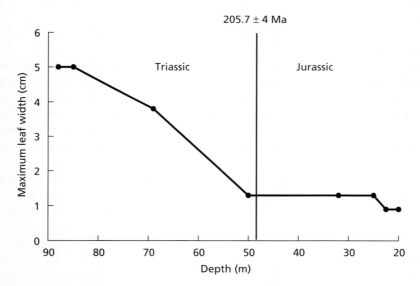

Figure 13.8 Trend in maximum width of fossil leaves, or leaf lobes, through the Triassic–Jurassic boundary. (Data from McElwain *et al.* 1999.)

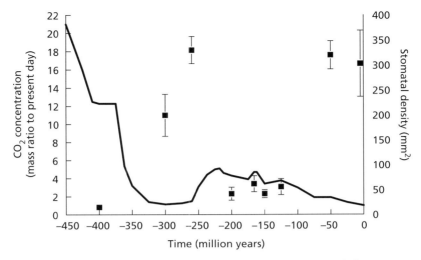

Figure 13.9 Atmospheric CO_2 concentration and stomatal density through the Phanerozoic. (Data from Beerling & Woodward 1997.)

pheric CO_2 concentration (Woodward 1987; Woodward & Kelly 1995), a response that is demonstrated by growing plants at different CO_2 concentrations (Woodward & Bazzaz 1988). Indeed observed reductions in the stomatal densities of leaves over the last 100 years rise in atmospheric CO_2 (Woodward 1987) provided the first evidence that plants in nature have been responding to changes in CO_2 concentration.

Stomatal density and index (Fig. 13.10) of fossil plants across the T–J boundary indicate very large reductions with time, to remarkably low densities and indices (McElwain et al. 1999). These low densities would have strongly limited the capacity for transpirational cooling but they also indicate that atmospheric CO_2 concentrations must have risen through this period of extinction. A combination of the estimated CO_2 concentration, from stomatal densities (McElwain et al. 1999) and the use of a basic equation relating atmospheric CO_2 concentration to temperature (Kothavala et al. 1999), indicates (Fig. 13.11) about a 4°C warming associated with more than a doubling of CO_2 concentration.

A further test of these simulations can now be achieved because the fossil leaf material has also been analysed for stable isotopic composition (the ratio of ^{13}C to ^{12}C, $\delta^{13}C$). The isotopic composition indicates the relative activities of transpiration and photosynthesis during the growing life of the leaf. This information can be used to constrain a model of the leaf energy balance (Beerling & Woodward 1997). The leaf model can then be used to investigate whether the projected changes in atmospheric CO_2 concentration and temperature over the T–J period (Fig. 13.11) would lead to lethally high temperatures and therefore selection for narrow leaves. The results of the simulation are shown in Fig. 13.12. They indicate

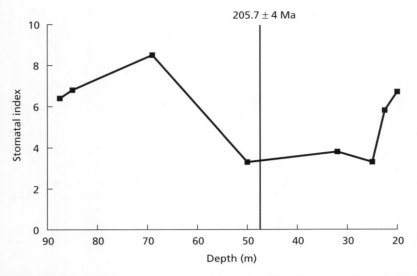

Figure 13.10 Stomatal index through the Triassic–Jurassic boundary. (Data from McElwain et al. 1999.)

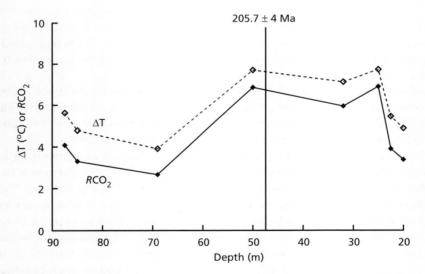

Figure 13.11 Simulated trends in atmospheric CO_2 (RCO_2, the mass ratio to the present day) and associated change in temperature through the Triassic–Jurassic boundary. (Data from McElwain et al. 1999.)

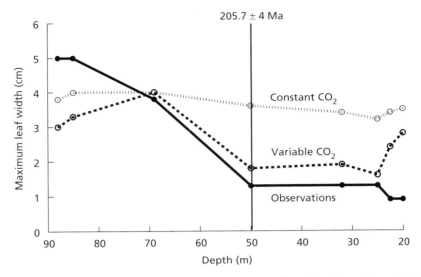

Figure 13.12 Simulated trends, through the Triassic–Jurassic boundary, in maximum leaf width assuming CO_2 and temperature are both constant (······) or variable (– – –) and in comparison with observations (———).

that the marked changes in leaf width are only predicted when both temperature and CO_2 increase through the T–J period. Further, quite independent support for the increase in CO_2 is provided from evidence that the Pangean continent was breaking up through this time, with very extensive volcanic activity (Marzoli *et al.* 1999). It is calculated that the CO_2 required to increase atmospheric CO_2 (see Fig. 13.11) during the T–J period is well within that estimated to be released by volcanic activity during this time at the break-up of Pangea (Marzoli *et al.* 1999).

A major achievement of this research has been the capacity to apply contemporary understanding of plant ecophysiology to the distant geological past and to find that this understanding works sufficiently well to predict measurements from alternative and independent data sources. This has two important consequences, the first is that the current challenge of understanding likely future effects of global change can follow from contemporary understanding. The second point is that geological understanding has relevance to contemporary and future change. A recent paper (Kirchner & Weil 2000) has suggested that the recovery of animal diversity can take as long as 10 million years after both mass and background extinction rates. If this is the case then the recovery of diversity from the current human-induced extinctions is also likely to be millions of years. It is a major challenge to both test this conclusion and an even greater challenge to reduce current rates of extinction. Analyses of the T–J event tend to support such a slow recovery, at least from the perspective of plants.

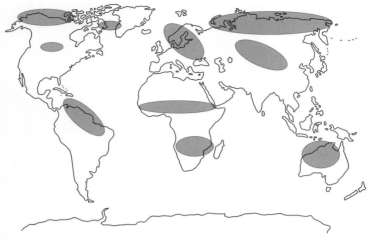

Figure 13.13 Modelled changes in vegetation type from the 20th century to the end of the 21st century. Modelling details as for Fig. 13.5 and indicating regions where vegetation type changes.

Modelling the responses of vegetation distribution to changes in climate over the next century is possible using a mixture of dynamic vegetation models and GCM simulations of climate (Cramer et al. 2001). These models can also be tested against recent and contemporary observational data and geological data (Beerling et al. 1999), again in combination with GCM simulations of past climates. Even though GCM simulations of past climates contain uncertain errors, there is nevertheless clear evidence that model simulations are consistent with geological data. Modelled vegetation changes for the end of the 21st century are quite extensive (Fig. 13.13), following 4°C of warming and an increase in CO_2 concentration to 800 p.p.m. In some cases these changes, such as the demise of high-latitude tundra, could lead to losses of diversity, although these areas are not highly diverse. However, the predicted vegetation changes in South America and Africa are from forest to either savannah or grassland and this will certainly entail very significant losses of diversity. The problem and challenge here is to understand how much diversity will be lost with such a change.

Vegetation, climate, CO_2 and diversity

Terrestrial plant diversity has been increasing since the arrival of plants on land (Niklas 1997; Fig. 13.14). It is particularly noteworthy that the increase in diversity is broadly opposite from the downward trend in atmospheric CO_2 concentration to the present day. Angiosperms in particular appear the most efficient plant group at reducing atmospheric CO_2 concentration by weathering (Volk 1987). The

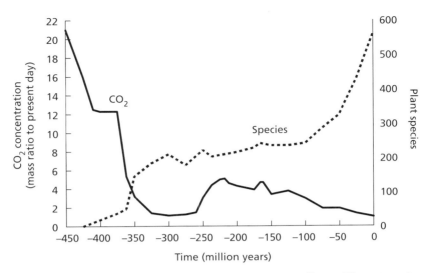

Figure 13.14 Opposing trends in terrestrial land plant diversity (from Niklas 1997 and primarily for North America and Europe) and atmospheric CO_2 concentration over the Phanerozoic.

marked increase in plant diversity over the last 100 million years, due to the evolution of the Angiosperms, is certainly strongly correlated with a downward trend in CO_2 and an increase in cold periods (see Fig. 13.2), although it is thought that the spread of Angiosperms was enhanced by warm global climates (Frakes et al. 1992). It is seen, not for the first time, that plants deplete CO_2 to concentrations close to global compensation points and the threshold for major extinction, only to be rescued by other factors. Yet the relationships between the origins of major plant groups and atmospheric CO_2 (Fig. 13.15) indicate that new plant groups often emerge when CO_2 concentrations are low, or declining and cold intervals become more frequent (see Fig. 13.2), as may also areas of increased aridity. It remains a major challenge to determine whether this relationship is anything other than purely correlational.

The Phanerozoic trends in atmospheric CO_2 concentration (see Fig. 13.1) can be used with a simple energy balance model of the Earth's surface (North et al. 1981) and estimates of the increasing solar constant (5% over this period, Caldeira & Kasting 1992) to calculate approximate trends in global surface temperatures. The reconstructions may only be approximate as the trends in CO_2 concentration are also approximate and it is also assumed that no other greenhouse gases have significantly influenced temperature. This simple simulation of temperature (Fig. 13.16) shows, apart from the very cold Carboniferous excursion, that the Earth's terrestrial temperatures have been decreasing with time, broadly in line with independent, geological estimates of climate (Frakes et al. 1992). This has occurred in

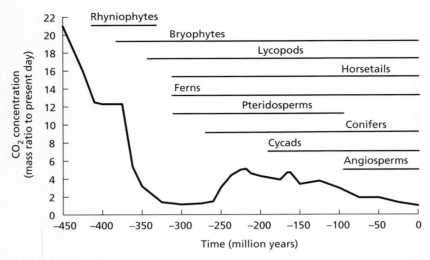

Figure 13.15 Timings of the origins of major plant groups with the trend in atmospheric CO_2 through the Phanerozoic. (Data from Woodward 1998.)

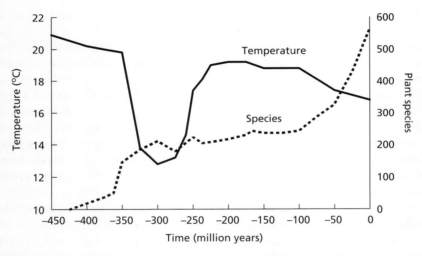

Figure 13.16 Modelled trends in surface temperature (——) through the Phanerozoic, in comparison with species diversity (– – –).

spite of an increase in the solar constant and is a measure of the effectiveness of atmospheric CO_2 as a greenhouse gas.

Over the last 250 million years diversity has climbed but the global temperature is predicted to have slowly declined (Fig. 13.16). This negative correlation is quite contrary to the contemporary situation (Fig. 13.17), where plant species diversity

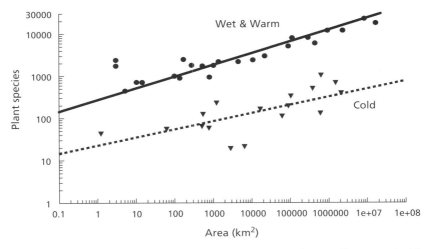

Figure 13.17 Plant species diversity with sample area for wet and warm climates and cold climates. (Data from Williams 1964.)

increases with temperature, at any spatial scale of observation. Of course the geological trends (see Fig. 13.16) are also a measure of rates of evolution and so the two graphs are not compatible. However, this difference does illuminate some of the problems in diversity research. As exposed in the chapters by Grime (Chapter 7), Lawton (Chapter 8) and Tilman (Chapter 9) there is uncertainty and disagreement about the relationship between productivity and diversity—which depends on which, or is it a duality? Small-scale experiments with variable resource supply (Hobbie et al. 1993) often show that diversity can decline, or show a more complex dynamic, with productivity. However, at the global scale, simulated increases in net carbon exchange over the 20th century are positively correlated with plant species diversity (Fig. 13.18). This chapter does not aim to enter into the debate on diversity and vegetation function (Grime 1997) but it is clear that the nature of this interaction remains an important challenge for ecology.

It is instructive to investigate the relationship between NCE and diversity at the global scale (Fig. 13.18). There is considerable interest in investigating these data, as areas with large positive values of NCE are significant parts of the terrestrial sink for anthropogenic releases of CO_2. Although there is a significant relationship between diversity and NCE ($r^2 = 0.44$) there is considerable scatter, indicating that regions with similar diversity have very different carbon balances, at least as estimated by modelling. Contemporary estimates of NCE and its components (net primary production, heterotrophic respiration and carbon losses due to disturbance) agree well with independent data, which provides some confidence in the simulations (Cramer et al. 2001). The two data sources, plant diversity (from Barthlott et al. 1996) and the simulations of NCE are quite independent, as the model of NCE does not account for species diversity, although it does account for

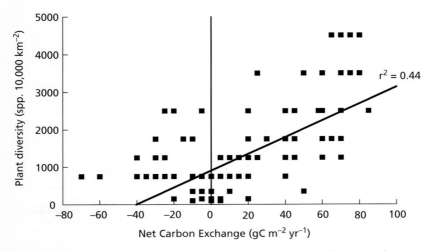

Figure 13.18 Relationship between plant diversity (from Barthlott et al. 1996) and simulated changes in net carbon exchange from the beginning to the end of the 20th century.

functional diversity (i.e. deciduous and evergreen trees, narrow- and broad-leaf trees, grasses and shrubs, Cramer et al. 2001), in a very simple fashion.

Myers et al. (2000) provide an independent assessment of areas of very high species diversity and which are also being subjected to significant species losses, primarily through human impacts. These areas contain as many as 44% of the vascular plant species but only occupy 1.4% of the Earth's surface (Myers et al. 2000) and are located between about 45°N and 40°S. The question here is whether changes in climate over the 20th century have also had a negative impact on these areas. These negative impacts are measured as a negative NCE for the 20th century as a whole (Fig. 13.18). Notable correlations between negative NCE and hotspots (Myers et al. 2000) are seen for a number of areas, the Atlantic Forest and Cerrado of Brazil, the western edge of Mesoamerica, West African Forests, Kenya, the Mediterranean, South-East Asia and west India. These sensitive areas, in terms of negative NCE, do not occur on or close to the equator.

It seems most surprising and perhaps purely coincidental that so many areas of high diversity are also responding negatively to the regional-scale 20th century trends in climate. All of the hotspots have been greatly reduced in area through various human impacts and, on average, only 12% of the primary vegetation remains (Myers et al. 2000). This combination of changing climate and reduction in area of vegetation can itself cause a climatic impact. Hayden (1998) reviews evidence for the impacts of vegetation on climate. For example, at least 50% of the precipitation over the Amazonian rain forest is recycled evapotranspiration. Therefore a drying of climate, for example, and a reduction in the area of this forest could cause a reduction in precipitation and, for example, an increased risk of

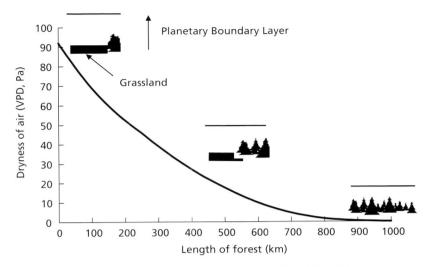

Figure 13.19 Modelled influence of vegetation cover on the humidity of the air (vapour pressure deficit, VPD) at the top of the planetary boundary layer, after 1000 km of travel over different extents of grassland followed by forest.

drought and fire, both of which would reduce NCE. A simulation of this effect is presented (Fig. 13.19) and clearly describes the effect. In the simulation, evapotranspiration from vegetation is captured within the planetary boundary layer (PBL) above the vegetation. The height of the PBL may be as much as 2 km and its height is dominantly controlled by surface convection. Thus, if the vegetation is only transpiring a small amount then convection will be high, as will the PBL. In the simulation there is a variable amount of grassland and tropical forest, along a transect of 1000 km. If the whole transect is actively transpiring forest then the PBL is low and very humid, due to the forest. Even at the top of the forest the air is saturated with water, and condensation nuclei in the air will cause rain. At the other extreme of grassland only, transpiration is less and convection greater. The PBL is higher and drier, with a much lower probability of rain.

The example (Fig. 13.19) indicates that very large areas of forest are required to maintain the humid and wet climate which is characteristic of rain forest and a necessary component of maintaining diversity. Dry the vegetation and species will become extinct (Turner 1996). However, the message from the simulation is that to in order to conserve species, considerable areas of contiguous land must be made into reserves, if the vegetation's climate—its bioclimate—is necessary for survival.

Future challenges

Ecology is often presented as a broad mixture of component approaches. However, one value of ecology is that some of these approaches can be combined to provide a new vision. The integration of geological aspects of ecology, as presented here, with contemporary understanding and modelling provides new understanding of both past and present. A continual challenge is the assurance that today's understanding of processes applies to the past; however, the continued development of new techniques has strengthened this assurance and provided new insight into both a turbulent past and likely future.

This chapter has indicated likely interactions between vegetation and climate. Unfortunately the robustness of these results can rarely be tested with precision, particularly over geological time scales. This is a problem, as a limited understanding of the past also suggests uncertainty in our understanding of the future. Understanding from the science of ecology is increasingly drawn into issues of policy and economics on pollution, land-cover change and climatic change, and so it is critical that this understanding is robust. Past achievements in ecology have established a rigorous outlook on a diverse science but important challenges remain. The following five challenges are identified as important for developing the fields of ecology identified in this chapter.

1 To expand our understanding and application of contemporary ecology to the past and future.
2 To identify the importance of evolutionary history on contemporary and future ecology.
3 To quantify the nature, scale and magnitude of feedbacks from vegetation on climate.
4 To explain the geological time series of plant (and animal) diversity.
5 To determine the importance of species diversity on ecosystem functioning and the importance of ecosystem functioning on species diversity.

As a final point it must be identified that a major contemporary challenge for ecology is to preserve the functioning of global ecosystems with their unique bioclimates and diversity, although allowing for inevitable changes in these latter components. This preservation will only be achieved from the basis of understanding how diversity and climate interact, a challenge which remains to be resolved.

Acknowledgements

I am grateful to David Beerling and Jenny McElwain for their comments on the manuscript.

References

Barthlott, W., Lauer, W. & Placke, A. (1996) Global distribution of species diversity in vascular plants. *Edrkunde* 50, 317–327.

Beerling, D.J. & Woodward, F.I. (1997) Changes in land plant function over the Phanerozoic: reconstructions based on the fossil record. *Botanical Journal of the Linnean Society* 124, 137–153.

Beerling, D.J., Woodward, F.I. & Valdes, P. (1999) Global terrestrial productivity in the mid-

Cretaceous (100 Ma): model simulations and data. In: *Evolution of the Cretaceous Ocean-Climate System* (eds E. Barrera & C.C. Johnson), pp. 385–390. Geological Society of America Special Paper 322, Boulder, Colorado.

Berner, R.A. (1994) GEOCARB II: a revised model of atmospheric CO_2 over Phanerozoic time. *American Journal of Science* **291**, 339–376.

Berner, R.A. (1997) The rise of plants and their effect on weathering and atmospheric CO_2. *Science* **276**, 544–.

Berner, R.A. (1998) The carbon cycle and CO_2 over Phanerozoic time: the role of land plants. *Philosophical Transactions of the Royal Society of London, Series B* **353**, 75–82.

Betts, R.A., Cox, P.M., Lee, S.E. & Woodward, F.I. (1997) Contrasting physiological and structural vegetation feedbacks. *Nature* **387**, 796–799.

Caldeira, K. & Kasting, J.F. (1992) The life span of the biosphere revisited. *Nature* **360**, 721–723.

Cao, M. & Woodward, F.I. (1998) Dynamic responses of terrestrial ecosystem carbon cycling to global climate change. *Nature* **393**, 249–252.

Charney, J.G., Quirk, W.J., Chow, S.H. & Kornfield, J. (1977) A comparative study of the effects of albedo change on drought in semi-arid regions. *Journal of Atmospheric Science* **34**, 1366–1385.

Cramer, W., Bondeau, A., Woodward, F.I. *et al.* (2001) Global response of terrestrial ecosystem structure and function to CO_2 and climate change: results from six dynamic global vegetation models. *Global Change Biology* in press.

Edwards, D. (1998) Climate signals in Palaeozoic plants. *Philosophical Transactions of the Royal Society of London, Series B* **353**, 141–157.

Erwin, D.H. (1998) The end and the beginning: recoveries from mass extinctions. *Trends in Ecology and Evolution* **13**, 344–349.

Frakes, L.A., Francis, J.E. & Syktus, J.I. (1992). *Climate Models of the Phanerozoic*. Cambridge University Press, Cambridge.

Gauslaa, Y. (1984) Heat resistance and energy budget in different Scandinavian plants. *Holarctic Ecology* **7**, 1–78.

Graham, A. (1999) *Late Cretaceous and Cenozoic History of North American Vegetation North of Mexico*. Oxford University Press, New York.

Grime, J.P. (1997) Biodiversity and ecosystem function: the debate deepens. *Science* **277**, 1260–1261.

Hayden, B.P. (1998) Ecosystem feedbacks on climate at the landscape scale. *Philosophical Transactions of the Royal Society of London, Series B* **353**, 5–18.

Heimann, M. (1997) A review of the contemporary global carbon cycle and as seen a century ago by Arrhenius and Högbom. *Ambio* **26**, 17–24.

Hobbie, S.E., Jensen, D.B. & Chapin, F.S. III (1993) Resource supply and disturbance as controls over present and future plant diversity. In: *Biodiversity and Ecosystem Function* (eds E.D. Schulze & H.A. Mooney), pp. 385–408. Springer-Verlag, Berlin.

Houghton, J.T., Callander, B.A. & Varney, S.K., eds (1992). *Climate Change 1992: the Supplementary Report to the IPCC Scientific Assessment*. Cambridge University Press, Cambridge.

Houghton, R.A., Davidson, E.A. & Woodwell, G.M. (1998) Missing sinks, feedbacks, and understanding the role of terrestrial ecosystems in the global carbon balance. *Global Biogeochemical Cycles* **12**, 25–34.

Kirchner, J.W. & Weil, A. (2000) Delayed biological recovery from extinctions throughout the fossil record. *Nature* **404**, 177–180.

Kothavala, Z., Oglesby, R.J. & Saltzman, B. (1999) Sensitivity of equilibrium surface temperature of CCM3 to systematic changes in atmospheric CO_2. *Geophysical Research Letters* **26**, 209–212.

Marzoli, A., Renne, P.R., Piccirillo, E.M., Ernesto, M., Bellieni, G. & De Min, A. (1999) Extensive 200-million-year-old continental flood basalts of the central Atlantic magmatic province. *Science* **284**, 616–618.

McElwain, J.C., Beerling, D.J. & Woodward, F.I. (1999) Fossil plants and global warming at the Triassic–Jurassic boundary. *Science* **285**, 1386–1390.

Mitchell, J.F.B., Johns, T.C., Gregory, J.M. & Tett, S.F.B. (1995) Climate response to increasing levels of greenhouse gases and sulphate aerosols. *Nature* **376**, 501–504.

Mora, C.I., Driese, S.G. & Colarusso, L.A. (1996) Middle to late Paleozoic atmospheric CO_2 levels from soil carbonate and organic matter. *Science* **271**, 1105–1107.

Myers, N., Mittermeler, R.A., Mittermeler, C.G., da Fonseca, G.A.B. & Kent, J. (2000) Biodiversity hotspots for conservation priorities. *Nature* **403**, 853–858.

New, M., Hulme, M. & Jones, P. (2000) Representing twentieth century space-time climate variability. II: Development of 1901–96 monthly grids of terrestrial surface climate. *Journal of Climate* **13**, 2217–2238.

Niklas, K.J. (1997). *The Evolutionary Biology of Plants*. University of Chicago Press, Chicago.

North, G.R., Cahalan, R.F. & Coakley, J.A. (1981) Energy balance climate models. *Review of Geophysics and Space Physics* **19**, 91–121.

Ramankutty, N. & Foley, J.A. (1999) Estimating historical changes in global land cover: croplands from 1700 to 1992. *Global Biogeochemical Cycles* **13**, 997–1027.

Retallack, G.J. (1997) Early forest soils and their role in Devonian global change. *Science* **276**, 583–585.

Robinson, J.M. (1990) Lignin, land plants, and fungi—biological evolution affecting Phanerozoic oxygen balance. *Geology* **18**, 607–610.

Sarmiento, J.L. & Lequere, C. (1996) Oceanic carbon dioxide uptake in a model of century-scale global warming. *Science* **274**, 1346–1350.

Shukla, J. & Mintz, Y. (1982) Influence of land-surface evapotranspiration on the Earth's climate. *Science* **215**, 1498–1501.

Turner, I.M. (1996) Species loss in fragments of tropical rain forests: a review of the evidence. *Journal of Applied Ecology* **33**, 200–209.

Volk, T. (1987) Feedbacks between weathering and atmospheric CO_2 over the last 100 million years. *American Journal of Science* **287**, 763–779.

Watson, A.J. & Liss P.S. (1998) Marine biological controls on climate via the carbon and sulphur geochemical cycles. *Philosophical Transactions of the Royal Society of London, Series B* **353**, 41–51.

Williams, C.B. (1964). *Patterns in the Balance of Nature*. Academic Press, London.

Woodward, F.I. (1987) Stomatal numbers are sensitive to increases in CO_2 from pre-industrial levels. *Nature* **327**, 617–618.

Woodward, F.I. (1998) Do plants really need stomata? *Journal of Experimental Botany* **49**, 471–480.

Woodward, F.I. & Bazzaz, F.A. (1988) The responses of stomatal density to CO_2 partial pressure. *Journal of Experimental Botany* **39**, 1771–1781.

Woodward, F.I. & Kelly, C.K. (1995) The influence of CO_2 concentration on stomatal density. *New Phytologist* **131**, 311–327.

Woodward, F.I., Lomas, M.R. & Betts, R.A. (1998) Vegetation-climate feedbacks in a greenhouse world. *Philosophical Transaction of the Royal Society of London, Series B* **353**, 29–38.

Part 4
Ecosystems, management and human impacts

Chapter 14
Lost linkages and lotic ecology: rediscovering small streams

J. L. Meyer* and J. B. Wallace*†

Introduction

The goal of these proceedings is to discuss the achievements of the past and the challenges facing us tomorrow. To explore this topic we will briefly examine the concepts that have shaped ecological research in flowing waters over the past quarter century and the opportunities for future advances in the field. We consider this time period because a seminal paper in stream ecology was published by Noel Hynes in 1975. The paper's title, 'The stream and its valley', conveys its message that organisms and processes in a stream ecosystem are shaped by the surrounding landscape. The linkage between a stream and its catchment was one of the earliest concepts in stream ecology (e.g. Ross 1963) and continues to engage stream ecologists today (e.g. Roth *et al*. 1996).

Advances in stream ecology over the past quarter century have occurred as ecologists explored the linkages that shape lotic ecosystems. Our conceptual models of flowing water ecosystems incorporate longitudinal (e.g. Vannote *et al*. 1980; Minshall *et al*. 1985), lateral (e.g. Junk *et al*. 1989) and subsurface (e.g. Boulton *et al*. 1998; Jones & Mulholland 2000) connections. The tragedy is that while ecologists have been carefully documenting the importance of these linkages, human society has been severing them by building dams and levees, cutting riparian forests, removing woody debris, mining groundwater, altering channel morphology and stream hydrology, and filling or piping entire channels. When these destructive practices are combined with anthropogenic alterations in water quality and spread of exotic species, it is not surprising that freshwater taxa are far more threatened than terrestrial taxa in North America (Ricciardi & Rasmussen 1999). Inadequate recognition of flowing waters as human-dominated ecosystems was a gap in much of 20th century stream research. Hence a challenge for ecology in the future lies in developing a predictive understanding of the connections between ecological conditions in flowing waters and the human attitudes, institutions and policies that dominate these ecosystems (Naiman *et al*. 1995). Linking ecological, ethical, economic and legal analyses will be essential to sustain the integrity of lotic

---

**Institute of Ecology, University of Georgia, Athens GA 30602, USA*
†*Department of Entomology, University of Georgic, Athens GA 30602, USA*

ecosystems (Meyer 1997). We explore these ideas here by focusing on small streams. The literature is much richer than can be presented in a short paper, so the citations are intended as introductions to an extensive body of research.

Headwater streams

Ecological studies in headwater streams have advanced our understanding of detrital food webs (Meyer 1994; Wallace et al. 1997), biogeochemistry (Meyer et al. 1988; Stream Solute Workshop 1990), linkages between ecosystems (Vannote et al. 1980; Boulton et al. 1998), the role of disturbance in structuring ecosystems (Resh et al. 1988), and ecosystem consequences of exotic species introductions (Huryn 1998).

Headwater streams in forested catchments are heterotrophic with food webs dependent on inputs of organic matter from surrounding ecosystems (Fisher & Likens 1973). In these complex detrital food webs (Fig. 14.1), fungi and bacteria are agents of leaf and wood decomposition (Harmon et al. 1986; Webster & Benfield 1986) as well as food resources for benthic invertebrates (Meyer 1994). Fungi develop higher biomass and have higher rates of productivity than bacteria on decaying leaves (Weyers & Suberkropp 1996); yet for many aquatic insects, ingestion of bacteria provides a significant source of carbon (Hall & Meyer 1998). Bacterial growth is supported by dissolved organic carbon (DOC) from the catchment and stream channel (Meyer et al. 1998), and bioavailability of DOC can be related to its chemical composition (Sun et al. 1997). Production of both primary and secondary consumers in headwater streams is tightly linked to the supply of leaf litter from the riparian forest and its retention in the channel, and invertebrate predators consume most of the secondary production (Wallace et al. 1997, 1999; Hall et al. 2000). Woody debris enhances organic matter and nutrient retention as well as providing a stable, stairstep profile that dissipates stream energy and reduces downcutting (Bilby & Likens 1980; Harmon et al. 1986; Wallace et al. 1995). Small forest streams are sites of input, storage, transformation and subsequent export of detritus to downstream reaches. They provide refugia for species that serve as colonists when downstream ecosystems are disturbed (Doppelt et al. 1993). Insectivorous birds may depend on emerging insects from small streams for a significant portion of their food (Gray 1993). These small streams are vulnerable to human alteration of the catchment, riparian zone, and channel.

Nutrient dynamics in streams can be described as a spiral that can be quantified (Newbold et al. 1981). Removal of nutrients from the water column increases with detrital standing stock, biological activity, and exchanges with the hyporheic zone, which is that part of the stream bed where there is exchange between ground- and surface water (Mulholland et al. 1985, 1997; Morrice et al. 1997). Intact riparian zones modulate nutrient inputs from catchments (Hedin et al. 1998). Both sources and spiralling of nutrients in headwaters have been altered by humans.

Headwater streams are tightly linked with the larger landscape and have served

LOST LINKAGES AND LOTIC ECOLOGY

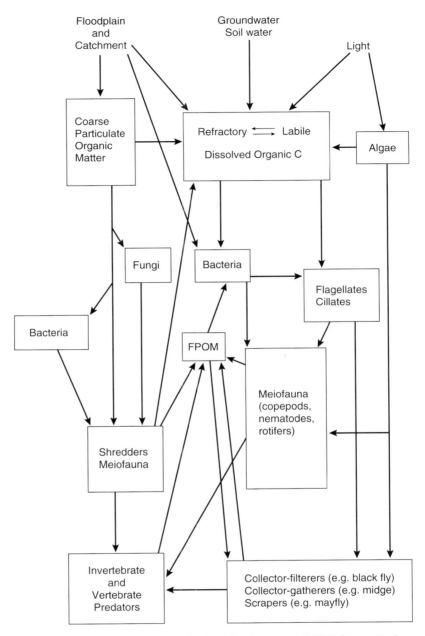

Figure 14.1 Simplified diagram of a food web in flowing waters. FPOM, fine particulate organic matter. (Reproduced with permission from Meyer 1994.)

as harbingers of ecosystem change resulting from acid deposition and increasing nitrogen deposition (Likens *et al.* 1977). Annual thermal regimes of small streams are potentially useful for detecting long-term global warming trends because groundwater entering small streams is usually within 1°C of the mean annual temperature of a region (Vannote & Sweeney 1980).

Headwater streams are a crucial part of the larger river network; hence, protection of headwater streams is the first line of defence for conservation and recovery of downstream ecosystems (Doppelt *et al.* 1993). A basic tenet of stream ecology is the river continuum concept, which views stream ecosystems as longitudinally changing gradients of hydrology, geomorphology and organic matter sources to which biological communities have evolved (Vannote *et al.* 1980). Subsequent conceptual developments emphasized lateral interactions with the flood plain (Junk *et al.* 1989) and with the hyporheic zone (Ward 1989), but upstream–downstream connections are dominant. Headwater streams are intimately linked with ecosystems downstream (e.g. Pringle 1997), and alteration of one part of the drainage network will have repercussions throughout the network. Inadequate attention has been paid to quantifying the effects of disturbances on various scales to entire basins and identifying direct and indirect mechanisms contributing to basin-wide degradation.

Human society and lotic ecosystems

Although studies in headwater streams have and will continue to advance ecological science, these ecosystems are threatened by human activities and are inadequately protected. It is essential that ecological science recognize human actions and institutions as forces structuring ecosystems just as geology and climate have long been recognized. As an example of human impact, consider earth-moving activity. Hooke (1999) estimates that rivers move 1 Gt (10^9 tons) of sediments annually in the conterminous United States whereas humans move 8.3 Gt annually. This is a conservative estimate of human earth-moving because it is based solely on housing construction, mining, agriculture and paved roads. The estimate ignores logging roads, river and harbour dredging, beach replenishment, and construction of large buildings. Even though this is an underestimate, the mass of earth moved by humans is over eight times the mass of sediment contributed to the oceans or interior basins by fluvial transport (Hooke 1999). The discrepancy between human and fluvial earth-moving activity is not evenly distributed across the landscape (Fig. 14.2). The largest peaks in fluvial earth-moving activity are west of the Mississippi, whereas the biggest peaks in human activity are east of the Mississippi, with mountain-top removal mining areas of central Appalachia providing particularly noticeable peaks (Hooke 1999). Humans are literally reshaping the landscape, and excess sediment is listed as the leading cause of impairment of streams in the United States (Waters 1995). Hence, a change in economic conditions or a shift in mining policy may have more impact on processes in a stream ecosystem than a flood, something more frequently considered by ecologists.

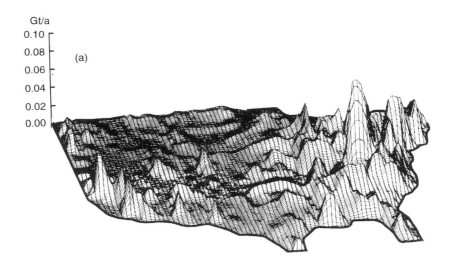

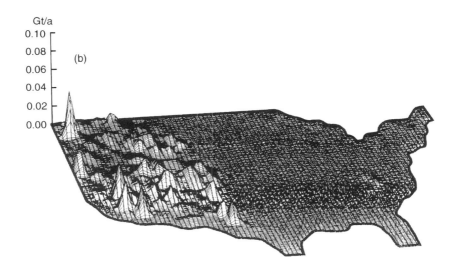

Figure 14.2 Rate of earth moved (a) by humans and (b) by rivers. Peak heights are proportional to 10^9 tons yr^{-1} moved in each 1° grid cell. (Reproduced with permission from Hooke 1999.)

Understanding the impact of 'the valley' on stream populations and processes requires us to consider both present and past landscapes. As ecologists we are trained to consider the evolutionary history of the species we study, but there has been less emphasis on the impact of human history on lotic ecosystems. Studies of historical legacies of human activity can alter our perceptions of ecosystem processes; for example, extirpation of beaver-altered stream ecosystems over much of the North American landscape long before there were any studies of flowing water ecosystems (e.g. Naiman *et al.* 1986). Similarly, the removal of large woody debris from rivers was one of the earliest activities of European immigrants to North America, and loss of snag habitat probably altered habitat availability and reduced secondary productivity in flowing waters long before there were ever scientific analyses of these phenomena (Sedell & Froggatt 1984). Legacies of past human abuse of the landscape live with us today in excessive sediment deposits from earlier agriculture that remains in the channel, altering composition of fishes and benthic invertebrates (Harding *et al.* 1998).

Humans impact lotic ecosystems by building dams, which eliminate and alter lotic ecosystems both up- and downstream of the reservoir (Pringle 1997; Rosenberg *et al.* 1997). Of the 5.2 million miles of US rivers (based on 1:500000 scale maps), only 2% of that length are free flowing for at least 200 km, and only 0.3% are protected as Wild and Scenic Rivers (Benke 1990). This is strikingly less than the 16.5% of public and private forests in the United States that are protected (Heinz Center 1999). Although dams and other federally licensed projects cannot be constructed on Wild and Scenic Rivers, the protection afforded by that designation does not extend to protection of their headwaters (Palmer 1993). Considerable opportunities for exciting ecological research are provided by the accelerating pace of dam removal in the United States as obsolete, unsafe, uneconomical or environmentally destructive dams are removed (Born *et al.* 1998). These offer opportunities for analyses of ecosystem recovery that extend upstream to evaluate impacts of restoring riverine network connectivity. Such studies could provide useful stimulus and guidance for removal of other dams.

Humans have greatly altered nutrient cycles in rivers and streams. Nitrate transport in rivers of the world is related to human population density in the catchment (Peierls *et al.* 1991), and nitrogen transport by temperate rivers is from two- to 20-fold higher now than in preindustrial times (Howarth *et al.* 1996). Much of the variation in phosphorus export in large rivers can be accounted for by human sewage inputs and fertilizer run-off (Caraco 1995). Increased nutrient transport in flowing waters is a consequence of both excess loading of nutrients and reduced capacity for removal of added nutrients. Ecology has made more progress in determining changes in loading than in developing a predictive understanding of nutrient removal in altered streams. Recent advances linking hydrodynamics with nutrient removal (e.g. Morrice *et al.* 1997) and application of stable isotope techniques (e.g. Wolheim *et al.* 1999; Mulholland *et al.* 2000) offer considerable promise for future progress, particularly if such studies are extended to heavily impacted streams.

In addition to excess nutrient loading, humans also add toxins such as heavy metals and synthetic organic compounds to rivers and streams. Most analyses of the consequences of those additions has been performed by toxicologists, and the regulations that govern discharge of these compounds reflect a reductionist approach. Considerable challenges for ecological research lie not only in exploring the synergistic effects of contaminants on populations and processes in streams (e.g. Clements 1999) but also in investigating the ecological impacts of newly discovered compounds (e.g. Committee on Hormonally Active Agents in the Environment 1999). Ecological science has made remarkable progress in the use of ecological indicators to detect impaired streams and rivers (e.g. Rosenberg & Resh 1993; Wright 1995; Karr & Chu 1999), and these methods are being incorporated into government monitoring programmes (Barbour et al. 1999). Abandoning exclusive reliance on chemical water quality indicators and incorporating ecologically meaningful measures of ecosystem impairment into regulatory decisions should enhance lotic ecosystem protection.

Cataloguing small streams

Human actions affect all sizes of rivers and streams (Naiman et al. 1995); here we explore human impact on small streams, which begins with a failure to recognize their extent. It is estimated that first- and second-order streams represent 95% of the stream channels and about 73% of the estimated stream channel length in the United States (Leopold et al. 1964). However, these estimates are conservative, since channels that are mapped on the ground have many more perennial streams than appear on standard 1:24 000 US Geological Survey maps (Leopold 1994). The catalogues of extent of rivers are based on maps drawn at a scale of 1:100 000 or 1:500 000, and analyses of length of free-flowing river (Benke 1990) or models of nutrient transport in channels of different size (Smith et al. 1997; Alexander et al. 2000) are based on networks drawn at this coarse scale; hence small streams are not included in these analyses.

A comparison of length of stream channel derived from different map scales in a well-studied southern Appalachian catchment illustrates the problem. The Coweeta Creek watershed is a 16.3 km^2 catchment comprising a large part of the Coweeta Hydrologic Laboratory in western North Carolina. Only 0.8 km of streams are indicated on a 1:500 000 scale map, 24.4 km are shown on a 1:24 000 scale map, and 56 km of stream are shown on a 1:7200 scale map (N. Gardiner, Institute of Ecology, University of Georgia, personal communication). Even the most detailed map available does not include all stream channels that can be found on the ground. Hence, at the very least, a map drawn to a 1:500 000 scale underestimates true stream channel length in this fifth-order basin by more than 70-fold. It is humbling to recognize that the 185 ecological research papers about streams that have been published in the past 25 years at Coweeta Hydrologic Laboratory have been done on stream channels that do not exist according to most national accountings of stream networks. This is not unique to the Coweeta basin: 1:24 000 scale maps identify only 21% of the stream channel length in the 728 km² Chattooga River basin (Hansen 2001).

The problem becomes more apparent when we recognize how these maps are drawn. Leopold (1994) notes that 'the blue lines on a map are drawn by nonprofessional, low-salaried personnel. In actual fact, they are drawn to fill a rather personalized aesthetic.' Headwater streams shown on a map meet no clearly defined statistical characteristic of the extent of streamflow (Leopold 1994).

Intermittent and ephemeral channels pose an even greater challenge for both representation and definition. Intermittent streams flow when the streambed is below the water table but not during dry periods. Ephemeral streams carry run-off from rainfall and flow only for short periods during and after storms. Some States have added their own twist to definitions. For example, West Virginia defines an intermittent stream as one that drains a watershed of at least one square mile; it also has a biological criterion for intermittent streams, which is that they do not support aquatic life whose life history requires residence in flowing waters for a continuous period of at least 6 months. Aquatic insect assemblages requiring 9–18 months of flow can be found in many channels represented by dotted blue lines on a topographic map. There is no ecologically sound, nationally consistent definition or accounting of intermittent and ephemeral stream channels. Society considers them of such little value that they are ignored. Although their inhabitants have been described (e.g. Williams 1996), quantification of the significance of intermittent and ephemeral streams in the landscape and of the goods and services they provide has not been explored.

Loss of headwater streams

Despite the ecological significance of small streams and the role that studies of them have played in developing stream ecological theory, they are largely overlooked and are being lost from the landscape at an alarming rate. The ecosystems that have formed a basis for our science are neither adequately recognized nor protected. Their abundance in the landscape may be a factor in their demise. Because they are small and numerous, they have been viewed as unimportant, insignificant or a general nuisance for agriculture, forestry, urban and residential development, and mining operations.

In the 19th and early 20th centuries, small streams were channellized to increase drainage and add farm acreage. For example, a striking decrease in drainage density in agricultural areas of Sweden can be seen on maps from 1820 compared with 1950 (Fig. 14.3). In this catchment, surface water area was reduced to 3.4% of its original extent. Similar practices were common over the same period in agricultural and urban areas of the United States and continued throughout the 20th century.

Small streams are also impacted by forestry operations. In areas with little topographic relief, extensive channellization and diking as seen in Fig. 14.3 is a common silvicultural practice. In all areas, removal of vegetation from a catchment alters

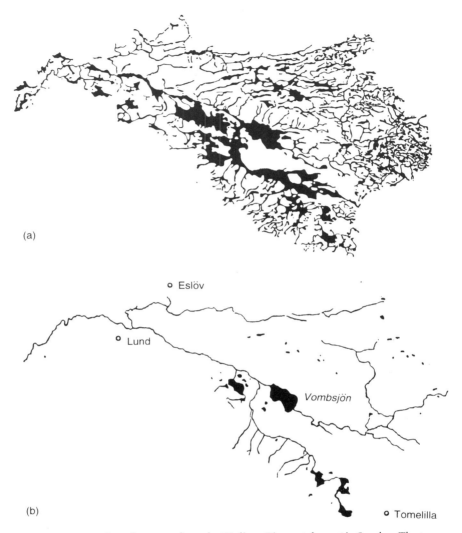

Figure 14.3 Loss of small streams from the Kävlinge River catchment in Sweden. The top panel (a) is a map from 1812 to 1820, whereas the bottom panel (b) shows the same area in 1950–53 after extensive diking and channellization. (Reproduced with permission from Wolf 1956.)

stream hydrology and increases nutrient export (e.g. Swank 1988) with marked effects on stream biota and ecosystem processes (e.g. Webster *et al.* 1992). Removal of forests leads to narrowing and downcutting of stream channels and consequent loss of headwater habitat area, for example first- and second-order stream channels in forests in eastern Pennsylvania are 2.5 times wider than those in grasslands (Sweeney 1993). Increased solar radiation reaching the stream channel after forest removal increases primary productivity, alters food-web structure, and increases stream temperatures (Webster *et al.* 1992). Small changes in temperature alter life-history characteristics of stream insects including adult size and fecundity (Sweeney 1993) and can provide growth advantages to one species over another (Edington & Hildrew 1973). One consequence of increased ecological understanding of the impacts of forestry practices on streams has been stricter regulations on streamside management zones, which require retention of forested riparian buffers along stream channels. In addition to problems of enforcement of buffer regulations, little protection is provided for intermittent or ephemeral stream channels, and little is known of the downstream impact of alteration of these channels during forestry operations. Small headwater streams are easily obstructed by woody debris, and these debris dams are important sites of organic matter and nutrient retention (Bilby & Likens 1980; Wallace *et al.* 1995). Deforestation results in a prolonged reduction in inputs of large woody debris, and until the successional forest matures, overall stream retentiveness is reduced (Hedin *et al.* 1988; Webster *et al.* 1992). Although the direct effects of forestry practices are on adjacent headwater streams, their impacts extend to downstream ecosystems.

When watersheds are developed for residential, commercial and industrial use, headwater streams are often filled and piped resulting in a decrease in drainage density. A classic example of the loss of headwaters with urbanization is Rock Creek in Maryland. During a period of rapid urbanization in Washington DC, 59.5 km of stream channel were eliminated in the catchment (Department of Interior 1968). The length of stream per unit area of catchment is termed drainage density, and in 1913 the Rock Creek basin had a drainage density of $1.53 \, \text{km} \, \text{km}^{-2}$. In 1966 its drainage density was $0.64 \, \text{km} \, \text{km}^{-2}$ (Department of Interior 1968). Drainage density was reduced by 58% as a consequence of urban development.

Urbanization in metropolitan Atlanta has also resulted in a reduction in drainage density in Upper Chattachoochee River catchments. Drainage density in four primarily forested and agricultural basins was $1.35 \pm 0.09 \, \text{km} \, \text{km}^{-2}$ (mean ± SE) whereas drainage density in six urban and suburban catchments was $0.91 \pm 0.03 \, \text{km} \, \text{km}^{-2}$ (E.A. Kramer, Natural Resources Spatial Analysis Laboratory, University of Georgia, unpublished data). One-third of the stream length has been lost in these catchments, primarily from small headwater streams, which are most vulnerable to filling and piping. Although this estimate of stream loss is alarmingly large, it is undoubtedly an underestimate because it is based on 1:24000 maps, which do not adequately display the smallest streams.

Road density has been increasing as drainage density has been decreasing in these urbanizing catchments around metropolitan Atlanta. In 14 catchments in the Upper Chattahoochee River basin, length of stream channel is negatively related to length of roads (Fig. 14.4, $r^2 = 0.68$, $P < 0.0003$). On average, 13.7 km of road have been added for every kilometre of headwater stream that has been lost because of the construction made possible by roads. Networks of headwater stream channels slow the flow of water and are also highly retentive of nutrients and organic matter. These retentive networks have been lost, and surface run-off now flows along

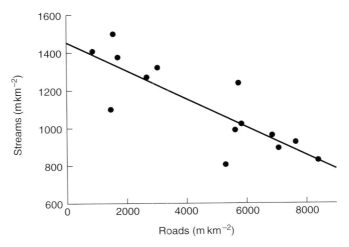

Figure 14.4 Relationship between length of streams and length of roads in catchments in the Upper Chattahoochee River basin near Atlanta, Georgia, USA. Land use in the catchments ranges from primarily forested to primarily residential and commercial. The regression line is: $Y = 1456 - 0.073X$, $r^2 = 0.68$, $P = 0.0003$, where Y is stream length (m km^{-2}) and X is road length (m km^{-2}). Data are from 1:24 000 maps interpreted by E. A. Kramer, Natural Resources Spatial Analysis Laboratory, University of Georgia, Athens, Georgia.

networks of impervious roadways with no capacity for retention of water, nutrients or organic matter. Small streams serve as capillaries of the drainage network, and their interaction with the surrounding landscape has been drastically altered.

Groundwater withdrawal for irrigation and other human uses has resulted in significant lowering of the water table in many regions of the world (Postel 1999). Overpumping of groundwater can profoundly affect headwater streams by making perennial streams ephemeral, and channels without water can extend far downstream. For example, a channel of the Santa Cruz River near Tucson, Arizona, was dry for several decades because of groundwater pumping (Grimm *et al.* 1997).

Mining operations are also responsible for the elimination of headwater streams. In the central Appalachian region in West Virginia, south-western Virginia, eastern Kentucky, and eastern Tennessee, coal is mined by a method known as mountain-top removal and valley fill (MTR/VF). This method of mining fills streams with overburden covering low sulphur coal seams (Fig. 14.5). From 1986 through 1998, MTR/VF coal mining has buried nearly 1450 km of streams in central Appalachia (USFWS 1998). In West Virginia alone over 750 km of streams have been buried (USFWS 1998). This estimate is very conservative since many kilometres of streams have gone uncounted because streams lengths were measured on 1:24 000 scale maps. MTR/VF is profitable for the coal mining industry but devastating for stream networks. Headwater streams that are entombed with 120 m of overburden are completely destroyed, and downstream reaches often

Figure 14.5 Mountain-top removal/valley fill mining operations in Kentucky. The upper photograph shows a filled valley; note the trucks on the road at the top of the fill for a sense of scale. The middle photograph shows excessive sedimentation in a stream below a mountain-top removal/valley fill mining operation. The lower photograph is a close-up of the sediment-choked stream. (Photographs by J. B. Wallace.)

suffer excess sedimentation (Fig. 14.5). Despite the prevalence of MTR/VF and its impact on streams, little ecological research has been done to document the impacts of MTR/VF on downstream ecosystems.

The observed reductions in drainage density resulting from human activities are largely a consequence of loss of small streams. Why are they not adequately protected? The US Army Corps of Engineers has been given the authority to regulate filling and piping of streams; no special permission is currently required for filling in or piping stream lengths up to 150 m because these activities are covered under a general nation-wide permit. Nation-wide permits are only supposed to be used for activities with no cumulative effects. In this case, there does seem to be a cumulative effect, namely the reduction in drainage density that is a common feature of urbanized catchments. Yet the nation-wide permit is still applied, although it is currently being revised. The Corps also has some regulatory authority relating to mining. In addition to filing for a permit under the Surface Mining Control and Reclamation Act, mining companies must receive a permit from the Corps for discharge of wastes (i.e. overburden) into streams. Such permits were perfunctory under a Corps nation-wide permit program. A recent judgement in a federal lawsuit has ruled that valley fills are not legal in intermittent or perennial streams, and the parties in the suit are trying to reach a settlement. In addition, the Corps nation-wide permit system is undergoing revision. Insights from ecological science on the significance of small streams in a drainage network are relevant to these deliberations.

Impacts of loss of small streams on ecosystem processes and stream biota

Elimination of small streams in a drainage network has numerous ecological consequences (Table 14.1). As small headwater channels are lost, flood frequency in the basin increases. The hydrograph is 'flashier' with the stream equalling or exceeding bankfull at 10–20 times its previous frequency (Dunne & Leopold 1978). This trend is exaggerated when stream loss is combined with increased impervious surfaces, which deliver water from the basin to downstream channels much more rapidly than intact headwater streams. For example, in Watts Branch, a stream draining an urbanizing watershed in Maryland, the number of floods increased from two to seven per year, and the magnitude of the average annual flood increased by 23% over three decades of accelerating residential development (Leopold 1994). Increased flood frequency increases bank erosion, channel widening and incision, and other changes in channel form (Arnold *et al.* 1982). In San Diego, California, extensive channel erosion contributed two-thirds of the in-stream sediment load and resulted in loss of valuable urban land (Trimble 1997). An increase in flood frequency and magnitude negatively impacts the stream biota, particularly when this is combined with increasing sediment transport (e.g. Waters 1995).

Elimination of headwater streams impacts ecosystem processes because small

Table 14.1 Ecological consequences of the alterations caused by loss of headwater streams.

Alteration	Consequence
Loss of hydrologic retention capacity	Increased frequency and intensity of flooding downstream and lower base flows
Increased downstream channel erosion	Increased sediment transport and reduced habitat quality
Reduced retention of sediments	Excess sedimentation downstream
Reduced retention and transformation of nutrients and contaminants	Increased nutrient and contaminant loading to downstream ecosystems
Reduced retention and mineralization of organic matter	Increased loading downstream
Reduced processing of allochthonous inputs	Reduced supply of fine particulate organic matter to downstream food webs
Reduced secondary production in headwaters	Less drift supplied to food webs downstream and less emergence production subsidizing riparian food webs
Loss of unique habitats	Increased extinction vulnerability of aquatic species (invertebrates, amphibians, fishes)
Altered thermal regime	Altered growth and reproduction in aquatic insects and fishes
Loss of thermal refuges and nursery areas	Increased mortality of fishes

streams are effective in mineralization of organic matter and in nutrient transformation and retention. Organic carbon turnover length (S) is a measure of the efficiency with which the stream mineralizes organic carbon and can be thought of as the average distance travelled by an organic carbon molecule from the time it enters the stream until it is respired to CO_2 (Newbold *et al.* 1982). Webster & Meyer (1997) compiled data on organic matter budgets for streams around the world and used these data to calculate S in 25 streams. S (km) is computed as organic matter in transport per unit stream width/respiration. Using values of S from Webster and Meyer (1997), we determined the following relationship between turnover length (S) and stream order (OR):

$$\log(S) = 0.008 + 0.496 \, OR \; (r^2 = 0.64, P < 0.0001)$$

We can then use this relationship to explore the impact of elimination of low-order streams on average turnover length in a basin. Using the length of stream in different orders at Coweeta Hydrologic Laboratory (Wallace 1988) and the above equation relating turnover length to stream order, the average turnover length in this fifth-order basin is 16 km. If a third of the stream length in the basin were eliminated, as has been observed in urban Atlanta catchments, and if all the streams eliminated were first-order streams, turnover length would increase to 22 km. If all first-order streams were eliminated from the basin, average turnover length would double to 33 km. Hence, elimination of headwater streams reduces the efficiency with which inputs of organic carbon are mineralized in the ecosystem. Mineralization of organic matter is an ecosystem service performed by flowing waters and valued by humans for wastewater purification. Reducing the length of headwater streams in the network significantly reduces the capacity of the ecosystem to provide this service.

Nutrients are removed from stream water by biotic and abiotic processes, and the average distance travelled by a molecule before being removed from the water column is called its uptake length (Newbold *et al.* 1981). Uptake lengths have been measured for ammonium, nitrate and phosphate in many streams using experimental additions of the nutrient and a conservative tracer (Stream Solute Workshop 1990). The shortest measures of uptake length are from small headwater streams, and as stream size (and discharge) increases, so does nutrient uptake length (Stream Solute Workshop 1990). Uptake length for both phosphorus and ammonium are less than 20 m in headwater streams at Coweeta Hydrologic Laboratory (e.g. Webster *et al.*, in press). Thus, an average nutrient molecule travels less than 20 m downstream before being removed from the water column in a small shallow stream, where there is extensive contact between the water column and benthic algae and microbes in surface sediments and the hyporheic zone. Some of the nitrogen removed from the water is ultimately lost via denitrification. This capacity for nutrient retention and transformation reduces the loading of nutrients to downstream ecosystems. When headwater streams are eliminated, floodwaters are delivered more rapidly, and more of the nutrients being applied to farm fields or lawns are delivered to receiving systems downstream. Therefore, as drainage density decreases because of loss of headwater streams, average nutrient uptake length in the basin increases, thereby reducing the ability of the stream network to provide a valued ecosystem service.

A simple model of soluble reactive phosphorus (SRP) removal from the water column based on numerous experiments done at Coweeta Hydrologic Laboratory illustrates the consequences of loss of headwater streams on nutrient export. As water moves downstream, SRP is added via subsurface water inputs from the surrounding catchment and removed by biotic and abiotic processes. This model ignores SRP input from leaching of materials stored in the streambed. Data from an instrumented hillslope at Coweeta (Yeakley *et al.* 1994) can be used to provide a reasonable estimate of phosphorus inputs via subsurface water of $0.1\,\text{kg P ha}^{-1}\,\text{yr}^{-1}$ or $11\,\text{mg P (m stream channel)}^{-1}\,\text{d}^{-1}$. Based on measurements of SRP removal from the water column, we use an uptake rate of $9\,\text{mg P m}^{-2}\,\text{d}^{-1}$ and a width of 0.5 m for first-order streams (Webster *et al.*, in press), and $13\,\text{mg P m}^{-2}\,\text{d}^{-1}$ and 1 m width for second-order streams (Mulholland *et al.* 1997). The relative length of stream in first- and second-order channels is from Wallace (1988). The model moves SRP downstream and the appropriate amounts of SRP are added and removed for each metre travelled. When the model is run with first-order streams intact, 63% of added phosphorus is retained in the stream network. When first-order streams are replaced with pipes (i.e. no SRP removal), only 34% of added SRP is retained, and the total amount of phosphorus exported increases 179%. This model illustrates that removal of retentive headwater streams can have consequences for nutrient transport to downstream ecosystems.

A recent evaluation of the sources of nitrogen being supplied to the Gulf of Mexico concluded that sources closest to large rivers were the greatest contributors to nitrogen even if they were at a great distance from the Gulf (Alexander et al. 2000). This result is a consequence of the relatively slower rate of nitrogen removal from larger, deeper rivers, and the relative efficiency with which nitrogen is removed from smaller streams. The analysis is based on a drainage network derived from 1 : 500 000 scale maps (Smith et al. 1997; Alexander et al. 2000), which do not include most first- to third-order streams. The model is calibrated against nutrient transport data from a series of large river stations, so the nitrogen removal that is occurring in small streams is subsumed in the model parameters. Although not explicitly considering small streams in the calculations probably does not alter the conclusions reached in this paper because of the way the model was calibrated, if total nutrient removal were apportioned to streams of different sizes, the smallest streams would be responsible for most of the nutrient removal. In fact, drainage density is an important predictor in the model from which these calculations were derived; as drainage density decreases (e.g. smaller streams are eliminated), nutrient transport increases (Smith et al. 1997).

In addition to impacting nutrient and organic matter transport, destruction of headwater streams represents a potential threat to biodiversity. Many threatened insect species are located in small headwater streams or springbrooks (Morse et al. 1997). The southern Appalachians have outstanding aquatic diversity, but 19 mayfly species, seven dragonfly species, 17 stonefly species and 38 caddisfly species are reported as vulnerable to extirpation at present (Morse et al. 1993, 1997). This may be an underestimation of the numbers of threatened and endangered species because many of the rare species are known from only one or two locations in springbrooks or seepage areas. This problem is not unique to the southern Appalachians; many of the candidates for threatened and endangered caddis flies in California are found in clear, cold, rapidly flowing water or in small spring streams (Erman & Nagano 1992). In western Oregon, the number of invertebrate taxa in intermittent streams exceeded that of permanent headwaters, and several undescribed species were associated with intermittent streams (Dieterich & Anderson 2000). Although insects are generally considered to have strong dispersal abilities because of their aerial adult phase, genetic analyses of peltoperlid stone flies in small streams of the northern Rocky Mountains demonstrated genetic isolation even among streams that are close geographically (Hughes et al. 1999). The absence of significant gene flow between such isolated populations indicates that recolonization after major disturbances may be limited and that these populations in small streams are vulnerable to elimination for extended periods. This vulnerability combined with the very limited range of many of these species increases their risk of extinction.

Estimates of the numbers of threatened and endangered invertebrate species are very conservative because of insufficient knowledge of the fauna, inadequate numbers of practising taxonomists able to identify new species, and absence of recent comprehensive surveys (Morse et al. 1997; Strayer 2000). Correcting these shortcomings offers a challenge for ecology in the future. It is likely that there are many new species and unrecognized ecological relationships in small streams, especially those tightly linked with groundwater (Strayer 2000). These are the very ecosystems that are being lost.

Headwater streams are also critical habitats for amphibians. Lungless plethodontid salamanders are believed to have originated and diversified in eastern North America, especially Appalachia (Beachy & Bruce 1992). Their lungless condition appears to be an adaptation for their principle larval habitat, small low-order streams, where they spend from a few months to 5 years (Beachy & Bruce 1992).

Headwater streams provide essential breeding habitat for some fish species. For example, the slackwater darter *Etheostoma boschungi* breeds in tiny streams, many of which are now small ditches flowing through pastures (Mettee *et al.* 1996). To the casual observer, a small ditch in a pasture hardly seems worth saving. That attitude is likely in part responsible for the fact that the slackwater darter is a federally listed endangered species. Another species dependent on small streams is the trispot darter *Etheostoma trisella*, which attaches its eggs to submerged blades of grass in tiny rivulets that flow from ephemeral ponds in fields (Ryon 1986). Headwater streams provide habitat for several endangered fish species in the south-eastern United States. Etnier (1997) identified 16 fish taxa occurring in first- and second-order south-eastern streams, and a quarter of the species are jeopardized because of non-point source pollution or extremely limited range. It is likely that this is an underestimate of imperilment since knowledge of species distributions are based on limited data collected two decades ago, and the human population in the region has grown rapidly since then (Burkhead & Jelks 2000). In addition, new species are being discovered and described that have very restricted ranges, diminishing their chance of long-term survival (Burkhead & Jelks 2000).

Small streams serve as nursery habitat or as a thermal refuge for species that spend most of their lives in larger systems. For example young-of-the-year brook trout *Salvelinus fontinalis* that were spawned in a lake migrated into small, groundwater-fed, inlet streams and spent the summer there (Curry *et al.* 1997). Stream temperatures, sustained by groundwater, were consistently cooler than in the littoral zone of the lake during summer. Groundwater is often warmer than stream water during winter, so small spring-fed streams provide a refuge from freezing for stream fishes (Power *et al.* 1999). Given the climatic extremes of continental North America, access to thermal refuges such as those provided by small spring-fed streams is an important aspect of survival for stream fishes (Power *et al.* 1999). Hence, isolating portions of the drainage and restricting access to critical temperature refuges could lead to extirpation or extinction (Power *et al.* 1999).

Challenges for the future

While ecologists have been carefully documenting the importance of longitudinal, lateral and hyporheic connectivity, human institutions and practices have been severing those connections. Hence, the greatest challenge for the future lies in more effectively linking ecology with humans and their institutions. What does it mean to incorporate human actions into our research? At the very least it means understanding the impacts of changing human institutions, policies and values on the ecosystems we study. This obviously requires interdisciplinary collaboration. In his seminal paper, Hynes (1975) noted '... one could say that some of our most important recent discoveries have been of the existence of hydrologists, foresters and soil scientists; which perhaps says something of our innocence'. The same can be said in the year 2001 about the interactions of ecologists with economists, anthropologists, political scientists, policy analysts and lawyers. Strengthening these interactions will enable ecologists to more effectively understand the impacts of human actions and institutions on ecological systems and the ways in which ecological systems influence human actions and institutions.

The ecological impacts on small streams that we have been discussing represent the cumulative impact of individual decision making. Ecological science has contributed to development of government policies for natural resource management; however, ecological science has had less impact on individual decision makers (e.g. developers, property owners) whose cumulative impact on stream networks has been significant. Understanding the individual decision-making process and its institutional context is fundamental to the disciplines of anthropology, economics and political science. Expanding collaborations with these disciplines will be fruitful.

Systems that are most vulnerable to the cumulative effects of seemingly insignificant individual decisions are those that are small and temporary. Those are the ecosystems that often have little legal protection. But they are also the ecosystems that can be protected by individual property owners. Ecological science can contribute by providing ecologically meaningful and legally enforceable definitions of these ecosystems, by cataloguing their extent and condition, and by quantifying the contribution of the seemingly insignificant small streams to the larger riverine ecosystem. The public cares about the stream in their back yard and the river in their town. To capitalize on this interest, the challenge for ecologists is to quantify the ecological goods and services provided by these ecosystems and valued by the public (Wilson & Carpenter 1999).

Over the past decades, ecology has assumed a global perspective, and we have made significant contributions to understanding the environmental issues that face our species. But not all ecologists are interested in or capable of working on a global scale, and the ecological challenges of the future are not all at that scale. Ecology is both local and global, and our science will prosper to the extent we continue to pursue questions and solutions on all relevant scales.

It is particularly crucial that lotic ecologists work at multiple scales. Rivers have been analysed as linear ecosystems with unidirectional flow, and hence ecologists have studied a longitudinal continuum composed of one stream in each order. Variability among streams of similar size has not been adequately sampled. The fact that a river is a network has been largely overlooked by lotic ecologists and is a fruitful direction for future research (Fisher 1997). The role of network structure or position in the network on population and ecosystem processes has not been adequately explored (Fisher 1997). The impact of multiple streams of different sizes on downstream processes has not been quantified. As small streams are lost from the network, and as other aspects of network structure and connectivity are changed by human actions, conducting lotic research with a network perspective is essential if we are to be able to quantify the ecosystem goods and services provided by a complex network of diverse streams and rivers.

Acknowledgements

This research has been supported by grants DEB-9632854 and DEB-9629268 from NSF and R82477-01-0 from EPA. B. Malmquist, C. Dahm and our graduate students provided helpful comments on this manuscript.

References

Alexander, R.B., Smith, R.A. & Schwarz, G.E. (2000) Effect of stream channel size on the delivery of nitrogen to the Gulf of Mexico. *Nature* **403**, 758–761.

Arnold, C.L., Boison, P.J. & Patton, P.C. (1982) Sawmill Brook—An example of rapid geomorphic change related to urbanization. *Journal of Geology* **90**, 155–166.

Barbour, M.T., Gerritsen, J., Snyder, B.D. & Stribling, J.B. (1999) Rapid bioassessment protocols for use in streams and wadeable rivers. *Periphyton, Benthic Macroinvertebrates and Fish*, 2nd edn. US Environmental Protection Agency; Office of Water, Washington DC.

Beachy, C.K. & Bruce, R.C. (1992) Lunglessness in plethodontid salamanders is consistent with the hypothesis of a mountain stream origin: a response to Ruben and Boucot. *American Naturalist* **139**, 839–847.

Benke, A.C. (1990) A perspective on America's vanishing streams. *Journal of the North American Benthological Society* **9**, 77–88.

Bilby, R.E. & Likens, G.E. (1980) Importance of organic debris dams in the structure and function of stream ecosystems. *Ecology* **61**, 1107–1113.

Born, S.M., Genskow, K.D., Filbert, T.L., Hernandez-Mora, N., Keefer, M.L. & White, D.A. (1998) Socioeconomic and institutional dimensions of dam removals: the Wisconsin experience. *Environmental Management* **22**, 359–370.

Boulton, A.J., Findlay, S., Marmonier, P., Stanley, E.H. & Valett, H.M. (1998) The functional significance of the hyporheic zone in streams and rivers. *Annual Review of Ecology and Systematics* **29**, 59–82.

Burkhead, N.M. & Jelks, H.L. (2000) Diversity, levels of impairment, and cryptic fishes in the southeastern United States. In: *Freshwater Ecoregions of North America: a Conservation Assessment* (eds R.A. Abell, D.M. Olson, E. Dinerstein et al.), pp. 30–32. Island Press, Washington DC.

Caraco, N.F. (1995) Influence of human populations on P transfers to aquatic systems: A regional scale study using large rivers. In: *Phosphorus in the Global Environment* (ed. H. Tiessen), pp. 235–244. John Wiley & Sons, New York.

Clements, W. (1999) Metal tolerance and predator–prey interactions in benthic macroinvertebrate stream communities. *Ecological Applications* **9**, 1073–1084.

Committee on Hormonally Active Agents in the Environment (1999). *Hormonally Active Agents in the Environment*. National Academy Press, Washington DC.

Curry, R.A., Brady, C., Noakes, D.L.G. & Danzmann, R.G. (1997) Use of small streams by young brook trout spawned in a lake. *Transactions of the American Fisheries Society* **126**, 77–83.

Department of Interior (1968). *The Nation's River*.

The Department of Interior Official Report on the Potomac. US Government Printing Office, Washington DC.

Dieterich, M. & Anderson, N.H. (2000) The invertebrate fauna of summer-dry streams in western Oregon. *Archiv für Hydrobiologie* **147**, 273–295.

Doppelt, B., Scurlock, M., Frissell, C. & Karr, J. (1993) *Entering the Watershed.* Island Press, Washington DC.

Dunne, T. & Leopold, L.B. (1978) *Water in Environmental Planning.* W.H. Freeman, New York.

Edington, J.M. & Hildrew, A.H. (1973) Experimental observations relating to the distribution of net-spinning Trichoptera in a stream. *Proceedings of the International Association of Theoretical and Applied Limnology* **18**, 1549–1558.

Erman, N.A. & Nagano, C.D. (1992) A review of the California caddisflies (Trichoptera) listed as candidate species on the 1989 Federal 'Endangered and threatened wildlife and plants—animal notice of review'. *California Fish and Game* **78**, 45–56.

Etnier, D.A. (1997) Jeopardized southeastern freshwater fishes: a search for causes. In: *Aquatic Fauna in Peril: the Southeastern Perspective* (eds G.W. Benz & D.E. Collins) pp. 87–104. Special Publication 1, Southeastern Aquatic Research Institute. Lenz Design and Communications, Decatur, Georgia.

Fisher, S.G. (1997) Creativity, idea generation, and the functional morphology of streams. *Journal of the North American Benthological Society* **16**, 305–318.

Fisher, S.G. & Likens, G.E. (1973) Energy flow in Bear Brook, New Hampshire: an integrative approach to stream ecosystem metabolism. *Ecological Monographs* **43**, 421–439.

Gray, L.J. (1993) Response of insectivorous birds to emerging aquatic insects in riparian habitats in a tallgrass prairie stream. *American Midland Naturalist* **129**, 288–300.

Grimm, N.B., Chacon, A., Dahm, C.N. et al. (1997) Sensitivity of aquatic ecosystems to climatic and anthropogenic changes: the basin and range, American Southwest and Mexico. *Hydrological Processes* **11**, 1023–1041.

Hall, R.O. Jr & Meyer, J.L. (1998) The trophic significance of bacteria in a detritus-based stream food web. *Ecology* **79**, 1995–2012.

Hall, R.O. Jr, Wallace, J.B. & Eggert, S.L. (2000) Organic matter flow in stream food webs with reduced detrital resource base. *Ecology* **81**, 3445–3463.

Hansen, W.F. (2001) Identifying stream types and management implications. *Forest Ecology and Management* **143**, 39–46.

Harding, J.S., Benfield, E.F., Bolstad, P.V., Helfman, G.S. & Jones, E.B.D. (1998) Stream biodiversity: The ghost of land use past. *Proceedings of the National Academy of Sciences of the USA* **95**, 14843–14847.

Harmon, M.E., Franklin, J.F., Swanson, F.J. et al. (1986) Ecology of coarse woody debris in temperate ecosystems. *Advances in Ecological Research* **15**, 133–302.

Hedin, L.O., Mayer, M.S. & Likens, G.E. (1988) The effect of deforestation on organic debris dams. *Proceedings of the International Association of Theoretical and Applied Limnology* **23**, 1135–1141.

Hedin, L.O., Von Fischer, J.C., Ostrum, N.E., Kennedy, B.P., Brown, M.G. & Robertson, G.P. (1998) Thermodynamic constraints on nitrogen transformations and other biogeochemical processes at soil–stream interfaces. *Ecology* **79**, 684–703.

Heinz Center. (1999) *Designing a Report on the State of the Nation's Ecosystems: Selected Measurements for Croplands, Forests and Coasts & Oceans.* The H. John Heinz Center, Washington DC.

Hooke, R.L. (1999) Spatial distribution of human geomorphic activity in the United States: Comparison with rivers. *Earth Surface Processes and Landforms* **24**, 687–692.

Howarth, R.W., Billen, G., Swaney, D. et al. (1996) Regional nitrogen budgets and riverine N & P fluxes for the drainages to the North Atlantic Ocean: natural and human influences. *Biogeochemistry* **35**, 75–139.

Hughes, J.M., Mather, P.B., Sheldon, A.L. & Allendorf, F.W. (1999) Genetic structure of the stonefly, *Yoraperla brevis*, populations: the extent of gene flow among adjacent montane streams. *Freshwater Biology* **41**, 63–72.

Huryn, A. (1998) Ecosystem-level evidence for top-down and bottom-up control of production in a grassland stream. *Oecologia* **115**, 173–183.

Hynes, H.B.N. (1975) The stream and its valley. *Proceedings of the International Association of Theoretical and Applied Limnology* **19**, 1–16.

Jones, J.A. & Mulholland, P.J. (2000) *Streams and Ground Waters*. Academic Press, San Diego.

Junk, W.J., Bayley, P.B. & Sparks, R.E. (1989) The flood pulse concept in river-floodplain systems. In: *Proceedings of the International Large River Symposium* (ed. D.P. Dodge), p. 110–127. Canadian Special Publications in Fisheries and Aquatic Sciences 106. Fisheries and Oceans. Government of Canada. Ottawa, Canada.

Karr, J.R. & Chu, E.W. (1999) *Restoring Life in Running Waters. Better Biological Monitoring*. Island Press, Washington DC.

Leopold, L.B. (1994). *A View of the River*. Harvard University Press, Cambridge, MA.

Leopold, L.B., Wolman, M.G. & Miller, J.P. (1964) *Fluvial Processes in Geomorphology*. W.H. Freeman, San Francisco.

Likens, G.E., Bormann, F.H., Pierce, R.S., Eaton, J.S. & Johnson, N.M. (1977) *Biogeochemistry of a Forested Ecosystem*. Springer-Verlag, New York.

Mettee, M.F., O'Neil, P.E. & Pierson, J.M. (1996) *Fishes of Alabama*. Oxmoor House, Birmingham, Alabama.

Meyer, J.L. (1994) The microbial loop in flowing waters. *Microbial Ecology* **28**, 195–199.

Meyer, J.L. (1997) Stream health: incorporating the human dimension to advance stream ecology. *Journal of the North American Benthological Society* **16**, 439–447.

Meyer, J.L., McDowell, W.H., Bott, T.L. *et al.* (1988) Elemental dynamics in streams. *Journal of the North American Benthological Society* **7**, 410–432.

Meyer, J.L., Wallace, J.B. & Eggert, S.L. (1998) Leaf litter as a source of dissolved organic carbon in streams. *Ecosystems* **1**, 240–249.

Minshall, G.W., Cummins, K.W., Peterson, R.C. *et al.* (1985) Developments in stream ecosystem theory. *Canadian Journal of Fisheries and Aquatic Sciences* **42**, 1045–1055.

Morrice, J.A., Valett, H.M., Dahm, C.N. & Campana, M.E. (1997) Alluvial characteristics, groundwater–surface water exchange and hydrologic retention in headwater streams. *Hydrological Processes* **11**, 253–267.

Morse, J.C., Stark, B.P. & McCafferty, W.P. (1993) Southern Appalachian streams at risk: Implications for mayflies, stoneflies, caddisflies, and other aquatic biota. *Aquatic Conservation: Marine and Freshwater Ecosystems* **3**, 293–303.

Morse, J.C., Stark, B.P., McCafferty, W.P. & Tennessen, K.J. (1997) Southern Appalachian and other southeastern streams at risk: implications for mayflies, dragonflies and damselflies, stoneflies and caddisflies. In: *Aquatic Fauna in Peril: the Southeastern Perspective* (eds G.W. Benz & D.E. Collins), pp. 17–42. Special Publication 1, Southeastern Aquatic Research Institute. Lenz Design and Communications, Decatur, Georgia.

Mulholland, P.J., Newbold, J.D., Elwood, J.W. & Webster, J.R. (1985) Phosphorus spiraling in a woodland stream: seasonal variations. *Ecology* **66**, 1012–1023.

Mulholland, P.J., Marzolf, E.R., Webster, J.R. & Hart, D.R. (1997) Evidence that hyporheic zones increase heterotrophic metabolism and phosphorus uptake in forest streams. *Limnology and Oceanography* **42**, 443–451.

Mulholland, P.J., Tank, J.L., Sanzone, D.M. *et al.* (2000) Nitrogen cycling in a forest stream determined by a ^{15}N tracer addition. *Ecological Monographs* **70**, 471–493.

Naiman, R.J., Magnuson, J.J., McKnight, D.M. & Stanford, J.A. (1995) *The Freshwater Imperative*. Island Press, Washington DC.

Naiman, R.J., Melillo, J.M. & Hobbie, J.E. (1986) Ecosystem alteration of boreal forest streams by beaver (*Castor canadensis*). *Ecology* **67**, 1254–1269.

Newbold, J.D., Elwood, J.W., O'Neill, R.V. & Van Winkle, W. (1981) Measuring nutrient spiraling in streams. *Canadian Journal of Fisheries and Aquatic Sciences* **38**, 860–863.

Newbold, J.D., Mulholland, P.J., Elwood, J.W. & O'Neill, R.V. (1982) Organic carbon spiraling in stream ecosystems. *Oikos* **38**, 266–272.

Palmer, T. (1993) *The Wild and Scenic Rivers of America*. Island Press, Washington DC.

Peierls, B.L., Caraco, N.F., Pace, M.L. & Cole, J.J. (1991) Human influence on river nitrogen. *Nature* **350**, 386–387.

Postel, S. (1999) *Pillar of Sand*. W.W. Norton, New York.

Power, G., Brown, R.S. & Imhof, J.G. (1999) Groundwater and fish—insights from northern North America. *Hydrological Processes* **13**, 401–422.

Pringle, C.M. (1997) Exploring how disturbance is transmitted upstreams: going against the flow. *Journal of the North American Benthological Society* **16**, 425–438.

Resh, V.H., Brown, A.V., Covich, A.P. *et al.* (1988) The role of disturbance in stream ecology. *Journal of the North American Benthological Society* **7**, 433–455.

Ricciardi, A. & Rasmussen, J.B. (1999) Extinction rates of North American freshwater fauna. *Conservation Biology* **13**, 1220–1222.

Rosenberg, D.M. & Resh, V.H., eds (1993). *Freshwater Biomonitoring and Benthic Macroinvertebrates*. Chapman & Hall, New York.

Rosenberg. D.M., Berkes, F., Bodaly, R.A., Hecky, R.E., Kelly, C.A. & Rudd, J.W.M. (1997) Large-scale impacts of hydroelectric development. *Environmental Reviews* **5**, 27–54.

Ross, H.H. (1963) Stream communities and terrestrial biomes. *Archiv Fur Hydrobiologie* **59**, 235–242.

Roth, N.E., Allan, J.D. & Erickson, D.E. (1996) Landscape influences on stream biotic integrity assessed at multiple spatial scales. *Landscape Ecology* **11**, 141–156.

Ryon, M.G. (1986) The life history and ecology of *Etheostoma trisella* (Pisces: Percidae). *American Midland Naturalist* **115**, 73–86.

Sedell, J.R. & Froggatt, J.L. (1984) Importance of streamside forests to large rivers: the isolation of the Willamette River, Oregon, U.S.A., from its floodplain by snagging and streamside forest removal. *Proceedings of the International Association of Theoretical and Applied Limnology* **22**, 1828–1834.

Smith, R.A., Schwarz, G.E. & Alexander, R.A. (1997) Regional interpretation of water-quality monitoring data. *Water Resources Research* **33**, 2781–2798.

Strayer, D.L. (2000) North America's freshwater invertebrates: A research priority. In: *Freshwater Ecoregions of North America: a Conservation Assessment* (ed. R.A. Abell, D.M. Olson, E. Dinerstein *et al.*), p. 104. Island Press, Washington DC.

Stream Solute Workshop. (1990) Concepts and methods for assessing solute dynamics in streams. *Journal of the North American Benthological Society* **9**, 95–119.

Sun, L., Perdue, E.M., Meyer, J.L. & Weis, J. (1997) Using elemental composition to predict bioavailability of dissolved organic matter in a Georgia river. *Limnology and Oceanography* **42**, 714–721.

Swank, W.T. (1988) Stream chemistry response to disturbance. In: *Forest Hydrology and Ecology at Coweeta* (eds W.T. Swank & D.A. Crossley), pp. 339–357. Springer-Verlag, New York.

Sweeney, B.W. (1993) Effects of streamside vegetation on macroinvertebrate communities of White Clay Creek in eastern North America. *Proceedings of the Academy of Natural Sciences of Philadelphia* **144**, 291–340.

Trimble, S.W. (1997) Contribution of stream channel erosion to sediment yield from an urbanizing watershed. *Science* **278**, 1442–1444.

United States Fish & Wildlife Service (1998) *Permitted stream losses due to valley filling in Kentucky, Pennsylvania, Virginia, and West Virginia: A partial inventory.* Pennsylvania Ecological Services Field Office, State College, PA.

Vannote, R.L. & Sweeney, B.W. (1980) Geographic analysis of thermal equilibria: a conceptual model for evaluating the effect of natural and modified thermal regimes on aquatic insect communities. *American Naturalist* **115**, 667–695.

Vannote, R.L., Minshall, G.W., Cummins, K.W., Sedell, J.R. & Cushing, C.E. (1980) The river continuum concept. *Canadian Journal of Fisheries and Aquatic Sciences* **37**, 130–137.

Wallace, J.B. (1988) Aquatic invertebrate research. In: *Forest Hydrology and Ecology at Coweeta* (eds W.T. Swank & D.A. Crossley), pp. 257–268. Springer-Verlag, New York.

Wallace, J.B., Webster, J.R. & Meyer, J.L. (1995) The influence of log additions on physical and biotic characteristics of a mountain stream. *Canadian Journal of Fisheries and Aquatic Sciences* **52**, 2120–2137.

Wallace, J.B., Eggert, S.L., Meyer, J.L. & Webster, J.R. (1997) Multiple trophic levels of a stream linked to terrestrial litter inputs. *Science* **277**, 102–104.

Wallace, J.B., Eggert, S.L., Meyer, J.L. & Webster, J.R. (1999) Effects of resource limitation on a detrital-based ecosystem. *Ecological Monographs* **69**, 409–442.

Ward, J.V. (1989) The four dimensional nature of lotic ecosystems. *Journal of the North American Benthological Society* **8**, 2–8.

Waters, T.F. (1995) *Sediment in Streams: Sources, Biological Effects and Control*. American Fisheries Society Monograph 7, Bethesda, Maryland.

Webster, J.R. & Benfield, E.F. (1986) Vascular plant breakdown in freshwater ecosystems. *Annual Review of Ecology and Systematics* **17**, 567–594.

Webster, J.R., Golladay, S.W., Benfield, E.F., Meyer, J.L., Swank, W.T. & Wallace, J.B. (1992) Catchment disturbance and stream response: An overview of stream research at Coweeta Hydrologic Laboratory. In: *The Conservation and Management of Rivers* (eds P.J. Boon, P. Calow & G.E. Petts), pp. 231–253. John Wiley & Sons, New York.

Webster, J.R. & Meyer, J.L., eds (1997) Stream organic matter budgets. *Journal of the North American Benthological Society* **16**, 3–168.

Webster, J.R., Tank, J.L., Wallace, J.B. *et al.* (in press) Effects of litter exclusion and wood removal on phosphorus and nitrogen retention in a forest stream. *Proceedings of the International Association of Theoretical and Applied Limnology* **27**.

Weyers, H.S. & Suberkropp, K. (1996) Fungal and bacterial production during the breakdown of yellow poplar leaves in 2 streams. *Journal of the North American Benthological Society* **15**, 408–420.

Williams, D.D. (1996) Environmental constraints in temporary fresh waters and their consequences for the insect fauna. *Journal of the North American Benthological Society* **15**, 634–650.

Wilson, M.A. & Carpenter, S.R. (1999) Economic valuation of freshwater ecosystem services in the United States: 1971–97. *Ecological Applications* **9**, 772–783.

Wolf, Ph. (1956). *Utdikad Civilisation*. [Drained Civilization] Gleerups. Malmo, Sweden.

Wolheim, W., Peterson, B., Deegan, L. *et al.* (1999) A coupled field and modeling approach for the analysis of nitrogen cycling in streams. *Journal of the North American Benthological Society* **18**, 199–221.

Wright, J.F. (1995) Development and use of a system for predicting the macroinvertebrate fauna in flowing waters. *Australian Journal of Ecology* **20**, 181–197.

Yeakley, J.A., Meyer, J.L. & Swank, W.T. (1994) Hillslope nutrient flux during near-stream vegetation removal. A multi-scaled modeling design. *Water, Air and Soil Pollution* **77**, 229–246.

Chapter 15

Plant–mammal interactions: lessons for our understanding of nature, and implications for biodiversity conservation

R. Dirzo

Introduction

Over the last few millennia, natural ecosystems have been transformed, polluted or degraded to varying degrees, as a result of human activities. With the advent of increasingly more sophisticated remote-sensing technologies and with the use of models and geographical information systems, we can have revealing pictures, assessments and even projections of the degree of transformation of the natural ecosystems of the Earth (e.g. Alcamo *et al.* 1998; Sala *et al.* 2000). Evaluations of the contemporary patterns of land-use and land-cover change at global, regional and national levels have been undertaken by several national and international projects, and the images and statistics they provide are appalling. This is particularly evident and currently very much publicized in the case of tropical forests. For example, a recent assessment at the global scale indicates that the current rate of tropical deforestation may be of the order of 15.4 million hectares per year, most of it concentrated in the tropics of the New World, where the most extensive tracts of forest still remain (Whitmore 1997).

One response of societies to this global environmental change has been the establishment of natural protected areas, the restoration of degraded ecosystems, and the consideration of management plans for the remaining wildlands located outside the formally protected areas. However, I would like to argue here that the remaining wildlands, including some protected parks, in many cases are far from being representative of the ecosystems where they have been established and are likely to be less so in the future. The reason for this is that they do not consider another significant threat and perturbation—the loss of animals, particularly the loss of mammals of medium and large size. Such loss of animals is a threat invisible to the 'eyes' of the remote sensing tools but I submit that it can be of such a magnitude that it may deserve to have an equivalent term to that of deforestation. I will refer to defaunation as the spasm of faunal loss that natural ecosystems are currently experiencing (see Dirzo & Miranda 1991).

Departamento de Ecología Evolutiva, Instituto de Ecología, UNAM, Ap. Post. 70–275, Mexico 04510 DF.
e-mail: urania@miranda.ecologia.unam.mx

A quantification of this phenomenon is very difficult to achieve. Yet there is scattered information suggesting that such a threat to the integrity of natural ecosystems is taking place in many localities of the world and that it deserves a serious analysis. For example, in a recent review Happold (1995) used several sources of information, some anecdotal and some quantitative, which led him to conclude that human–mammal conflicts in Africa constitute a serious threat to the large animals of that continent (Table 15.1). He found that the types of interaction could be classified as antagonistic, neutral or positive and that these types of interaction depended on animal attributes such as body size and reproductive rate. The emerging patterns were that large mammals show marked population declines. The smallest animals, in contrast, seem to be favoured by humans.

In another assessment, based on a review of the current threats to tropical forests, particularly of the New World, Phillips (1996) speculated on the impacts of several anthropogenic environmental threats on carnivorous and herbivorous animals. He classified the animals into two categories, vaguely defined as large and small, and assigned scores of positive or negative impact of each of the threats. Considering those threats for which the degree of uncertainty is less marked, he speculated that both carnivores and herbivores of large size should be negatively

Table 15.1 A summary of conflicts between humans and mammals in Africa. The table shows the type of interactions, the causes of conflict, the characteristics of animals for each type of interaction and the consequences on animal populations. (Modified with permission from Happold 1995.)

Nature of interaction	Reasons for interaction	Species characteristics	Animal response
Antagonistic	High value to humans High trophy value Cause damage High monetary value Competition for land Host for diseases	Large size Low reproductive rate Compete directly with humans/stock for resources	Populations decrease (elephants, large/medium artiodactyls and carnivores, most primates)
Neutral	Less value to humans Limited value for trophies Low monetary value Limited competition for land Unknown as hosts for diseases	Variable size Moderate reproductive rate Limited competition with humans/stock for resources	Populations static (many medium-sized mammals, many orders)
Positive	No value to humans Limited competition with humans for resources; good at exploiting human resources Frequently favoured by habitat modification	Usually small size Usually high reproductive rate	Populations increase (small rodents, shrews, bats, other small species)

impacted by such threats. In contrast, he concluded that small animals, but particularly the herbivorous species, are likely to be benefited by the environmental threats considered, especially by the hunting of large animals and by forest fragmentation. This speculative analysis arrived at similar patterns as those detected in the more analytical assessment for the African mammals: human perturbations to natural ecosystems and direct animal exploitation lead to population declines of the large mammals, while the smallest animals seem to be favoured. These qualitative patterns may constitute a working hypothesis of potential global interest. Next I will try to analyse to what extent these reviews can be supported by more detailed or quantitative information, and to what extent this is a more widespread rather than an isolated phenomenon.

Is mammalian defaunation another global environmental change?

Case studies from several regions, particularly in the tropics, indicate that defaunation takes place due to two major types of causes: direct and indirect. An analysis of such case studies may give a first insight to answering the question above.

Indirect causes of defaunation

The most prevalent indirect cause of defaunation in the tropics is the destruction and fragmentation of the habitat. Such habitat conversion, illustrated for example at a local level in the case of the region of Los Tuxtlas, Mexico (Dirzo & García 1992), leads to the reduction of populations or, in some species, complete eradication of many mammals of medium and large size. In this region large vertebrates such as the jaguar, tapir and spider monkey, among the mammals, are locally extinct, and so is the harpy eagle, among the large birds (Dirzo & Miranda 1991).

At a larger, regional scale, such as East Africa, habitat destruction and fragmentation lead to the reduction of the geographical range of several species (Happold 1995). Examples of this situation include the contraction of the geographical range of the black rhino, which in a 50-year period has been reduced to a few relatively large patches and numerous small fragments. The quantitative information on numbers of black rhino that should accompany the destruction of its habitat is not available, but the corresponding inferences should be readily obvious. Moreover, a similar situation occurs with other large-size species such as the elephant in East Africa. This species also shows a significant, although not as marked, contraction of its range in East Africa. Although in some African countries or regions elephant populations seem to remain stable or even to be increasing (e.g. in some parts of Zimbabwe, or in Botswana), in many others they are declining. Yet, despite the variation in population trends depending on the locality, overall elephant population in Africa is reported to have declined from 1.3 to 0.5 million during the 1980s.

Finally, some evidence indicates that habitat fragmentation leading to defaunation may affect species differentially. For example, the formation of Lake Guri, associated to the development of a dam in Venezuela, created several dozen islands of tropical forest, from what used to be a topographically complex but continuous forest up to 1986. The islands range in size from 650 hectares to less than 0.1 hectare. John Terborgh and colleagues have been surveying these islands and a mainland control site located less than 2 km from the islands, comparing the presence and absence of vertebrates and invertebrates and studying some ecological processes (Terborgh et al. 1997). A preliminary account of the presence and absence of mammals in this system highlights the impact of forest fragmentation on mammalian defaunation. The survey shows that the mainland and the largest islands contain the full complement of species typical of the region, while smaller and smallest islands lack several components of the mammalian fauna, particularly the medium and large species, such as the carnivores, the ungulates and two of the monkeys. In contrast, the small rodents are present in most of the islands, including the smallest ones.

Direct causes of defaunation

Hunting for local subsistence and for commercial purposes and illegal trading of animals are the predominant direct causes of defaunation. Quantitative documentation of these causes is very difficult to obtain, partly because of the illegal nature of some of them. However, information derived from studies by Kent Redford and John Robinson, particularly in the Neotropics, provide evidence that suggests that direct causes constitute a major pulse of defaunation.

Animal exploitation for commercial purposes

An investigation of numbers of animals and animal skins exported from Iquitos, Peru, indicates that in the period 1962–67, over 1.6 million animals were exported (Redford & Robinson 1991). This activity concentrated in the medium- to large-size mammals, including the two species of peccary, which together with the mazama deer were the most heavily attacked. The exportation of live monkeys amounted to a dramatic figure of almost 184 000. Carnivorous top predators were also heavily exported, including the largest of them, the jaguar, with a total of over 5300 animals killed for the exportation of their skin.

Subsistence hunting

Native communities of many localities in the tropics carry out a continuous extraction of game for subsistence. In this activity, mammals are preferred over birds, and birds are preferred over reptiles. Thus, it is reasonable to assume that the degree of defaunation via hunting reflects such preferences. Redford's (1992) study of hunting by indigenous and colonist people of the Brazilian Amazon provides an insight of the magnitude of this phenomenon.

A wide variety of species are hunted in this area, ranging from the *Cebus apella* monkeys, which are killed at a rate of 2.5 animals per person per year, to the tapirs, which are killed at a rate of 0.05 animals per person per year. Taking into consideration the number of people living in the region and the rates of killing per person, Redford estimated a dramatic number of ≈ 14 million animals killed per year. This figure does not consider the number of animals fatally wounded per killed animal, or the number of additional animals that die in the absence of those killed (e.g. infants without their mothers). Another interesting aspect of this study is that this heavy hunting is directed to the herbivorous mammals while the large carnivores are hunted to a lower extent, although those animals are heavily extracted for commercial purposes.

The vulnerability to population decline due to hunting
Given the high magnitude of hunting it is instructive to analyse the response of the different mammal species to hunting. Changes in animal abundance were analysed by Bodmer *et al.* (1997) to assess if demographic parameters characteristic of the species were correlated to population decline in 16 species of mammals with body weight >1 kg. The study was carried out in the north-western Peruvian Amazon. For this analysis, Bodmer *et al.* (1997) evaluated population decline due to hunting by comparing the absolute differences in abundance between two adjacent sites, one infrequently hunted and one heavily hunted. The differences in abundance were assessed from comparable transect censuses in the two sites. The species' demographic parameters analysed were the intrinsic rate of natural increase (r_{max}), longevity (inferred from age to last reproduction), and generation time (inferred from age of first reproduction).

The results showed that declines in the abundance of mammals between the infrequently hunted site and the heavily hunted site were negatively correlated with r_{max}. Mammals with lower rates of increase had greater declines in abundance, while species with high growth rates declined less or even increased in their abundance. This result also indicates that r_{max} is a critical variable influencing the vulnerability to local extinction due to hunting.

Decline in abundance due to hunting was also significantly related to longevity. The relationship was positive, indicating that long-lived species had greater declines in abundance than short-lived species, and that the former are more vulnerable to local extinction than short-lived species.

Finally, there was a significant positive relationship between decline in abundance, comparing the infrequently hunted and the heavily hunted site, and generation time. Mammals with longer generation times exhibited greater declines and vulnerability to extinction than mammals with short generation times.

The mammals detected as being more vulnerable to local extinction (due to hunting) because of their low rates of growth, long-lived individuals and long generation times, included the 'large' species such as the tapir, the primates and the carnivores. The less vulnerable species, those with high growth rates, short-lived and with short generation times, included the smallest species of the spectrum analysed, such as the large rodents, the peccaries and the brocket deer. Although this study did not consider mammals <1 kg, given the results I find it reasonable to argue that the small granivorous rodents in tropical forests are not likely to be affected, or are likely to even be benefited by hunting.

Defaunation at the global level

The 1996 IUCN Red List of Threatened Animals (Baillie & Groombridge 1996) offers some interesting insights regarding the situation of mammals at the global scale and complements the case studies described above. The assessment indicates that the number of threatened mammals has reached a comparable level to that of birds and these two groups currently constitute the most seriously threatened animals among the vertebrates. The number of mammal species in all categories of threat presently is close to 1100.

The total number of threatened species of mammals can be further analysed with different criteria. A taxonomic disaggregation of the data indicates that the degree of threatening varies markedly with the order of mammals (Fig. 15.1). The Proboscideans, constituted by two species of elephant, are critically endangered. Proboscideans are followed by the Peryssodactyla, in which only 40% of the species are under lower risk, the Primates, with close to 50% of the species from vulnerable to critically endangered, and the Dermoptera (lemurs), with 50% of vulnerable species. At the other extreme, the less endangered are the rodents, with a great majority of the species, close to 85%, under lower risk, and the Lagomorphs, with close to 80% of the species under lower risk. Surprisingly, only some 30% of the Carnivora are from vulnerable to critically endangered, but a closer analysis would reveal that the high proportion of species under lower risk are small species such as foxes. In sum, the overall tendency of this taxonomic disaggregation is that the groups of species with characteristics that could be grossly described as large- and

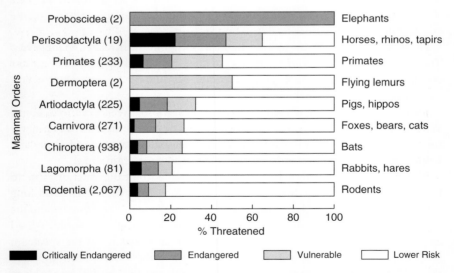

Figure 15.1 The proportion of species in four categories on threat according to the IUCN 1996 Red List, for a selection of nine orders of mammals. The orders have been arranged in a descending order according to the degree of threat. (Reproduced with permission from Baillie & Groombridge 1996.)

medium-size species are the most heavily threatened, while the least threatened are the mammals that could be grossly defined as animals of the smallest size.

The IUCN data can also be organized geographically to search for spatial patterns of animal threatening at a global scale. The most evident aspect of the geographical distribution of threatening is that many of the countries with the largest numbers of endangered species are located in tropical or subtropical regions. The top of the list is Indonesia, with 128 threatened species, India and China with 75, Brazil with 71, Mexico with 68, and Australia with 58. Other countries of relatively smaller size, also located in tropical and semitropical regions follow, including Papua New Guinea, Philippines, Madagascar, Kenya and Malaysia, with numbers between 45 and 55. Although the tropical trend is fairly evident, some non-tropical countries or regions such as China, parts of Mexico and Australia also harbour a large number of endangered species.

Finally, it is instructive to look at the temporal changes in the numbers of species in the IUCN Red List of threatened animals. The time course of the numbers of threatened animals has been analysed for years 1990, 1994 and 1996 (Table 15.2). It is immediately obvious that amphibians and mammals have more than doubled their numbers of threatened species in just 6 years. Although the case of the amphibians has been widely publicized, the increase in risk of the mammals is not so appreciated. However, Smith *et al.* (1993) used the data on changes in status of risk between 1986 and 1990 to estimate the characteristic extinction times (CET, the number of years in which 50% of the species may become extinct—see details in Smith *et al.* 1993). Quite remarkably, the estimated CET for mammals forecasts the extinction of half the species of this group within a period as short as 2.5 centuries.

The previous review of available information and the scenarios I discussed reveal a situation of alarm from the biodiversity point of view. I would argue that such a spasm of potential defaunation in the planet constitutes another major

Table 15.2 Temporal changes in the numbers of species in the IUCN Red List of threatened animals in different groups of organisms for the period 1990–96 (Baillie & Groombridge 1996). The last column shows the estimated characteristic extinction times (CET) for the period 1986–90. (Reproduced with permission from Smith *et al.* 1993.)

	1990	1994	1996	CET 86–90 (years)
Mammals	535	741	1096	250
Birds	1026	970	1107	350
Reptiles	169	316	253	2000
Amphibians	57	169	124	3000
Fishes	713	979	734	900
Invertebrates	1977	2754	1891	–
Total	4477	5929	5205	–

global environmental change that society may want to address on the basis of ethical, aesthetic, scientific and even utilitarian reasons. I will address some concerns from the scientific perspective, by exploring the ecological and biodiversity consequences of defaunation.

Ecological and biodiversity consequences of defaunation in natural ecosystems

Beyond our concern that defaunation might bring about the loss of a large number of populations, species or evolutionary lineages, including some close relatives of *Homo sapiens*, as well as the loss of a large number of charismatic decorations of natural ecosystems, from the conservation biology point of view a major concern should be the disruptions or local/global extinctions of ecological processes. Evidently mammals are involved in several ecological processes ranging from their functioning as ecosystem engineers (Jones et al. 1994), to agents of seed and fruit dispersal (e.g. Fragoso 1997; Julliot 1997) and plant pollinators (Janson et al. 1981; Carthew 1993). The ecological consequences of the loss of mammals involved in such roles have been investigated to some extent, or can be deduced from studies of their ecology (e.g. Goldingay et al. 1991), or can be inferred from studies with other animals that play such roles (Chapman & Onderdonk 1998; Howe 1990). However, the role of mammals as herbivores, particularly in the case of the meso and mega mammals, has been poorly investigated and thus the potential consequences of their loss from ecosystems are not appreciated.

Following Paine's (1966) influential paper on the consequences of the removal of trophic components on food-web complexity and species diversity and community structure, I find it of interest to analyse the potential consequences of the loss of mammalian herbivores, from the point of view of the disruption of trophic cascades. Most available studies on trophic cascading involving mammals are pertinent to animals that operate as top predators (see Terborgh 1992), and most of the available evidence corresponds to aquatic ('wet') trophic webs (e.g. Estes et al. 1998). Recent research, however, has shown that trophic cascades are also present in terrestrial ecosystems, including diverse ones (Pace et al. 1999). Here I explore trophic cascading in terrestrial ecosystems in the light of mammalian defaunation involving herbivores in tropical ecosystems (see also McLaren & Peterson 1994).

Studies on mammalian herbivory in natural ecosystems: an analysis of direct and indirect effects on multitrophic interactions

Among the vertebrates, mammalian herbivores, although a minor fraction of the global species diversity, affect vegetation in several ways and may potentially play an important role in structuring plant–insect–natural enemy trophic interactions. Mammalian herbivores, a global contingent of about 1000 species, includes organisms as varied as a 60 g *Heteromys* granivorous mice, 250 kg tapirs in neotropical forests, to 5 ton vegetarian elephants in African or Indian forests. The most evident aspects of plants/vegetation affected by mammalian herbivores are: (i) effects on plant species diversity and composition; (ii) spatial and temporal patterns of plant resource abundance; (iii) plant growth form/architecture; (iv) plant chemistry, both defensive and nutritional.

Figure 15.2 explains the essence of my argument on the role of mammals in multitrophic interactions. In this diagrammatic model, the box of herbivores includes mammals and insects, and the arrows indicate the first-order, second-order and third-order effects. This diagrammatic model also presents the basic general question I address in this section: to what extent do mammals affect vegetation in such a way that this in turn affects phytophagous insects and how does this in turn scale up to the herbivore's natural enemies? Moreover, to what extent do such effects generate feedback loops down to the herbivorous insects that scale down to the plants? Several of these direct and indirect components of the multitrophic interaction have been documented in elegant studies with insects, but information is very scarce as to how mammalian herbivores drive such indirect effects and loops via their effects on the plant resource matrix. Next I will illustrate these triggers using examples from temperate and tropical systems. Although

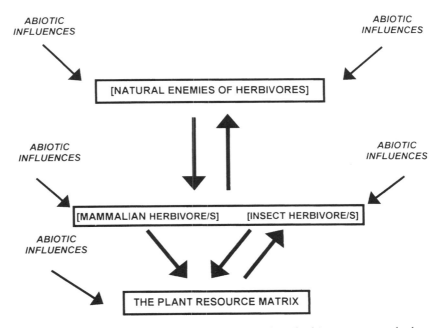

Figure 15.2 A diagrammatic model of interactions involving herbivorous mammals, the plant resource base, phytophagous insects, and natural enemies of the phytophagous insects. The arrow between mammalian herbivores and the plant resource base corresponds to first-order interactions. The ascending arrow between the plant resource base and insects indicates second-order interactions, and the arrows between phytophagous insects and natural enemies correspond to third-order interactions.

my emphasis in this paper is on tropical ecosystems, I will use a temperate example to highlight the great complexity we should expect to see, given the considerably higher diversity of tropical ecosystems, when knowledge on the latter is improved.

A temperate example: beaver herbivory on cottonwoods in riparian habitats

The temperate example concerns the effects of mammalian herbivores on plant growth and chemistry, illustrated by recent studies by Martinsen et al. (1998) in the riparian habitats of the south-western United States, where the beaver *Castor canadensis* is an important herbivore of cottonwood trees. When cut down by beavers, most of the cottonwoods resprout producing trees with different growth form and with new foliage, which is different from non-resprout foliage, both morphologically and chemically. In this site, the beetle *Chrysomella confluens* is a specific herbivore on the foliage of cottonwoods, skeletonizing these plants' leaves and occasionally causing extensive defoliation. Beetles are particularly abundant on resprouting trees. For example, the size of the patches of beaver activity significantly explained most of the variation in the number of beetle larvae clutches/plot. Thus, an indirect (second-order) effect of beavers on beetles is apparent and it was hypothesised that chemical changes in the resprout foliage may be the driving mechanism.

The analysis of leaf phenolic glycosides in cottonwood foliage indicated that their concentration was about twice as high in the resprout foliage, in comparison to non-resprout tissue. These glycosides, when sequestered by larvae, break down to salicin. Salicin is transported to dorsal glands of the beetles, where it is converted to salicylaldehyde, a defensive compound. Upon threatening by a predator, the larvae exert their glands and expose the predators to salicylaldehyde. At the same time, the conversion of salicin to salicylaldehyde liberates glucose, which can be used for beetle nutrition. In addition, it was found that resprout foliage is about 20% richer in nitrogen. As a consequence, when fed resprout foliage, beetles showed a lower time to complete their development and higher adult weight. Of greater importance in terms of cascading effects, beetle defence against predatory ants (*Formica propinqua*) was also enhanced. Beetles developing on resprout tissue survived for a significantly longer time before they were attacked and dragged by *F. propinqua* ants into their mound. Because the defensive mechanisms of the larvae are effective against several predators, besides ants, cottonwood defoliation by beetles can be so intensive that it may effectively reduce the abundance and diversity of other phytophagous insects that feed on cottonwood. Moreover, because the activity of beavers is very patchy, their presence determines a complex spatial structure of the abundance and diversity of the arthropod community in these riparian habitats.

In synthesis, herbivory by beavers causes significant changes in the chemistry and growth form of cottonwood plants, which in turn affect an unrelated phytophagous insect, the activities of which in turn affect other arthropods in the system, including predatory ants, and other phytophagous insects. At any rate, the image that these studies provide shows a complex array of multitrophic effects and feedback's triggered by mammalian herbivory. Clearly, any perturbations that either increase or reduce the abundance of beavers in these riparian systems may have serious consequences on several components of the multitrophic interactions of such keystone ecosystems.

A tropical example: mammalian herbivory in the forest understorey

My tropical example addresses the effects of mammalian herbivory on plant species diversity and structure in neotropical forests.

A distinguishing aspect of the temperate example I just presented is that the effects were generated by a predominant or keystone herbivore. Such a situation is unlikely in the neotropical forests, where the guild of mammalian herbivores is composed of a very diverse assemblage of animals ranging from pacas to agouties, peccaries, deer, tapirs, etc. The logistic problem then arises as to how to study the role of such a diverse array of mammalian herbivores on the vegetation and other components of the trophic web?

The contemporary massive defaunation of several tropical forests provides an unfortunate, yet useful, opportunity to see what happens to the forest, particularly the understorey, when the medium to large mammals are reduced or eradicated. My approach to the study of the consequences of defaunation has been to carry out comparative analyses of sites with contrasting level of conservation of the mammalian fauna. In this example, the forest of Los Tuxtlas (LT), south-east Mexico, is a site predicted to have a fauna under poor conditions of conservation, due to hunting, illegal trading and especially habitat reduction and fragmentation (Dirzo & Miranda 1991). This site is being compared with Montes Azules (MA), a site where we expect the fauna to be in an excellent conservation situation, given that MA is considerably larger, non-fragmented and under official protection as a Biosphere Reserve (Dirzo & Miranda 1991).

An analysis of the fauna has been carried out for several years in both sites using a strictly comparable system of detection of footprints and sightings (Dirzo & Miranda 1991). Results have been obtained in terms of: (i) numbers of species detected by both methods, and (ii) an index of their occurrence. From these studies it is evident that LT shows a reduction in the number of species detected, and those still detectable show a significantly reduced occurrence in the forest understorey.

Associated with this recent defaunation, the levels of mammalian herbivory on individually marked plants and leaves contrast dramatically. In Monte Azules, where the fauna is intact, around 30% of the seedlings and saplings are damaged by mammalian herbivores. This contrasts with the undetectable damage by mammals in the largely defaunated site, Los Tuxtlas. In addition, many of these mammals operate as seed predators (particularly of large seeds) and may do considerable disturbance in the understorey by trampling (e.g. by large herds of peccaries, tapirs). This loss of the expected load of herbivory by vertebrates correlates with a significant first-order effect: a change in the structure and diversity of the understorey

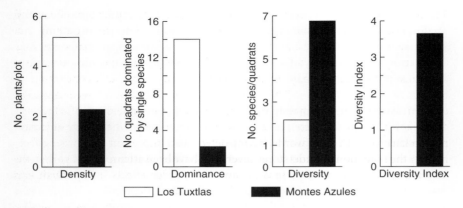

Figure 15.3 Structure and diversity of the forest understorey in a defaunated site (Los Tuxtlas) and a site with the mammalian fauna intact (Montes Azules). Density corresponds to the number of individuals per square metre. Dominance corresponds to the number of plots (out of 20) dominated by one single species; a dominant species was defined as that in which at least 50% of the plants in the plot corresponded to a single species. Diversity indicates the number of different species per plot. The diversity index is Shannon's Index. All data derived from 20 plots in each site. (Reproduced with permission from Dirzo & Miranda 1991.)

(Fig. 15.3). The defaunated site presents a significantly higher density of seedlings, a significantly reduced number of different species per plot and, in general, a substantial reduction, over threefold, in its diversity index. In synthesis, while the understorey at Montes Azules shows a variety of species and density is low, Los Tuxtlas displays a defaunation or 'empty forest' syndrome (cf. Redford 1992), characterized by the presence of dense, monoculture carpets of seedlings, locally dominated by one or a few species.

The information I presented so far is correlative and may not necessarily imply a cause-and-effect relationship between defaunation and the tremendous modification of the understorey. However, a series of experimental manipulations consisting of the comparison of exclosures and control plots support the observations, described above, given that exclosures developed a less diverse plant community than control plots in MA. In contrast, at LT no difference was observed between exclosures and control plots (R. Dirzo, unpublished results). Thus, I am led to conclude that mammalian herbivory plays a fundamental role in the structure and diversity of the tropical forest understorey, and that the absence of mammalian herbivores leads to dramatic changes in the understorey plant community. These findings pose a number of intriguing questions that may deserve further work. One of them is the untangling of the proximal mechanisms responsible for the generation of these mono-dominated seedling carpets. This is an aspect that warrants further work.

The existence of possible mechanisms that might compensate for the lack of mammalian herbivory in tropical forests is an additional important question. In particular, it could be expected that the development of mono-dominated seedling carpets may have an effect on the patterns of herbivory and population dynamics of the insects that feed on these plants, along the lines of the resource concentration hypothesis (e.g. Root 1973). The monospecific concentration of seedlings may cause an increased level of herbivory by phytophagous insects (many of which are specialists in tropical forests), and this might compensate for the lack of mammalian herbivory. In other words, having detected an important first-order effect, namely the reduction of understorey diversity, I have been attempting to gain some information on the possible second- and third-order effects, using particular insect–plant interactions in Los Tuxtlas, the defaunated site.

I will address this issue using the system of the moth *Urania fulgens* (Lepidoptera: Uraniidae) and its host plant, *Omphalea oleifera* (Euphorbiaceae) in Los Tuxtlas. The advantages of this system include, for the purposes of the hypothesis, the following.

1 The interaction is very specific. *Urania* rejects and will die of starvation rather than accept any other plant tissue. *Omphalea* is not eaten by any other phytophagous insect. I presume that previous to defaunation of this forest, *Omphalea* was eaten and/or trampled by mammals.

2 *Urania* is a migratory species and its feeding period is concentrated to a few months (February to September) during which, in some years, the moth undergoes major population explosions.

3 Herbivory, especially during the years of population explosion, can be very high.

4 We still know very little about the interaction of *Urania* and its natural enemies, but we do know that wasps and flies parasitize eggs and larvae, respectively. This offers some potential to explore third-order effects.

To explore second- and third-order effects, I developed monospecific (30 seedlings planted plus the natural understorey) and diverse (10 seedlings plus the natural understorey) experimental plots. These plots consisted of 1 m² quadrats established at the edge of the vertical projection of the crowns (20 m from the trunk) of reproductive *Omphalea* trees at LT. There were four plots adjacent to an isolated tree; two of them were monocultures and the other two were diverse. The monoculture plots represent the current situation of the understorey, while the diverse plots are meant to represent the situation of the understorey before defaunation took place in Los Tuxtlas.

I present preliminary data on three response variables: herbivory, larval density, and the magnitude of larval parasitism by Tachiniid flies.

Herbivory
The overall time course of herbivory (Fig. 15.4) reflected the natural history of the moth. Soon after the year's immigration, around September, herbivory increased and by the time the moths left LT, around February, the levels had fallen to a low of about 8% leaf area damaged. This low level remained throughout the year.

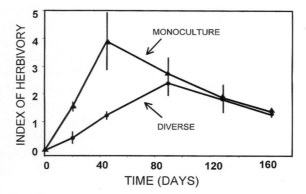

Figure 15.4 The time course of the mean levels of herbivory per plant for saplings of *O. oleifera* in monoculture and diverse plots. The herbivory index was calculated as $IH = [\Sigma (n_i \times i)]/N$, where n_i is the number of leaves in each of five categories of damage, i denote the categories of damage (0 = 0 leaf area damaged, 1 = 1–6%, 2 = 6–12%, 3 = 12–25%, 4 = 25–50%, 5 = 50–100%) and N is the total number of leaves per plant.

This general pattern contrasted between the two treatments: in the monoculture plots there was a steep increase and a major and earlier peak followed by a drastic fall. The plants in the diverse plots showed a steady increase that converged with the fall of the herbivory in the monoculture plots. This convergence suggests that insect herbivory does not compensate for the lack of mammalian herbivory. Since no other herbivore feeds on this plant, the observed pattern may be determined, to a large extent, by the density and feeding capability of the *Urania* larvae.

Larval density
In conjunction with the herbivory, larval density (Fig. 15.5) reached an early and very high peak in the monoculture plots, followed by a very dramatic fall after 50 days. The larvae in the diverse plots exhibited a more gradual initial increase that remained more or less constant until the end of the season, when *Urania* left the area. No larval predation by vertebrates was observed during this period, so it is possible that the contrast in caterpillar density is influenced by parasitoids.

Larval parasitism
Larval parasitism, leading to larval death, showed a correlated response with the previous two variables, but the contrast between the two types of plots was even more marked (Fig. 15.6). Only a few larvae were attacked and killed in the diverse plots, while in the monoculture plots parasitism was high from the beginning and remained high until almost the end of the season.

These data suggest some potentially significant linkages: while the monoculture plots supported a greater density of larvae per plant, the magnitude of herbivory

PLANT–MAMMAL INTERACTIONS

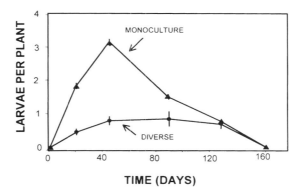

Figure 15.5 The time course of the mean number of larvae per plant for saplings of *O. oleifera* in monoculture and diverse plots.

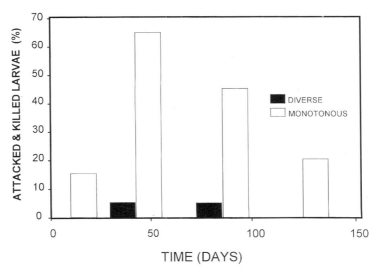

Figure 15.6 The incidence of parasitoid attack on plants of *O. oleifera* in monoculture plots. The bars indicate the percentage of larvae that were parasitized and killed by Tachiniid flies.

was significantly higher only for the first 45 days. The drastic fall of herbivory might have been determined by the tremendous incidence of larval parasitism, which took a heavy toll right from the beginning and throughout the season. It would appear that the level of leaf damage by the specialist insect was influenced from the top, by the herbivore's natural enemies, which in turn appear to respond to the structure and diversity of the vegetation. These results were obtained under the experimental simulation of what happens in the presence and absence of mammalian herbivores, and so allow me to speculate, tentatively, that mammalian herbivory may trigger profound changes in several components of the trophic structure of the *Urania–Omphalea* system.

In summary, given the findings of these two case studies, I suggest that mammals, through their effects on plant chemistry or growth form (as was the case of the beaver–cottonwood temperate example), or through their effects on diversity and abundance of plants (as in the case of Los Tuxtlas tropical forest), may play a significant role in the patterns of insect herbivory, and may affect the dynamics of trophic interactions (see Fig. 15.2). Undoubtedly we are far from having an adequate understanding of such mammal-driven complexity, but I think this is a promising avenue of scientific inquiry.

Conclusions

To conclude, I would like to contend that, at least for some systems, our current scientific descriptions of multitrophic interactions might be quite distorted, given the increasing absence of mammalian herbivores due to anthropogenic disturbance in temperate and tropical systems.

In the light of my contention that defaunation seems to be another global environmental change, and given our primitive knowledge of the role of mammalian herbivores in natural ecosystems, I argue that we need to dedicate more efforts to the conservation and rigorous study of these organisms. We need to ecologically understand these animals rather than, or in addition to, simply view them as charismatic decorations of the natural world. Conservation biology needs to go beyond the traditional approach of conservation of taxa, to the conservation of ecological and evolutionary processes.

Acknowledgements

I am grateful to the organizers of the BES/ESA Symposium for their invitation to present a paper in it. An earlier draft of this paper was read and constructively criticized by Nancy Huntly, Néstor Mariano, and two anonymous reviewers. Karina Boege and Guillermina Gómez offered fruitful discussions during the preparation and design of this paper. Raúl I. Martínez helped in several stages of the preparation of the manuscript. The studies on defaunation in Mexican tropical forests were supported by a CONACyT grant (31856-N).

References

Alcamo, J., Leemans, R. & Kreileman, E. (1998). *Global Change Scenarios of the 21st Century*. Elsevier, Oxford.

Baillie, J. & Groombridge, B. (1996) 1996 IUCN Red List of Threatened Animals. IUCN The World Conservation Union, London.

Bodmer, R.E., Eisenberg, J.F. & Redford, K.H. (1997) Hunting and the likelihood of extinction of Amazonian mammals. *Conservation Biology* **11**, 460–466.

Carthew, S.M. (1993) An assessment of pollinator visitation to *Banksia spinulosa*. *Australian Journal of Ecology* **18**, 257–268.

Chapman, C.A. & Onderdonk, D.A. (1998) Forests

without primates: primate/plant codependency. *American Journal of Primatology* **45**, 127–141.

Dirzo, R. & García, M.C. (1992) Rates of deforestation in Los Tuxtlas, a Neotropical area in Southeast Mexico. *Conservation Biology* **6**, 84–90.

Dirzo, R. & Miranda, A. (1991) Altered patterns of herbivory and diversity in the forest understory: a case study of the possible consequences of contemporary defaunation. In: *Plant–Animal Interactions: Evolutionary Ecology in Tropical and Temperate Regions* (eds P.W. Price, T.M. Lewinsohn, G.W. Fernandes & W.W. Benson), pp. 273–287. John Wiley and Sons, New York.

Estes, J.A., Tinker, M.T., Williams, T.M. & Doak, D.F. (1998) Killer whale predation on sea otters linking oceanic and nearshore ecosystems. *Science* **282**, 473–476.

Fragoso, J.M.V. (1997) Tapir-generated seed shadows: scale-dependent patchiness in the Amazonian rain forest. *Journal of Ecology* **85**, 519–529.

Goldingay, R.L., Carthew, S.M. & Whelan, R.J. (1991) The importance of non-flying mammals in pollination. *Oikos* **61**, 79–81.

Happold, D.C.D. (1995) The interactions between humans and mammals in Africa in relation to conservation: a review. *Biodiversity and Conservation* **4**, 395–414.

Howe, H. (1990) Seed dispersal by birds and mammals: implications for seedling demography. In: *Reproductive Ecology of Tropical Forest Plants* (eds K.S. Bawa & M. Hadley), pp. 191–218, UNESCO, Paris.

Janson, C.H., Terborgh, J. & Emmons, L.H. (1981) Non-flying mammals as pollinating agents in the Amazonian forests. *Biotropica Supplement* **15**, 1–6.

Jones, C.G., Lawton, J.H. & Shachak, M. (1994) Organisms as ecosystems engineers. *Oikos* **69**, 373–386.

Julliot, C. (1997) Impact of seed dispersal by red howler monkeys *Allouata seniculus* on the seedling population in the understorey of tropical rain forest. *Journal of Ecology* **85**, 431–440.

Martinsen, G.D., Briebe, E.M. & Whitham, T.G. (1998) Indirect interactions mediated by changing plant chemistry: beaver browsing benefits beetles. *Ecology* **79**, 192–200.

McLaren, B.E. & Peterson, R.O. (1994) Wolves, moose and tree rings on Isle Royale. *Science* **266**, 1555–1558.

Pace, M.L., Cole, J.J., Carpenter, S., R. & Kitchell, J.F. (1999) Trophic cascades revealed in diverse ecosystems. *Trends in Ecology and Evolution* **14**, 483–488.

Paine, R.T. (1966) Food web complexity and species diversity. *American Naturalist* **100**, 65–75.

Phillips, O.L. (1996) The changing ecology of tropical forests. Biodiversity and Conservation **6**, 291–311.

Redford, K.H. (1992) The empty forest. *Bioscience* **42**, 412–422.

Redford, K.H. & Robinson, J.G. (1991) Subsistence and commercial uses of wildlife in Latin America. In: *Neotropical Wildlife Use and Conservation* (eds J.G. Robinson & K.H. Redford), pp. 6–23. University of Chicago Press, Chicago.

Root, R.B. (1973) Organisation of a plant–arthropod association in simple and diverse habitats: the fauna of collards (*Brassica oleracea*). *Ecological Monographs* **43**, 95–124.

Sala, O.E., Chapin, F.S., Armesto, J.J. *et al.* (2000) Global biodiversity scenarios for the year 2100. *Science* **287**, 1770–1774.

Smith, F.D.M., May, R.M., Pellew, R., Johnson, T.H. & Walter, K.S. (1993) Estimating extinction rates. *Nature* **364**, 494–496.

Terborgh, J. (1992) Maintenance of diversity of tropical forests. *Biotropica* **24**, 283–292.

Terborgh, J., Lopez, L., Tello, J., Yu, D. & Bruni, A.R. (1997) Transitory states in relaxing ecosystems of land bridge islands. In: *Tropical Forest Remnants* (eds W.F. Lawrance & R.O. Bierregaard), pp. 256–274. University of Chicago Press, Chicago.

Whitmore, T.C. (1997) Tropical forest disturbance, disappearance, and species loss. in: *Tropical Forest Remnants* (eds W.F. Lawrance & R.O. Bierregaard), pp. 3–12. University of Chicago Press, Chicago.

Chapter 16
Ecological economic theory for managing ecosystem services

J. Roughgarden and P. R. Armsworth

Introduction

The last 50 years in ecology have been heady times, and ecologists begin the new millennium enthusiastically. For those who see the glass as half full, the subject has developed experimental methods, mathematical theories, and empowering technologies. The status of ecology has risen from total obscurity to general awareness—one of us had to spell the word 'ecology' for her parents 30 years ago, and recently support for 'the ecology' has been affirmed by even US presidential candidates. For those who see the glass as half empty, the environment is withstanding daily abuse, species are going extinct, the sea and air are warming, and population growth is conquering the globe. Ecologists face spiralling information requirements from a society that makes increasing demands of its environment, and when the relevant scientific information is available, ecologists still struggle to see their recommendations implemented by policy-makers. The tranquillity of New Year's Eve 2000, belies great changes in the near future, both socially and environmentally.

Humans now dominate the biosphere (Vitousek *et al.* 1997), and their domination of it will increase through this new century. Ecologists have a responsibility to study the dramatic effects of anthrogenic impacts on ecosystems. This will require an understanding of the socioeconomic forces that motivate human actions, and an integration of ecological and social sciences.

The relationship between ecological science and society is a contested boundary. How should ecology connect with society? Indeed, who are ecologists: are they plain folk? Or are they scientists, specialists somehow apart from society? Should ecologists advocate public policy positions, or should they serve primarily as advisors?

The new, important subject of ecological economics is emerging from the meeting of ecological and social sciences. Some ecologists are appalled at the very idea of an ecological–economic collaboration—sleeping with the devil (see viewpoints discussed in Goulder & Kennedy 1997; O'Neill *et al.* 1998; Roughgarden 2001). Others see economics as a flawed discipline that should be subsumed within a new general systems theory to encompass ecology, economics, sociology, psy-

Department of Biological Sciences, Stanford University, Stanford CA 94305, USA

chology and the rest (Costanza 1996; Naveh 2000). Still others, such as ourselves, envision an interdisciplinary project. This interdisciplinary project borrows from theories in environmental and natural resource economics.

Normative theories in economics provide a systematic framework for societal decision making, one that builds on individuals' preferences. Environmental economics is a subdiscipline of economics that considers decisions relating to environmental commons, such as clean air and water (Kolstad 2000). Natural resource economics is a second economic subdiscipline, one that examines the extraction of exhaustible and renewable resources, such as mineral deposits, timber and fisheries (Dasgupta & Heal 1979; Kneese & Sweeney 1985; Clark 1990).

The interdisciplinary project of ecological economics extends environmental and natural resource economics. The concept of a resource is generalized from a few target species to a much larger set of ecological systems, and that of a service is generalized to account for the multitude of benefits that society derives from ecosystems (Ehrlich & Mooney 1983; Daily 1997). Ecological economic analyses must also acknowledge the limitations on our ability to control nature. These limitations necessitate a different style of economic analysis than has conventionally been performed: we provide an illustration below (pp. 345–349). There is a distinction between our capability to know about nature and our ability to control it. Both could be improved with greater investment, but we believe that it is the latter that will ultimately constrain management.

This chapter introduces mathematical theory that we have developed for managing a variety of ecosystem services using ecological economic principles. It is intended to provide a primer of economic ideas for ecologists. In the next section we begin with traditional harvesting theories, and show how these can be modified and extended to account for other types of ecosystem service, and for fluctuations in environmental conditions when only simple, inexpensive management policies can be implemented. We then look at multispecies communities and how one should manage extractive and non-extractive industries exploiting these. We conclude with a discussion.

For brevity, we cannot cover all technical frontiers of ecological economics. A number of important topics have therefore been omitted, such as economic game theory and incentive design. We have also elected to present simple theoretical models that offer some degree of generality. Policy-makers seeking to apply our results will need to tailor the models to their particular study systems.

Single-species models

Types of services
The ecosystem services considered in traditional resource economics are extractive, such as annual harvests of fish, trees, pelts and horns. Harvests are sold in markets, fetching prices that represent economic value.

Increasingly though, attention is focusing on '*in situ* services' that do not involve harvesting, and which are derived directly from the standing stock of organisms. A grove of trees that prevents soil erosion is one example. If the trees are removed, then a retaining wall must be constructed. Aesthetic values of organisms and ecosystems provide further important examples of *in situ* services (Krutilla 1967). These services earn rent in various ways. They might earn rent in the usual sense of people paying to watch an ecosystem (or its components) as tourists, or paying to live in the ecosystem on a patch of land leased from a park or national forest. They might also earn 'virtual rent' which would be the cost of replacing what the ecosystem does, such as the annualized cost of building a retaining wall plus its annual maintenance. Another way virtual rent is realized is when people vote to prevent commercial development in favour of open space. By voting for open space instead of development, people have, in effect, purchased the open space at the price of the annual income they are forgoing by not having the commercial enterprise there. Although *in situ* services are not bought and sold at fish and farmer's markets, they are still valuable economically because people have expressed a preference for these services whenever they choose them over alternatives that can be economically valued in a conventional way (Pearce 1993; Smith 1993).

Here we assume the services have been identified and valued in a particular cultural and geographical context, and ask what configuration a local ecosystem should have to deliver as much service as possible.

Targets and tactics
A 'production function' describes how much service is produced by an ecosystem as a function of how it is internally organized.

Let us begin with a simple extractive service. The production function for a single population being harvested sustainably should be familiar—it is the right-hand side of an equation for population dynamics, $\Delta n = f(n)$. This is called a production function, because the formula states how many animals may be sustainably harvested, which is the specific ecosystem service being considered here, as a function of the stock size, which describes the state of the ecosystem in this case.

For example, logistic growth is $\Delta n = rn(k-n)/k$ where n is population size, Δn is the change in population size between years, r is the (discrete-time) intrinsic rate of increase, and k the carrying capacity. The production function, as illustrated in Fig. 16.1, is a downward opening parabola whose peak is at $k/2$ and which intercepts the horizontal axis at 0 and k where the slopes are r and $-r$, respectively.

Production functions for many species from plants through invertebrates to vertebrates in aquatic, marine and terrestrial environments are hump-shaped curves, and the logistic is a middle-of-the-road model illustrating the key principles (Roughgarden 1997). They all have this shape because production initially increases from zero as the population grows, but as density dependence sets in production slows, peaking at the point of maximum sustainable yield, and thereafter drops to zero at the equilibrium population size.

A 'target' is a desired stock size. The population is held at this target by harvesting the biomass that this stock produces. The harvest rate is a 'control variable'—it is the policy instrument used to keep the ecosystem where we want it to be.

Various tactics are used to implement the control (Milner-Gulland & Mace 1998). A 'harvest quota' stipulates that a certain number, say 100 000 tons, can be harvested from some region until the next year's quota is set. An 'effort (or mortality) quota' stipulates that the harvesting season is say, 2 months long, based on the assumption that 50 000 tons are caught per month, so that after 2 months the same 100 000 tons end up being captured. A harvest quota is hard on the population because if the stock is lower than anticipated the 100 000 are caught anyway. An effort quota is softer because if the stock is low, then less than the 100 000 are caught by the time the 2 months are up, and so the population has not been hit as hard. But an effort quota is tougher to implement because catch per unit effort continually increases with improving technology, and the season length must continually shrink to compensate.

Any population size between 0 and k can be chosen as a target. The population is positioned on target either with a one-time special harvest to reduce it to the desired size, or by not harvesting at all to allow it to grow. Once on target, the net production is sustainably harvested using harvest or effort quotas (Intriligator 1971; Clark 1990).

The basic question then becomes: what is the best target?

Traditional best target

Elementary economic theory is clear about what the best target is when considering solely a harvesting industry (Plourde 1970). One compares the harvest earnings with what one could earn from an alternative investment in other parts of the economy, and traditionally the best target is that which maximizes the difference between these two. People are assumed to value earnings generated in the present more than those earned in the future, and future values are scaled by a social discount factor when calculating the optimal target.

The harvest earnings are given by the extractive production function. The alternative earnings are represented by a straight line whose slope is the discount rate. The point where the vertical distance between the production function and line is greatest is the optimal target predicted by this elementary theory (Fig. 16.1). This target earns the most money sustainably. That is, it does so under the assumptions so far.

This best target suffers from drawbacks, however. First it is low. It is always to the left of the peak of the extractive production function, which is a stock size less than that producing the maximum sustainable yield (when $n = k/2$). This low target stock size places the population at risk of extinction. The risk of losing the stock is not considered in this traditional analysis. So yes, if it were indeed feasible to maintain the stock indefinitely at a low abundance, the conventional economic recommendation would yield the most money for a single harvesting industry. But failing to consider the implications of a low stock size for population extinction gives a misleadingly rosy picture of projected earnings.

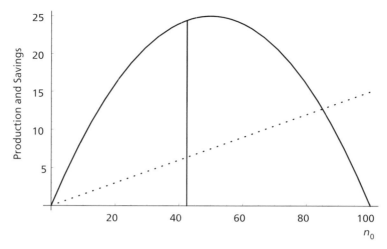

Figure 16.1 Earnings from extractive production according to the logistic model (solid dome-shaped curve) and from alternative investments (dotted diagonal line). (The earnings from alternative investments are called the 'opportunity cost'.) Vertical line shows target stock that maximizes the difference between these two. This is the most profitable target stock, assuming no risk to the ecosystem, and is to the left of the peak of the production function. The illustration is plotted using $r = 1$ and $k = 100$, and an interest rate of $\rho = 0.15$. This optimal target is the n_0 that maximizes $f(n_0) - \rho n_0$ where $f(n_0)$ is the ecosystem extractive production function and ρn_0 is earnings from alternative investments with interest rate ρ. Differentiating with respect to n_0 yields $f'(n_0) = \rho$. The optimal target is the root of this equation—it is to the left of the peak of f, because that is where the slope of f is positive. This formula is called the 'golden rule' in economics.

Second, the lower the target, the weaker the stability of the population—the less forgiving it is to human mistakes and to natural perturbations. In mathematical models, stability is measured by the 'eigenvalue' of the system at the steady state (see Roughgarden 1998 or any comprehensive theoretical ecology text). According to the way eigenvalues are defined, an eigenvalue of zero indicates the best stability; as the eigenvalue goes from zero to one, the stability weakens; and if the eigenvalue is greater than one, the steady state is unstable and extraordinary controls are needed to keep the system positioned there. Figure 16.2 shows that the stability of the target progressively weakens as the target is shifted from equilibrium k down to zero for both control tactics. Harvest quotas always produce weaker stability than effort quotas, but either way, the lower the target, the weaker the system's stability.

Thus, elementary economic theory for how best to manage a single stock for an extractive service jeopardizes the ecosystem by recommending a low target. The lowness itself is risky, and also the low target produces an ecosystem whose weak stability is unforgiving of human error and susceptible to natural perturbation. Is

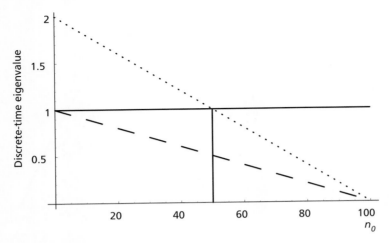

Figure 16.2 Eigenvalue of the dynamics of sustainable harvest as a function of target stock. Upper dotted line for catch quota and lower dashed line for mortality quota. The lower the target, the weaker the stability, and for catch–quota regulation, targets to the left of the peak of the extractive production function ($<k/2$ in the logistic model) are unstable.

this the inevitable recommendation of economic theory? Can an extended economic theory recommend higher targets with less risk of damage to nature?

Density-dependent prices

Economists say that dangers of following the elementary theory are overstated because costs of operation inherently favour a larger stock. When a stock is rare, then considerable effort must be expended in finding and catching another unit of harvest and the costs of doing so will be high. Harvesting costs will be much lower when the stock is abundant. When the net price (market price less costs) per organism increases with stock size, a higher target is always recommended than for a flat price (Clark 1990), as illustrated in Fig. 16.3. We acknowledge this possibility, but doubt its importance in predicting larger optimal targets for some situations. For example, for the Newfoundland cod fishery, this contention led to an optimal target only slightly larger than that based on a flat price (Roughgarden & Smith 1996).

In situ services

If we now include *in situ* services that a stock supplies, then the optimal target may be substantially higher than traditionally recommended. Summing the revenue from all extractive services plus the rent from all *in situ* services leads to a whole-system production function, as illustrated in Fig. 16.4. The optimal target using this whole-system production function is always higher than that predicted when considering harvest alone.

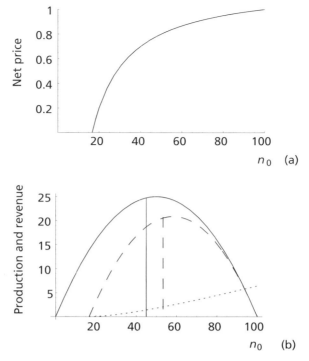

Figure 16.3 (a) Density-dependent net price. (b) Revenue, which is price times production, is lower dashed curve, and new opportunity cost, $\rho \int_{n_{min}}^{n_0} p(n)\, dn$ where n_{min} solves $p(n_{min}) = 0$, is dotted curve. Vertical lines: solid is optimal target from Fig. 16.1 and dashed is new optimal target. Density-dependent price raises optimal target.

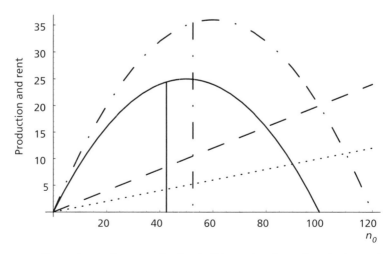

Figure 16.4 Extractive service or production (solid curve) and rent from *in situ* service (dashed diagonal line). Outer (dot–dash) curve is combined revenue from both services. Optimal target for both services combined is higher than for extraction alone

If the return from *in situ* rent is high enough relative to the return from harvest, the optimal target can even be k, indicating that no harvest is economically optimal. The same conclusion is reached by Hartman (1976), who finds that no timber industry should be permitted if the recreational value of old-growth forest is sufficiently large. Bulte and van Kooten (2000) show that the tourism revenue associated with whale watching might be sufficient to justify continuing the moratorium on commercial harvesting of minke whales in the North Atlantic. Indeed, nothing prevents the optimal target from being greater than k, as in a zoo, where the abundance of animals exceeds their natural abundance. In this case the price of admission paid by tourists watching giraffes is much greater than the price of giraffe meat at a supermarket.

Therefore, the recommendations of an economic analysis may commonly align with a conservationist agenda, provided *in situ* services are aggressively included in the business plan for the ecosystem's management.

Zonation vs. higher stock targets

What is the best method for managing extractive and *in situ* services provided by a single stock? One model for environmental preservation would have society fence in areas reserved for *in situ* services and deny access to extractive industries. Is this the best approach economically? Alternatively, could the entire area of the ecosystem supply both types of service throughout its range, provided the users of each service adhere to their allotments?

We have constructed a model for a population in a homogeneous stretch of habitat. It has two control variables: the location of a fence separating a reserve from an extractive zone, and the amount of harvest allowed outside the fence. We compared the optimal solution to this bivariate control problem with that for an ecosystem where both services were exploited simultaneously throughout the area. Relevant examples would be a coral reef fish stock or an old-growth forest that had both recreational and harvesting value, and where two types of harvesting strategy were available: a moderate harvesting strategy that did not cause recreational users to devalue the resource and an intensive harvesting strategy that devalued recreational services in the harvesting zone.

As Fig. 16.5 illustrates, the total revenue from both services combined is always higher if the services coexist than if the *in situ* services are cordoned off in a reserve and extractive services condemned to a mining zone. The target predicted using the whole-system production function above (p. 342) is optimal for this model, and all other strategies are less efficient.

The difference between the two curves can be interpreted in several ways. If society were unsure whether the regulatory authority had the requisite institutional power to control harvesting, then reserves would provide a precautionary enforcement tool to guarantee that the resource was not depleted. The difference between the curves then would represent a potential cost of conservation. A further interpretation arises in situations where a variety of harvesting techniques can be employed, some cheaper but more destructive than others. The difference between

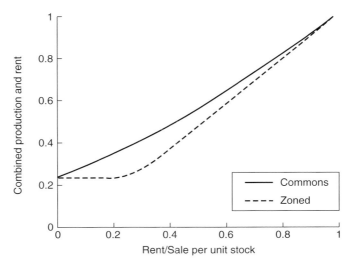

Figure 16.5 Discounted present value (scaled between 0 and 1) of investing in the optimal zonation between extractive and *in situ* services, and investing in a commons, which is the optimal spatially mixed combination of both extractive and *in situ* service. Horizontal axis is the *in situ* value per organism, relative to the sale price of an organism, which is its extractive value. The sale price is taken as fixed at $1 per organism and the *in situ* value varies. Optimal zonation generates less revenue than optimal spatial mixing in a commons regardless of the relative value of *in situ* services to extraction.

the two curves then describes an economic incentive for adopting less destructive techniques.

Under what circumstances might a reserve be economically justifiable? If the two industries cannot coexist, a multiple-use strategy with a reserve earns more revenue than either industry can in isolation. Also, certain areas of ecological importance, such as mating grounds, nurseries, watering holes and so forth, need fencing in any sensible plan of ecosystem use. A final important situation in which a reserve may be justifiable is discussed by Lauck *et al.* (1998) and Mangel (2000). These authors show that when implementation of harvesting policies is imperfect, a reserve could be advantageous, because it enforces an effort reduction without risking costly implementation errors. In their models, creating a reserve incurred no inefficiency, because they assumed perfect mixing across reserve boundaries.

Using an ecosystem's stability

The findings of conventional economic theory, which assumes a sophisticated degree of control over natural systems, are largely irrelevant for instances where authorities have limited institutional flexibility and cannot afford sophisticated management strategies that require much continuous updating. An alternative optimal target can be derived for such situations.

The approach we now introduce rests on the concept of a quasi-stationary distribution and associated probability of extinction (Nisbet & Gurney 1982; Renshaw 1991). Here's the idea.

The ecosystem is assumed to reside in a fluctuating environment. The fluctuations arise naturally as, for example, a population's r and k bounce up and down between good and bad years. Also, the fluctuations are introduced by people who imperfectly execute policy—harvesters who accidentally catch too many individuals, scientists who set mistaken quotas, and so forth. As a result, the population inevitably becomes extinct. Even unharvested populations eventually become extinct (as dinosaurs testify). Extinction is therefore said to be an 'absorbing boundary'. Still, if we are careful, we can plan so that the waiting time to extinction is very long, hundreds to thousands of years.

Imagine a statistical collection of replicates of our harvested population, called an ensemble. The ultimate distribution of population sizes throughout the ensemble, in the limit as time tends to infinity, results in everything being at zero—every population becomes extinct. But, the distribution of population sizes for the replicates *before* they become extinct is called the quasi-stationary distribution. The proportion of the non-extinct replicates that become extinct each year is the probability of extinction.

It is possible to calculate both the quasi-stationary distribution and the rate at which replicates are becoming extinct as a function of the target stock, as detailed in Box 16.1. The lower the target the higher the probability of extinction, as illustrated by the knee-shaped curve in Fig. 16.6.

A low target yields a high present-day profit but with a high extinction rate, so the expected profit over the planning horizon is low. A high target yields a low extinction rate but with a low present-day profit, so again the expected profit over the planning horizon is low. An intermediate target strikes the best balance

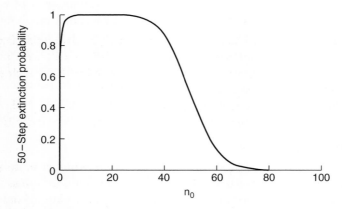

Figure 16.6 Probability of population extinction after 50 steps as a function of target stock, obtained by extending the one-step probability over 50 steps

BOX 16.1

The expected worth of the ecosystem after τ steps is the expected liquidation value of the population plus the sum of the compounded earnings from the expected harvest during the entire period (planning horizon),

$$E_\tau(n_0) = [1-\chi(n_0)]^\tau n_0 + \sum_{i=0}^{\tau-1}[1-\chi(n_0)]^{\tau-i}(1+\rho)^i f(n_0)$$

where $f(n_0)$ is the production extracted as a function of the target stock, n_0, and $\chi(n_0)$ is the probability of extinction per time step as a function of n_0. This assumes that probabilities of extinction at consecutive time steps are independent and that the price remains the same during the planning horizon. Meanwhile, the alternative worth that could be accumulated by investing the value of the target stock, n_0, elsewhere to earn interest at rate ρ compounded over the τ time steps (the opportunity cost) is

$$O_\tau(n_0) = (1+\rho)^\tau n_0$$

The optimal target, taking risk of extinction into account, is the n_0 that maximizes the difference between $E_\tau(n_0)$ and $O_\tau(n_0)$. The critical quantity to calculate for these formulas is the probability of extinction as a function of target, $\chi(n_0)$.

The probability of extinction during one time step as a function of target stock is

$$\chi(n_0) = \int_0^\infty x_{n_0}(n)\, dn$$

where $x_{n_0}(n)$ is the flux of populations becoming extinct from size n, assuming a specified target stock of n_0. This, in turn, is the product of the probability that the population is at size n times the conditional probability of going extinct given that its size is n. The extinction flux from a particular n, for a specified n_0, is

$$x_{n_0}(n) = q_{n_0}(n)\, z_{n_0}(n)$$

where $q_{n_0}(n)$ is the quasi-stationary distribution of population sizes (the distribution of population sizes among the populations that have not yet become extinct) and $z_{n_0}(n)$ is the conditional probability that a single population jumps to extinction from a population size of n.

The stochastic process that supplies formulas for $q_{n_0}(n)$ and $z_{n_0}(n)$ is a first-order autoregressive time series (Cox & Miller 1968; Karlin & Taylor 1975)

$$(n-n_0)_{t+1} = \lambda(n_0)(n-n_0)_t + \varepsilon_t$$

where $n - n_0$ is the deviation of the population size from the target stock, ε_t is an independent and normally distributed random variable with standard deviation, s, and $\lambda(n_0)$ is the discrete-time eigenvalue obtained by linearizing the dynamics of the sustainable production and harvest around the target

Continued p. 348

BOX 16.1 (*Continued*)

stock. Furthermore, zero is assumed to be an absorbing boundary representing extinction. Therefore, all populations eventually become extinct, but exist with a quasi-stationary distribution prior to extinction, and with an asymptotic probability of extinction per time step.

The quasi-stationary distribution is approximately

$$q_{n_0}(n) = \frac{(\sqrt{2/\pi})e^{-\left[\frac{(n-n_0)^2}{2\sigma^2(n_0)}\right]}}{\sigma(n_0)\left(1+\mathrm{Erf}\left[\frac{n_0}{\sqrt{2}\sigma(n_0)}\right]\right)}$$

This comes from the stationary distribution to the linear stochastic process without an absorbing boundary, which is truncated at $n=0$ and renormalized to unit area. The peak, which is approximately the mean, is at n_0. The standard deviation of the quasi-stationary distribution is approximately given by the familiar formula for the standard deviation of the stationary distribution of a first-order autoregressive process

$$\sigma(n_0) = \sqrt{\frac{s^2}{1-\lambda^2(n_0)}}$$

Given that a population's size is n, the probability that an ε_t appears which drops the population size to or below the absorbing boundary at 0 is

$$z_{n_0}(n) = (1/2)\left(1+\mathrm{Erf}\left[\frac{\lambda(n_0)(n_0-n)-n_0}{\sqrt{2}s}\right]\right)$$

The formulae for the quasi-stationary distribution and probability of extinction have been compared with numerical simulations, and appear exceedingly accurate.

Specializing, the harvest policy is assumed to be a mortality (or effort) quota. The dynamics of sustainable production and harvest then are

$$\Delta n = f - mn$$

Furthermore, the logistic production function is

$$f = \frac{rn(k-n)}{k}$$

The mortality coefficient then needed to attain a given target stock, n_0, is

$$m(n_0) = \frac{r(k-n_0)}{k}$$

The eigenvalue of these dynamics around the sustainable production–harvest equilibrium is

$$\lambda(n_0) = 1 + \partial(\Delta n)/\partial n\big|_{n \to n_0, m \to m(n_0)}$$

Continued

BOX 16.1 (Continued)

which evaluates to

$$\lambda(n_0) = 1 - \frac{rn_0}{k}$$

To summarize, the traditional optimal target is the root, n_0, in

$$\frac{d[f(n_0) - \rho n_0]}{dn_0} = 0$$

In contrast, the optimal target over τ time steps taking risk of extinction into account is the root, n_0, of

$$\frac{d[E_\tau(n_0) - O_\tau(n_0)]}{dn_0} = 0$$

between present-day profit from extraction vs. risk of extinction over the planning horizon, as illustrated in Fig. 16.7.

The long-term optimal target taking risk of extinction into account is higher than the conventional target. The optimal target is typically to the right of the peak of the production function, rather than to the left as traditionally recommended, and may be as high as $2k/3$.

This economic calculation can justify conservation under even the most pessimistic of assumptions—that the sole service is extractive and managers are not averse to taking risks. Any *in situ* services together with risk-averse management and density-dependent pricing would further strengthen an economic case for conservation.

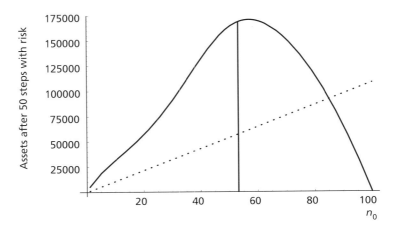

Figure 16.7 Expected assets from ecosystem extraction and from alternative investments after 50 steps as a function of target stock including risk of extinction. The optimal target is shown as a vertical line

349

Multispecies models

Weeding vs. biodiversity

Many threats to ecosystems result from open-access situations where exploitative activities are unregulated. But, will biodiversity be protected even if exploitation of the environment is tightly regulated? Not necessarily. One reason for this is the economic incentive for private landowners to weed, or selectively remove species from communities. This incentive results from asymmetries in prices and the fact that species can inhibit one another's growth.

Consider an ecosystem with two species that compete with each other, such as two species of trees. Suppose one is an expensive hardwood used for furniture, and the other a cheap softwood used only for charcoal. Both can be harvested, yet, if the competition between them is strong enough, then the total harvest revenue is maximized if the softwood is weeded out and hardwood is planted in its place, reducing the biodiversity from two species to one.

The amount of incentive for weeding depends on both the extent of asymmetry in price and the strength of the species interaction, as illustrated in Fig. 16.8. If the cheap species does not inhibit the growth of the expensive species, then there is no incentive to remove it. If the cheap species competes strongly for light and nutrients with the expensive species, then the economic incentive to remove it is high. On the other hand, if both species fetch the same price, say one from furniture makers and the other from boat builders, then again there is little incentive to weed, assuming other aspects of the trees' biology are similar.

The same argument applies to *in situ* services. Suppose one species generates considerably more rent than the other, and that the less valuable one greatly inhibits the valuable one's growth, then again there is an incentive for weeding. Thus, if some plants are beautiful and others ugly, then the economics of delivering *in situ* services favours a landscape sculpted with lovely plants and weeded of prickly, ugly and poisonous things.

We feel conservationists interested in saving biodiversity must somehow deal with this basic economic consideration. But how?

One might advocate that any present asymmetry in prices is only temporary. Tastes change, and with them prices. With the increasing popularity of sushi, fish are now sold that were valueless before. Aesthetic tastes for nature change too, from formal gardens to patches of pristine vegetation. The role of species in supplying ecosystem services can also change through time. So, biodiversity conservation may be advocated as a kind of inventory warehousing, so that nature's store is supplied to meet future demands. This justification is encapsulated in the economic concept of 'option values': these are the added value of decisions that leave future options open (Arrow & Fisher 1974; Dixit & Pindyck 1994).

One could also argue that biodiversity as such is valuable to society, that society derives benefits from biodiversity not enjoyed by private landowners. In this situation, institutional arrangements are needed to facilitate repayments to landowners who refrain from weeding for the social benefits that accrue as a result. Still, the

ECOLOGICAL ECONOMIC THEORY FOR MANAGEMENT

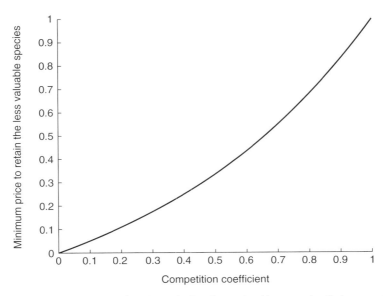

Figure 16.8 Minimum price of species-2 for it to be retained in an optimally harvested two-species ecosystem as a function of the competition coefficient between the species. The target stocks to maximize revenue from harvesting a two-component ecosystem, assuming no risk of extinction, are the n_1 and n_2 that maximize $p_1 f_1(n_1,n_2) + p_2 f_2(n_1,n_2) - \rho(p_1 n_1 + p_2 n_2)$ where p_i and f_i are, respectively, the market price per organism and production functions accounting for the species interaction for organisms of type i, and ρ is the return rate from an alternative investment. Differentiating with respect to n_1 and n_2, and rewriting yields an extended 'golden rule', $J'p = \rho p$ where p is a column vector of prices and J' is the transpose of the system's Jacobian (defined in Roughgarden 1998 or equivalent texts). For two ecologically symmetric Lotka–Volterra competitors, the production functions are $f_1(n_1,n_2) = rn_1(k - n_1 - an_2)/k$, $f_2(n_1,n_2) = rn_2(k - an_1 - n_2)/k$ where a is the competition coefficient ($0 < a < 1$). Label as species-2 the less economically valuable, and scale the price of the first species to 1, yielding a price vector of $(1; p)$ with $p < 1$. p must be higher than a threshold for species-2 to be harvested at all; otherwise it should be removed as a 'weed'. This threshold, which depends on the strength of competition between the species, is $p_{\min} = a/(2-a)$.

strongest case for biodiversity will emerge from our valuing each and every species for its own sake. The way to have a large party is to welcome each guest individually, not to advertise the total head count.

Positive feedback from ecosystem investment

How much conservation is enough? Will a hectare or two do the job, or do we need really large stretches of habitat? In traditional economics, investments typically yield diminishing marginal benefits, which means that each additional dollar's investment yields less and less incremental revenue. In contrast, investing in ecosystems may yield *increasing* marginal returns.

The idea is simple. According to the species–area curve (MacArthur & Wilson 1967), there are more species per hectare living in a large piece of habitat than in a small piece of habitat. The species–area curve itself does show diminishing return to scale—each hectare added yields fewer and fewer to the total species diversity. However, each species added yields economic benefits from all of the hectares already conserved. The overall effect then is like positive feedback. Each additional hectare makes all the other hectares better, so overall an increasing return to scale results, as illustrated in Fig. 16.9. This result would apply when setting aside species-rich habitats for biodiversity prospecting. This result is solely about the shape of the gross benefit function; the shape of the cost function would also need consideration.

An increasing return to scale also emerged in a model for an exploited metapopulation (Brown & Roughgarden 1997). An increasing return to scale introduces

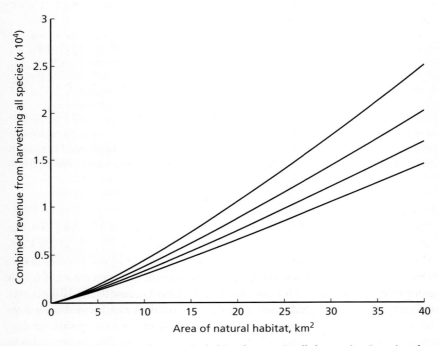

Figure 16.9 Revenue summed over entire habitat from optimally harvesting S species who compete with diffuse competition as a function of habitat area. Diffuse competition is a prototype single trophic level organization that can be solved for an arbitrary number of species. The production functions are, for $i = 1\ldots S$, $f_i(n_1,\ldots,n_S) = rn_i(k - \Sigma_{j=1}^{S} a_{ij} n_j)/k$. The matrix of competition coefficients, $a_{ii} = 1$, $a_{i,j} = a(i \neq j)$. If all the species fetch the same price, the price vector is a column of 1s. The number of species varies with habitat area as a power law, $S = bA^z$. Curves from top to bottom are for $a = 0, 0.01, 0.02$ and 0.03. Also, $r = 1, \rho = 0.05$, $k = 100, b = 10$ and $z = 1/4$.

a different intuition for investment in nature compared to investment in typical enterprises. Nature does not depreciate; nature appreciates.

Discussion

Sources of environmental preference

Why should we value nature? From an economic perspective it does not matter why, just as long as we do.

Some find that buying and selling nature integrates natural products and services seamlessly into our human economy, ensuring an efficient rather than wasteful use of nature (Krautkraemer 1995). Others see environmental stewardship as a moral responsibility (Leopold 1966). Still others see the environment as an extension of self (Allen 1990). People with each of these positions may feel uncomfortable with those of others. Yet in some sense it does not matter. So far as ecological economics is concerned, all reasons for environmental preferences are extensionally equivalent. We do not need ideological uniformity.

Economists use the phrase 'consumer sovereignty' to describe how economics takes consumer preferences as a primitive. Hopefully, consumers will value the good, but economic theory is indifferent to the moral standing of whatever the consumer preferences actually are. This leaves one free to advocate environmental preferences from any of these philosophical positions.

Economists have a tool-kit of sophisticated valuation techniques at their disposal. None of these techniques is perfect, and some are more appropriate for certain situations and services than others. For an introduction to relevant methods and their advantages and disadvantages consult Pearce (1993), Smith (1993) and Heal (2000).

Postmodern ecology and economics

The last century enjoyed continuing debates, yet suffered debilitating violence, between partisans of two rival economic systems, capitalism and socialism. Capitalism was not kind to the environment, to women or to the poor, because allocational inequities and restricted access to markets meant that preferences of certain groups were not accounted for. Socialism was bad too, because its need for a central authority to manage production excluded perspectives not focused on power. Today, one game remains in town. However, the absence of a competing ideology should not preclude debate over how that one game is played. For the next century, we believe an important challenge for society is to construct a postmodern economics, one founded on free-market exchange of goods and services yet containing new institutions that promote equity and environmental care.

To move beyond the oppositional stances of the recent past, we need new images of possibilities, images that include elements of nature, body, community, technology and markets. We envision a postmodern landscape with less intrusive models of management, with more commons, fewer partitions and broadly dispersed ecosystem use. We see more citizen participation and less peer review, more local and less global, and a respectful balance between particular and general. We will celebrate diversity, for in diversity lies freedom from control.

Summary

In this paper, we have attempted to illustrate how one might construct a theory of ecological economics to inform the management of ecosystem services. We began with traditional single-species models from natural resource economics and showed how these could be generalized to account for *in situ* values. Then we discussed how these models might relate to issues of reserve design. We concluded that section by introducing a theoretical method for determining optimal management strategies when constrained by imperfect implementation and institutional inflexibility. Importantly, this method costed out extinction risk. In the next section, we showed how the earlier theory could be extended to multispecies systems, and demonstrated that without institutional intervention even tightly regulated ecosystems are at risk. Our final model illustrated how the conventional economic wisdom of diminishing marginal benefits may not apply to certain ecosystem services.

It was our aim to produce a primer of economic ideas for ecologists, one that would provoke questions and stimulate research and creative thinking. To this end, we have addressed a diverse set of questions rather than any one question in detail, and have attempted to highlight instances where economic revisionism may be required.

Acknowledgements

We thank Sandy Andelman, Jordi Bascompte, Susan Holmes, Hugh Possingham and Ottar Bjornstad of the National Center for Ecological Analysis and Synthesis (NCEAS) Working Group in Theoretical Ecological Economics for discussion throughout this work, and Jim Reichman and the staff of NCEAS for the interest, trust and supportive environment that made this research possible. The NCEAS is a centre funded by the National Science Foundation (Grant DEB-94-21535), the University of California at Santa Barbara, and the State of California. Stephen Carpenter, Gretchen Daily, Jessica Hellmann, Paul Higgins, Jennifer Hughes and three anonymous referees provided valuable feedback on the manuscript. P.R.A. received additional support from the US–UK Fulbright Commission.

References

Allen, P.G. (1990) The woman I love is a planet. In: *Reweaving the World: the Emergence of Ecofeminism* (eds I. Diamond & G. Orenstein), pp. 52–57. Sierra Club Books, San Francisco.

Arrow, K.J. & Fisher, A.C. (1974) Environmental preservation, uncertainty, and irreversibility. *Quarterly Journal of Economics* **88**, 312–319.

Brown, G. & Roughgarden, J. (1997) A metapopulation with private property and common pool. *Ecological Economics* **22**, 65–71.

Bulte, E. & van Kooten, G.C. (2000) Economic science, endangered species and biodiversity loss. *Conservation Biology* **14**, 113–119.

Clark, C.W. (1990) *Mathematical Bioeconomics: the Optimal Management of Renewable Resources*, 2nd edn. Wiley-Interscience, New York.

Costanza, R. (1996) Ecological economics: reintegrating the study of humans and nature. *Ecological Applications* **6**, 978–990.

Cox, D.R. & Miller, H.D. (1968). *The Theory of Stochastic Processes*. John Wiley & Sons, New York.

Daily, G.C. (1997) *Nature's Services: Societal Dependence on Natural Ecosystems*. Island Press, Washington DC.

Dasgupta, P.S. & Heal, G.M. (1979) *Economic Theory and Exhaustible Resources*. Cambridge University Press, Cambridge.

Dixit, A.K. & Pindyck, R.S. (1994) *Investment Under Uncertainty*. Princeton University Press, Princeton, NJ.

Ehrlich, P.R. & Mooney, H.A. (1983) Extinction, substitution, and ecosystem services. *Bioscience* **33**, 248–254.

Goulder, L.H. & Kennedy, D. (1997) Valuing ecosystem services: philosophical bases and empirical methods. In: *Nature's Services: Societal Dependence on Natural Ecosystems* (ed. G.C. Daily), pp. 23–47. Island Press, Washington DC.

Hartman, R. (1976) The harvesting decision when a standing forest has value. *Economic Inquiry* **14**, 52–58.

Heal, G. (2000) Valuing ecosystem services. *Ecosystems* **3**, 24–30.

Intriligator, M.D. (1971) *Mathematical Optimization and Economic Theory*. Prentice-Hall, Englewood Cliffs, NJ.

Karlin, S. & Taylor, H.M. (1975) *A First Course in Stochastic Processes*, 2nd edn. Academic Press, New York.

Kneese, A.V. & Sweeney, J.L. (1985) *Handbook of Natural Resource and Energy Economics*, Vols 1–3. Elsevier, Amsterdam.

Kolstad, C.D. (2000) *Environmental Economics*. Oxford University Press, Oxford.

Krautkraemer, J.A. (1995) Incentives, development and population: a growth-theoretic perspective. In: *The Economics and Ecology of Biodiversity Decline: the Forces Driving Global Change* (ed. T.M. Swanson), pp. 13–23. Cambridge University Press, Cambridge.

Krutilla, J.V. (1967) Conservation reconsidered. *American Economic Review* **47**, 777–786.

Lauck, T., Clark, C.W., Mangel, M. & Munro, G.R. (1998) Implementing the precautionary principle in fisheries management through marine reserves. *Ecological Applications* **8**, s72–s78.

Leopold, A. (1966) *A Sand County Almanac with Essays from Round River*. Ballantine Books, New York.

MacArthur, R.H. & Wilson, E.O. (1967) *The Theory of Island Biogeography*. Princeton University Press, Princeton, NJ.

Mangel, M. (2000) On the fraction of habitat allocated to marine reserves. *Ecology Letters* **3**, 15–22.

Milner-Gulland, E.J. & Mace, R. (1998) *Conservation of Biological Resources*. Blackwell, Oxford.

Naveh, Z. (2000) The Total Human Ecosystem: integrating ecology and economics. *Bioscience* **50**, 357–361.

Nisbet, R.M. & Gurney, W.S.C. (1982) *Modelling Fluctuating Populations*. John Wiley & Sons, Chichester.

O'Neill, R.V., Kahn, J.R. & Russell, C.S. (1998) Economics and ecology: the need for détente in conservation ecology. *Conservation Ecology*, 2. [online] URL: http://www.consecol.org/vol2/iss1/art4.

Pearce, D.W. (1993) *Economic Values and the Natural World*. MIT Press, Cambridge, MA.

Plourde, C.G. (1970) A simple model of replenishable resource exploitation. *American Economic Review* **60**, 518–522.

Renshaw, E. (1991) *Modelling Biological Populations in Space and Time*. Cambridge University Press, Cambridge.

Roughgarden, J. (1997) Production functions from ecological populations: a survey with emphasis on spatially implicit models. In: *Spatial Ecology: the Role of Space in Population Dynamics and Interspecific Interactions* (eds D. Tilman & P. Kareiva), pp. 296–317. Princeton University Press, Princeton, NJ.

Roughgarden, J. (1998) *Primer of Ecological Theory*. Prentice-Hall, Upper Saddle River, NJ.

Roughgarden, J. (2001) Guide to diplomatic relations with economists. *Bulletin of the Ecological Society of America* **82**, 85–88.

Roughgarden, J. & Smith, F. (1996) Why fisheries collapse and what to do about it. *Proceedings of the National Academy of the Sciences* **93**, 5078–5083.

Smith, V.K. (1993) Nonmarket valuation of environmental resources: an interpretive appraisal. *Land Economics* **69**, 1–26.

Vitousek, P.M., Mooney, H.A., Lubchenco, J. & Melillo, J.M. (1997) Human domination of Earth's ecosystems. *Science* **277**, 494–499.

Chapter 17
Alternate states of ecosystems: evidence and some implications

S. R. Carpenter

Introduction

Ecosystems' components exhibit a great range of turnover rates, ranging from chemical reactions that reach equilibrium in milliseconds, to life cycles of organisms that span years or decades, to geomorphic processes that develop over millennia. The coupling of processes with diverse turnover rates creates the complex dynamics that we observe in ecosystems: gradual trends, apparently stable regimes, and sometimes surprising outbreaks and collapses. It is easy to understand that gradual change in the environment may lead to gradual change in biological constituents of ecosystems. From this, it may appear to follow that abrupt changes in ecosystem structure or processes are attributable to abrupt changes in the environment. However, ecologists have long recognized that gradual changes in environmental variables can sometimes cause abrupt changes in ecosystems. 'Big effects from small causes' (Ricker 1963) have long fascinated ecologists and challenged managers of natural resources.

Multiple stable states are one of the phenomena that can lead to massive changes in ecosystems from only minor changes in the environment (Holling 1973). Various mathematical models have been proposed to investigate multiple stable states and abrupt transitions among them (May 1977; Guckenheimer & Holmes 1983; Kuznetsov 1995; Ludwig et al. 1997; Scheffer 1998). Ecologists have also devoted extensive field work and empirical analysis to assessing and understanding multiple states. There is considerable diversity of opinion, however, about the general applicability of the non-linear dynamic models that may explain multiple states of ecosystems. Phenomena that appear to be caused by multiple states may instead be explainable by variables that were inadvertently left out of the analysis, or by completely novel factors such as species invasion or pollution by exotic chemicals. Thus, the unambiguous demonstration of multiple states requires careful characterization of the system and synthesis of diverse lines of evidence.

This paper is about the theory, empiricism and practical consequences of multiple states. Several examples are discussed to illustrate the diversity of ecological

Center for Limnology, University of Wisconsin, Madison, Wisconsin 53706, USA. E-mail: srcarpen@facstaff.wisc.edu

situations in which alternate states can arise. Some comments on the nature of evidence for multiple states are summarized, and approaches for assessing multiple states in ecological data are outlined. It appears that statistical tests for multiple states demand substantial bodies of data. While adequate data can be gathered in principle, and have been collected in some cases, in many management situations it may be impossible to conclude whether multiple states exist on the basis of extant data. Using economic optimization criteria, it can be shown that even if the probability of multiple states is low the implications for policy choice may be profound. The heavy impact of multiple states on policy choice results from the high costs of restoring the ecosystem if it is shifted into an undesirable stable state. These high costs are due to the hysteretic or irreversible dynamics that occur in systems with multiple states. Thus, even low probabilities of multiple states may lead to considerable caution in environmental policy, and support the need for careful experimentation to determine risks more accurately.

Theory

The idea of multiple stable states was introduced to ecology through theoretical papers (Lewontin 1969; Holling 1973; Austin & Cook 1974; Noy-Meir 1975; May 1977). These papers stimulated a considerable body of research aimed at demonstrating and understanding multiple stable states in the field (see following sections). This section summarizes the relevant theory.

Most textbooks of ecology give examples of ecological models that exhibit multiple states. As a pertinent example, consider an animal population X in which the adults are subject to harvesting and the juveniles are subject to predation by another species. These other predators are themselves controlled via consumption by the adults of X. A simple model of such a system is

$$X_{t+1} = X_t - kX_t + fX_t\{1 - mX_t - [c/(1+X_t q)]\} \tag{1}$$

Harvest mortality is k, fecundity is f, density-dependent mortality is m, and c is the maximum mortality rate due to predation on juveniles by the other species. This last mortality term is inversely related to adult density because the adults prey upon the other species. For certain combinations of parameter values, equation 1 has two equilibria (Fig. 17.1(a)). The upper equilibrium is stable, while the lower one is unstable leading to collapse of the population toward zero. A population near the upper stable point could be disturbed below the lower equilibrium, causing the population to fall toward zero. This phenomenon, called critical depensation, is an important consideration in the management of fish stocks (Walters 1986; Hilborn & Walters 1992; Lierman & Hilborn 1997).

Dynamics of X can change discontinuously, from persistence to multiple equilibria to collapse, as a parameter changes gradually and continuously (Fig. 17.1(b)). For example, if the juvenile predation parameter is 1 and harvest is increased gradually from zero, the population initially has one stable positive equilibrium, then has two equilibria (and could collapse if disturbed below the lower

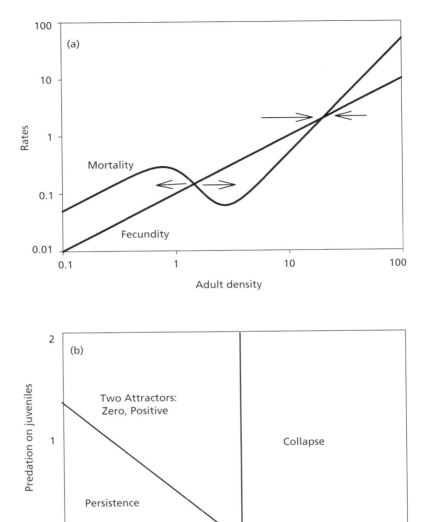

Figure 17.1 (a) Rates of fecundity and total mortality (harvest + density-dependent + predation on juveniles) for equation 1 showing two equilibria (parameter values: $f=0.5$, $m=0.01, c=1, k=0.4$). (b) Relationship of the equilibria of equation 1 to values of predation on juveniles (c) and harvest of adults (k) (other parameters: $f=0.5, m=0.01$).

unstable one), then has only an equilibrium at zero as harvest rate passes above ≈ 0.5. Sharp changes in dynamics from gradual changes in parameters are an important reason for ecologists' interest in alternate states. In ecological models, most parameters are not universal constants. Instead, they are quantities that are roughly constant for certain specified scales of time and space. Thus, parameters may change as the spatial or temporal extent of the problem changes. It is important to know whether such gradual changes can lead to sharp qualitative changes in ecosystem dynamics.

When coupled ecosystem components have very different turnover rates, the component that turns over slowly may control the types of dynamics possible for the faster component. We find coupled slow and fast processes in mineralization of soil and competition of plants, or in life cycles of fish and plankton, or trees and canopy insects, for example. In general, suppose two ecosystem components are coupled by the equations

$$X_{1,t+1} = X_{1,t} + f_1(X_{1,t}, X_{2,t}, B_1) \qquad (2)$$

$$X_{2,t+1} = X_{2,t} + cf_2(X_{1,t}, X_{2,t}, B_2) \qquad (3)$$

where functions f_1 and f_2 describe the interactions, B_1 and B_2 are vectors of parameters, and c is a parameter that controls the turnover rate of X_2. As c becomes small, X_2 becomes slow and will change only gradually in comparison with X_1. If X_1 is subject to sharp changes in dynamics as X_2 passes a threshold, then the slow changes in X_2 operate analogously to the changes in the parameter k in the example of Fig. 17.1(b). Thus, the hierarchy of turnover times in ecosystem components may lead to a hierarchy of controls in which slower components determine the qualitative dynamics of faster ones. Of course, a change in dynamics of X_1 feeds back to affect X_2 through equation 3. Because of these feedbacks, the taxonomy of possible dynamics is rich and complicated.

Some models of hysteretic shifts among multiple states are presented by Noy-Meir (1975), Peterman (1977), Ludwig et al. (1997) and Carpenter et al. (1999). Vandermeer and Yodzis (1999) discuss a general class of models that can lead to shifts among alternate states, and argue that these may explain a number of long-term changes seen in ecosystems.

The contrast between gradual, stable dynamics and sharp shifts among very different states is profound. It has deep implications for our understanding of ecosystem change, methods for predicting ecosystem change, and policy choice. Ludwig et al. (1997) provide an especially clear metaphor, which I can do no better than to quote in full:

If a weight is added suddenly to a raft floating on water, the usual response is for the weighted raft to oscillate, but the oscillations gradually decrease in amplitude as the energy of the oscillations is dissipated in waves and, eventually, in heat. The weighted raft will come to rest in a different position than the unweighted raft, but we think of the new configuration as essentially the same as the old one. The system is stable. If we gradually increase the weight on the raft, eventually the configuration will change. If the weight is hung below the raft, the raft will sink deeper and deeper into the water as more and more displacement is required to balance the higher gravitational force. Eventually, the buoyant force cannot balance the gravitational force and the whole configuration sinks: the system is no longer stable. On the other hand, if the weight is placed on top of the raft, the raft may flip over suddenly and lose the weight and its other contents long before the point at which the system, as a whole, would sink. This sudden loss of stability may be more dangerous than the gradual sinking, because there may be little warning or opportunity to prepare for it. We may think of the raft system as losing its resilience as more weight is placed on it. Suppose that we accept the 'balance of nature' and the steady flows of resources that it implies. As we demand more and more of the products of natural systems, and we load them with more and more of our waste products, are we likely to experience a gradual loss of stability or a sudden one?

The raft metaphor illustrates the risks of abrupt change, but multiple states also imply that recovery from unwanted changes may be slow and costly, or impossible.

Evidence and examples

In a prominent review paper, Connell and Sousa (1983) criticized case studies of alternate states and concluded that 'There was no evidence of multiple stable states in unexploited natural populations or communities' (p. 808). Their paper prompted considerable debate about the nature of evidence for stability in general and alternate states in particular (Peterson 1984; Sousa & Connell 1985; Sutherland 1990). More recently, several review papers or books have presented numerous examples of alternate or alternating stable states (Laycock 1991; Knowlton 1992; Wilson & Agnew 1992; Jeppesen et al. 1998; Scheffer 1998). There appears to be a sea change in the literature, for which several explanations are possible. Almost 20 years of additional research are available since the critique by Connell and Sousa (1983). It is likely that their criticisms have caused scientists to be more rigorous and thorough in adducing evidence for or against alternate states. While Connell and Sousa (1983) disallowed examples of alternate states in which the environment appeared to change, Peterson (1984) pointed out that environmental changes mediated by organisms could be an important mechanism leading to alternate states. In both terrestrial plant communities (Wilson & Agnew 1992) and shallow lakes (Jeppesen et al. 1998; Scheffer 1998), environmental changes caused by organisms are central to maintenance of alternate states and to shifts among them. Climate modellers recognize that feedbacks between the atmosphere and vegetation create alternate states in monsoonal systems (Broström et al. 1998). Much of the argument of Connell and Sousa (1983) is directed at how one could demonstrate stability in a stochastic world. Ecologists now have empirical tools for assessing stochastic stability (Ives 1995). Also, it is now clear that multiple attractors for ecosystem dynamics may be points or various kinds of cycles (Zimmer 1999). Alternate states are but a subset of the attractors that may arise in ecosystems.

This section presents selected recent examples of alternate states. The goal is to illustrate the variety of systems in which multiple states occur, as well as the diversity of mechanisms that are involved in multiple states. This paper considers cases of multiple stable states in the strict sense, as well as cases in which feedbacks between slow and fast variables create alternating regimes. Such systems may cycle over long time periods corresponding to many turnovers of the slow variables. Over shorter time periods that nevertheless correspond to many turnovers of the fast variables, such regimes may appear stable. Moreover, certain types of gradual changes or certain perturbations may cause rapid shifts among regimes.

Rapid climate change

Simulation models suggest that oceanic circulation can switch abruptly among attractors that create significantly different climates (Rahmstorf 1997). Thus, greenhouse warming may have rapid, unexpected effects on climate (Broecker 1997). The various regimes found in climate systems are temporary, because one attractor eventually gives way to another. Each regime, however, persists for a long time relative to ecological cycles, and shifts between regimes may have great ecological importance. Palaeoclimate reconstructions based on ice cores and marine sediments show that 'climate changes large enough to have extensive impacts on our society have occurred in less than 10 years' (Taylor 1999). In one case, during the recovery from the most recent glaciation, Europe's climate abruptly lurched back to ice-age conditions for about 400 years. This shift was most likely caused by a massive influx of meltwater to the North Atlantic which disrupted the oceanic currents that deliver heat to eastern North America, Europe and Scandinavia (Rahmstorf 1997). The increase in freshwater input to the oceans is comparable to increases that may occur as a result of greenhouse warming. Other simulation studies have demonstrated rapid, coupled shifts of vegetation and climate in Saharan Africa that resemble switches between alternate states (Broström *et al.* 1998; Braconnot *et al.* 1999; DeMenocal *et al.* 2000).

Phosphorus recycling and lake eutrophication

Excessive inputs of phosphorus to lakes cause eutrophication, characterized by blooms of noxious algae, periods of oxygen depletion, fish kills, loss of recreational amenities, and additional costs to purify water for drinking, industry or irrigation (Smith 1998). Eutrophy can be a self-sustaining state because of phosphorus recycling from sediments (Nürnberg 1984). At high phosphorus levels, anoxia develops near the sediment surface. Anoxia solubilizes iron–phosphorus complexes, and thereby accelerates phosphorus recycling, creating a positive feedback that sustains eutrophication. At low phosphorus levels, water near the sediment remains oxygenated, converting phosphorus to insoluble forms and thereby creating a positive feedback that sustains the clear-water condition.

Recycling causes sources of phosphorus to lake water change non-linearly with phosphorus mass or concentration in the water (Fig. 17.2). At low levels of water phosphorus, recycling is small, while at high levels of water phosphorus recycling is rapid. At an intermediate level of water phosphorus, duration of anoxia increases steeply leading to a steep increase in recycling rate. Loss rate of phosphorus from the water is roughly linear with phosphorus mass in the water. This system can have one or two stable states depending on the relative positions of the loss and input curves (Carpenter et al. 1999b). Several lines of evidence are consistent with this model of lake dynamics (Carpenter et al. 1999a):

1 chemical studies of phosphorus recycling demonstrate the underlying mechanism;

2 comparative studies of large numbers of lakes show that recycling rates can exceed loss rates from the water column (as required for the loss and input curves to cross in three places as depicted in Fig. 17.2);

3 all available case studies of lake restoration can be explained by the model, including cases in which reduction of phosphorus inputs failed to mitigate eutrophication because of excessive phosphorus recycling from sediments;

4 all common methods for restoring eutrophication can be represented as special cases of the model; and

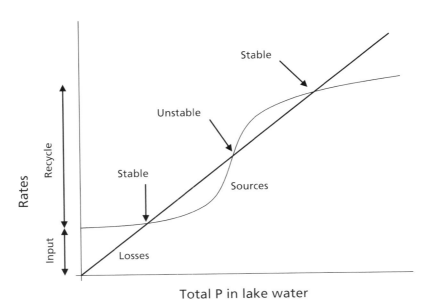

Figure 17.2 Rates of phosphorus flux in a lake ecosystem vs. mass of phosphorus in the water (Carpenter et al. 1999b). Straight line shows losses (outflow plus sedimentation), curved line shows sources (input plus recycling).

5 the model adequately fits time series from one of the world's longest-lasting limnological studies.

Succession in hemlock–hardwood forests

In forests of the Western Great Lakes region, degree of dominance by late successional shade-tolerant species depends on neighbourhood effects (the likelihood that a canopy tree will be replaced by a tree of the same species) and disturbance intensity (proportion of trees killed during a disturbance episode) (Frelich & Reich 1999). If neighbourhood effects are weak, then dominance by shade-tolerant species decreases monotonically as disturbances intensify. However, if neighbourhood effects are strong then forest composition has alternate stable states. Frelich and Reich present data on postdisturbance composition of stands vs. disturbance intensity, for stands of different initial tree composition (Fig. 17.3). Lightly disturbed sites remain dominated by shade-tolerant hemlock and hardwoods, and disturbance must become relatively intense to alter this situation. At sites where disturbance has been heavy, shade-intolerant aspen and paper birch predominate. As disturbance intensity is relaxed, aspen and paper birch remain dominant until disturbance intensity reaches a relatively mild threshold level. The degree of disturbance necessary to switch from shade-intolerant species to shade-tolerant ones is much milder than the degree of disturbance required for the reverse switch. Such hysteresis is characteristic of alternate or alternating states (Scheffer 1998). Wilson and Agnew (1992) review many examples of positive feedback systems that maintain particular vegetation types. Transitions between such feedback systems create alternations among stable states.

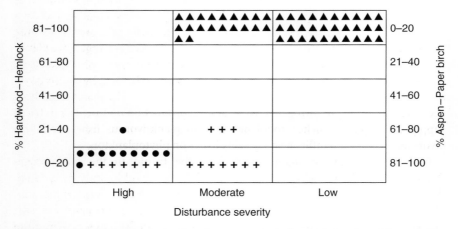

Figure 17.3 Dominant species vs. disturbance severity in hemlock–hardwood forests of the Western Great Lakes region. Closed circles denote stands that changed state after disturbance. Stands that did not change state after disturbance are denoted by triangles (hardwood–hemlock) or crosses (aspen–birch). (Reproduced with permission from Frelich & Reich 1999.)

African savannah, elephants and fire

Woodlands in certain areas of Africa are being replaced by grasslands. The explanations for the decline include fire alone (one steady state controlled by frequency of fire) and a combination of fires and herbivory (two steady states). Using measurements of fire frequency, grazing rate of various herbivores, and rates of woodland decline, Dublin *et al.* (1990) showed that fire is the only plausible explanation for grassland creation. Grazing by elephant and other browsers is insufficient to cause the shift to grassland. However, once the grassland is formed, increased elephant density is sufficient to maintain the grass-dominated state. Simulation studies suggested that hunting could reduce elephant densities sufficiently to allow the vegetation to return to the wooded state. Thus, Dublin *et al.* (1990) conclude that the system exhibits two stable states, woodland and grassland, the first maintained by low to moderate fire frequency and the second maintained by elephant browsing. They show that the shift from woodland to grassland is most plausibly explained by increased fire, and that the reverse shift could potentially be achieved by herbivore reduction. Numerous other cases of alternate and alternating states in grasslands are reviewed by Laycock (1991). Zimov *et al.* (1995) use palaeo-ecological data, experiments, and a simulation model to show that transitions between shortgrass steppe and tundra are alternate stable states controlled by climate and herbivore grazing. They suggest that megafaunal extinctions caused by human hunting explain the transition from steppe to tundra in Beringia at the end of the Pleistocene.

Phytoplankton vs. macrophyte dominance in shallow lakes

Shallow lakes (those that are completely mixed during the ice-free season) exist in turbid or macrophyte-dominated states (Jeppesen *et al.* 1998; Scheffer 1998). In the turbid state, shading by phytoplankton prevents the expansion of macrophyte beds. Wind-driven mixing resuspends sediment and recycles nutrients, sustaining phytoplankton production. Herbivorous zooplankton have no refuge from planktivores, are subject to heavy predation, and are unable to suppress phytoplankton. In the macrophyte-dominated state, sediments are stabilized by macrophyte roots. Nutrient recycling is low, and rapid denitrification in the macrophyte beds decreases availability of nitrogen. Phytoplankton production is limited by nutrient supply. Macrophytes shelter zooplankton from planktivorous fishes, and herbivorous zooplankton inflict heavy grazing losses on phytoplankton.

Shallow lakes can be manipulated from the turbid to the macrophyte-dominated state by reducing biomass of planktivorous and benthivorous fishes. Removal of planktivores allows herbivory to increase, thereby reducing phytoplankton biomass. Benthivorous fishes resuspend sediments and uproot macrophytes, so their removal decreases nutrient recycling and facilitates expansion of macrophyte beds. The reverse transition, from macrophyte dominance to turbidity, can be accomplished by fish reintroduction or by waterfowl grazing on macrophytes (Van Donk & Gulati 1995; Jeppesen *et al.* 1998).

Blindow *et al.* (1993) present cases of shifts between stable states in shallow lakes, using time series spanning nearly a century. Repeated shifts between the turbid and vegetated states were associated with hydrologic fluctuations.

Shifting predator dominance in marine benthos

Barkai and McQuaid (1988) present an example of alternate states maintained by predation off Western Cape, South Africa (Fig. 17.4). On Malgas Island, the rock lobster *Jasus lalandii* is sustained by several prey species, and consistently holds populations of the whelk *Burnupena papyracea* at low levels. On nearby Marcus Island, *Burnupena papyracea* maintains high population densities supported by mussel prey, and rock lobsters are absent. Experiments show that rock lobsters thrive in cages on Marcus Island, but uncaged rock lobsters are immediately attacked and consumed by whelks. Barkai and McQuaid note that rock lobsters were once abundant on Marcus Island, and their populations may have been reduced by a period of low oxygen, allowing the whelks to become dominant. Other cases of alternate states in marine benthos are discussed by Knowlton (1992) and Petraitis and Latham (1999).

Fish recruitment cycles and trophic cascades in lakes

Trophic cascades from fish to phytoplankton have now been demonstrated in many pelagic systems (Carpenter & Kitchell 1993; Hansson *et al.* 1998; Pace *et al.* 1999; Persson 1999). Cascades can lead to multiple states of fish community structure through depensation, as illustrated by equation 1 and Fig. 17.1 (Steele & Henderson 1984).

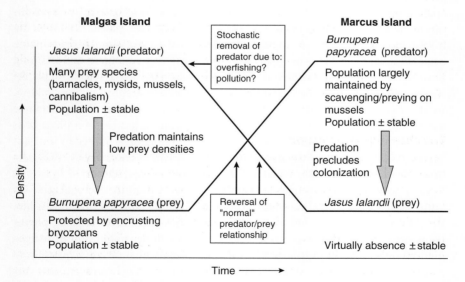

Figure 17.4 Density of rock lobster (*Jasus lalandii*) and whelk (*Burnupena papyracea*) and mechanisms structuring their interaction on Malgas Island and Marcus Island, South Africa. (Reproduced with permission from Barkai & McQuaid 1988.)

In simple fish communities, roughly periodic shifts among regimes of piscivore vs. planktivore dominance can arise from recruitment processes of top predators (Carpenter 1988). Even moderate densities of adult fish can suppress recruitment through predation (including cannibalism). As a dominant cohort of adults dies off, predation declines to the point where a new cohort can recruit. The new year class transforms the food web as the fishes grow and consume successively larger prey. Once the new cohort reaches the apex of the food web, it structures the community for the lifetime of the fishes. Thus, the system cycles through regimes of dominance by adult fishes, punctuated by intervals when recruitment is possible and planktivory by juvenile fishes may be intense (Carpenter 1988). Evidence for such cycles comes from models fit to data from whole-lake experiments (Carpenter 1988), palaeolimnological records (Carpenter & Leavitt 1991) and long-term studies of all trophic levels of unperturbed lakes (Post *et al.* 1997; Sanderson 1998; Sanderson *et al.* 1999). The fish recruitment phenomenon is a cycle (Carpenter 1988). Over intervals of a few years, or hundreds of plankton generations, the regimes are stable to modest perturbations (Carpenter *et al.* 1992). Thus, at time scales pertinent to the plankton, the system appears to have alternate states.

Alternate states of a haddock fishery

Models with single stable states and alternate stable states were explicitly compared to long-term data on haddock stocks by Spencer and Collie (1997). Although the alternate state model appeared to better represent the range of variability in the data, the two models could not be distinguished statistically (Fig. 17.5). However, Spencer and Collie (1997) pointed out that significant management miscalculations could result from ignoring the possibility of alternate states. Their analysis provides a concrete example of the two general points to be explored in the remainder of this chapter—that multiple stable states are not easy to verify using a single source of information, yet ignoring the possibility of multiple stable states invites serious errors in ecosystem management.

The challenge of detection

Several recent papers have considered the types of evidence necessary to test for the existence of alternate states. Scheffer (1998) describes the symptoms of hysteresis in an ecological process using plots of a state variable vs. a control factor (Fig. 17.6). Long-term observations or experimental data may show that state changes abruptly at a certain threshold level of the control factor (Fig. 17.6a). Reversal of the state occurs at a different level of the control factor (Fig. 17.6b). The difference in thresholds depending on direction of the state transformation is a key characteristic of hysteresis. Certain perturbations may cause the system to shift from one basin of attraction to another (Fig. 17.6c). Long-term studies and experiments are likely sources of evidence for the phenomena depicted in Fig. 17.6(a–c). Comparative data from many different ecosystems may also provide evidence for alternate

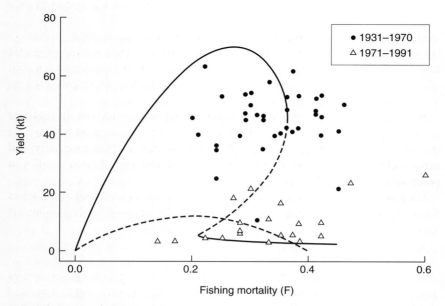

Figure 17.5 Haddock yield vs. fishing mortality for the George's Bank haddock stock. The dashed line is the Schaefer model (one stable point) and the solid–dashed line is the Steele–Henderson model (two stable points). Solid circles are 1931–70. Heavy fishing mortality occurred in the late 1960s. Open triangles are 1971–91, after the period of heavy exploitation. (Reproduced with permission from Spencer & Collie 1997.)

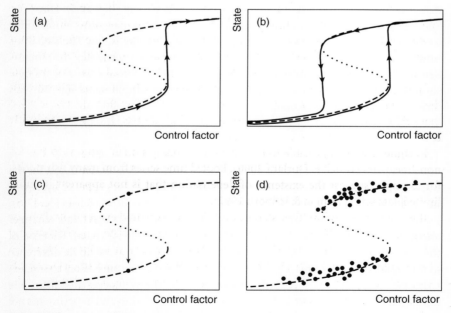

Figure 17.6 Characteristics of hysteretic systems. In each panel, the vertical axis is a state variable and the horizontal axis is a controlling factor. (Reproduced with permission from Scheffer 1998.)

states. In this case we would expect to see a bimodal distribution of states yet overlapping distributions of the control factor (Fig. 17.6d). It is possible, however, that data such as Fig. 17.6(d) could result from a second control factor that acts on one of the two clusters of observations, rather than from alternate states. This possibility could be tested by experimentation.

Petraitis and Latham (1999) discuss experimental designs that test for alternate states. They are especially interested in biotic mechanisms that generate alternate states of species assemblages. They point out that an especially informative type of experiment is one in which both the intensity of disturbance and the availability of potential alternative colonists can be manipulated.

Long-term observations and ecosystem experiments typically yield time series from one or a few ecosystems, which could potentially be tested for the presence of multiple states. One possibility is to fit a polynomial

$$X_{t+1} = X_t + f(X_t, X_2^t, U_t) \tag{4}$$

X is a state variable of interest, U is a control variable (or vector of control variables), and f is a function of X and U which contains cubic or higher powers of X. The polynomial will have either one or more stable states. The probability of multiple stable states can be obtained by bootstrapping from the parameter distribution to determine the proportion of parameter sets that yield multiple stable states.

Carpenter and Pace (1997) devised three different methods of testing for alternate states in ecological time series. One method was similar to the application of equation 4 described above. The second method was based on discriminating alternative polynomial models, one of which had alternate states. The third method was based on discriminating alternative autoregressive processes, one centred on a single mean and the other centred on two different means. Their goal was to test for alternate states in the interactions of chlorophyll and coloured dissolved organic carbon compounds in lake ecosystem experiments. Models with alternate states fit their data about as well as models with a single attractor. Therefore they concluded that evidence for alternate states was equivocal. Their approach may be applicable to a wide variety of ecological time series.

In some cases it is possible to pool data from many similar systems to test for non-linearities (Brock & Durlauf 2000). Pooled time series from many fish stocks, for example, suggest the existence of depensation that is not apparent in more limited data sets (Lierman & Hilborn 1997).

If the hypothesis of multiple stable points is true, we would expect to find one or more unstable equilibria marked by diverging dynamics. Both ecologists and economists have developed statistical methods for detecting instabilities (Dennis *et al.* 1995; Ellner & Turchin 1995; Brock *et al.* 1996; Whang & Linton 1999; Dechert & Hommes 2000). These might be adapted to test for divergences that could be explained by alternate states.

The case studies presented by Spencer and Collie (1997) and Carpenter and Pace (1997) suggest that rigorous detection of alternate states places heavy demands on the data. To explore this, I generated artificial time series for a system with two stable states (Appendix 1). In the examples shown in Fig. 17.7, the stable states occur at $X = 1$ and $X = -0.4$, which are separated by an unstable (repelling) state at $X = 0.4$. Deterministic simulations show convergence to the stable states (Fig. 17.7a). Process error causes small random shocks to the system at each time step (Fig. 17.7b). Process error arises whenever a model is incompletely specified or the

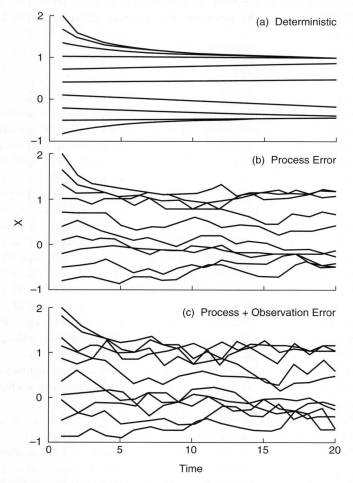

Figure 17.7 Simulations of a process with two alternate states at $X = 1$ and -0.4, separated by an unstable (repelling) state at $X = 0.4$. (a) Deterministic simulations; (b) simulations with process error ($s_v = 0.1$); (c) simulations with both process error ($s_v = 0.1$) and observation error ($s_w = 0.01$).

ecosystem is subject to external disturbances (Hilborn & Mangel 1997), which will always be the case for open ecological systems (Burnham & Anderson 1998). Process error complicates the trajectories, and can even cause trajectories near the unstable point to switch basins of attraction. Observation error cannot cause switches between basins of attraction, but it does obscure the dynamics (Fig. 17.7c). Most ecological time series are obtained by repeated sampling of a heterogeneous population, and such observations inevitably have some uncertainty which represents observation error (Hilborn & Mangel 1997). Both observation and process errors arise in most ecological time series.

Are alternate states detectable in a collection of time series like those of Fig. 17.7? To illustrate the prospects for detecting alternate states, I fit the polynomial

$$X_{t+1} = b_0 + (1+b_1)X_t + b_2 X_t^2 + b_3 X_t^3 \tag{5}$$

to two types of data: strictly observational data in which the system was sampled for a specified period of time, and experimental data in which the system was reset to a randomly selected point every 12 time steps (Appendix 1). I then determined whether the difference equation 5 had one or two stable points. This exercise was repeated 1000 times for a given number of time steps, and the proportion of analyses yielding two stable states was calculated. Because the data were simulated by a process with two stable states, the proportion of analyses yielding two stable states is a measure of the probability of correctly detecting the existence of alternate states when they are in fact present.

The numerical experiments show a wide range of outcomes for detection of alternate states (Fig. 17.8). With low process error and experimentation, alternate states were identified about 90% of the time if more than 100 time steps were observed (Fig. 17.8a). However, without experimentation the probability of identifying alternate states actually declined as more time steps were observed, if the process error was small. With small process errors and no experimental perturbations, there will be very few chance events that shift a given simulation from one attractor to another. Therefore it is unlikely that the system will occupy both attractors in the course of a single simulation. Therefore, the analysis of that simulation will find only a single attractor. The regressions become more rigorous as more time steps are observed, so the probability of detecting alternate states actually declines.

With moderate process error, the performance of the experimental approach decreases while that of the non-experimental approach improves (Fig. 17.8b). The greater process error increases the frequency of state shifts due to random shocks. Combined with the experimental perturbations, these random shocks lead to frequent state shifts which decrease the sensitivity of the regressions to the existence of two distinct states. In the non-experimental case, however, the random shocks increase the frequency of state shifts enough to improve the ability of the regressions to detect alternate states.

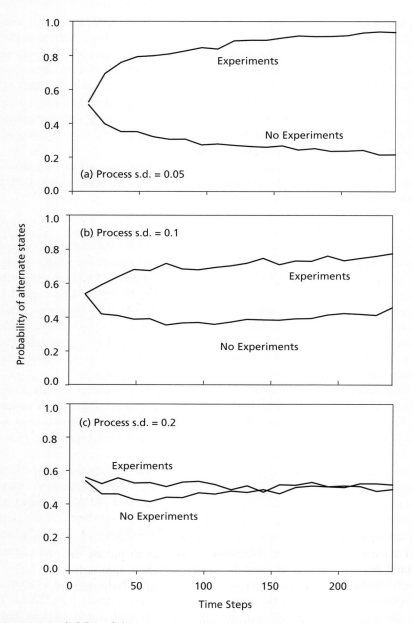

Figure 17.8 Probability of alternate states vs. length of time series for simulated data sets with and without experimental pertubations. (a) Process error $s_v = 0.05$; (b) process error $s_v = 0.1$; (c) process error $s_v = 0.2$. In all cases, observation error $s_w = 0.01$.

Both experimental and non-experimental approaches have a similarly low success rate for detecting alternate states if process error is large (Fig. 17.8c). With large process error, random shifts between states are common. These obscure the distinction between the states, and cause the regressions to identify the existence of alternate states about half the time. Note that the experiments are random manipulations in this exercise. In a more realistic simulation, the experiments would be carefully selected to maximize the chance of discriminating the alternative models (Burnham & Anderson 1998). Non-random experiments are likely to be more informative than the random ones depicted here.

Ives's (1995) study of stochastic stability of point attractors illuminates the patterns seen in Fig. 17.8. The apparent stability of an ecological time series can be measured using autoregressions similar to equation 4. Stability, as measured in this way, is proportional to the ratio of the slope of the attractor (as measured by the eigenvalue of the difference equation for the autoregressive process) to the process error variance (calculated from the residuals of the regression) (Ives 1995). Basins with steep attractors and small shocks appear stable, while systems with shallow attractors and large shocks appear unstable. Ives (1995) did not consider multiple attractors, and a full analysis of multiple attractors would have to consider the width as well as the steepness of the attractors (Holling 1973; Ludwig *et al.* 1997). Nevertheless, large process errors would tend to shift the system frequently among attractors, thereby making the attractors difficult to detect.

The general conclusion from these simulations is that alternate states should be detectable in long-term ecological data if observations are conducted for a sufficiently long period of time, careful experimental manipulations are performed, and process error is not too large. In addition, researchers must attend to interactions at several scales of time and space to understand mechanisms that can cause or suppress multiple attractors.

Perhaps the strongest arguments for multiple states will be based on multiple lines of evidence, blending long-term observation, experimentation, comparisons of contrasting ecosystems, and modelling tied closely to data. The examples presented in this paper, and many others, are in fact based on synthesis of multiple lines of evidence. We are not likely to develop a simple litmus test for alternate states. Future researchers who wish to pursue the phenomenon would be well advised to combine several different approaches and multiple types of information. In any given case, it seems likely that many years of careful investigation will be necessary to demonstrate alternate states at the probability levels typically demanded by working scientists. Ecologists should give high priority to such research, because of the conceptual importance of multiple attractors in basic ecology and their profound implications for ecosystem management. Managers, however, will often have to take action in situations where the probability of alternate states is moderate and inconclusive. How should one proceed in this situation? This question is addressed in the final section of the paper.

Probabilities, policy choice and precaution

How should the possibility of alternate states affect environmental policies? Questions of this type are usually addressed using decision theory (Lindley 1985). A decision analysis compares the future costs and benefits of alternative policy choices. For each possible choice of management action, future costs and benefits are calculated based on predictions derived from models of economic and ecological processes. These models involve uncertain parameters and stochastic inputs, so the predictions are probabilistic and comparisons are based on expected net benefits integrated over a range of possible outcomes. The models are dynamic: for each policy choice they yield a time stream of expected net benefits. Expected net benefits in the future are discounted, such that future benefits decline exponentially over time (Heal 1997). The practical consequence of discounting is that short-term benefits are weighted more heavily than long-term benefits. Also, discounting is convenient mathematically. Certain integrals which must be computed in decision problems would be infinite if the discount factor was 1 (i.e. long-term benefits had the same weight as short-term benefits). Economists rationalize discounting as a measure of people's preference for a sum of money now vs. later, as measured for example by interest rates paid on loans (Weitzman 1998).

There is considerable debate about applicability of economic cost–benefit analysis to environmental problems (Sagoff 1988; Bromley 1990; Brown 1991; Costanza 2000; Heal 2000; Ludwig 2000; Pritchard et al. 2000). A few questions that arise are: Is it ethical to monetize all aspects of nature, or are some aspects priceless? How can one appropriately combine the preferences of diverse individuals in a measure of aggregate value to society? Is the economic principle of efficiency an appropriate criterion for humanity's interaction with nature? Also, the discount factor is highly contentious (Heal 1997; Weitzman 1998). While any ecologist delving into economics should be aware of these important controversies, they are not the focus of this chapter. Instead, we will explore how decision making should be affected by a modest probability of alternate states, if there are severe penalties for shifting the ecosystem into an undesirable state.

Consider an ecosystem with two alternate states, one with great net value and the other with small net value. The high-value state can be exploited by extracting resources or discharging pollutants. The high-value state can be sustained under a moderate regime of resource extraction or pollutant discharge, but increasing levels of these stresses increase the risk that the ecosystem will shift to the low-value state. Once in the low-value state, a great deal of time and expense are required to recover the high-value state. Decision analysis can be used to show that resource extraction and pollutant discharge should be reduced as a precaution against a very expensive shift to the low-value state (Ludwig 1995; Carpenter et al. 1999a). How should management restrictions respond to uncertainty about the risk of a shift to the low-value state? The only way to prove its existence is to shift the ecosystem into the low-value state, an experiment to be avoided.

Uncertainty about alternate states should make managers more cautious. To illustrate this point, I have used the model of Carpenter et al. (1999b) to calculate the effect of uncertainty on policies for pollutant input to a lake (Appendix 2). The lake has two stable states that differ greatly in value. If excessive pollutant inputs shift the lake to the low-value state, the lake cannot be shifted back to the high-value state simply by reducing pollutant input to zero, because the pollutant is recycled effectively within the lake (Appendix 2). We call the low-value state 'irreversible' because recovery cannot be accomplished by reducing pollutant input alone. Economic benefits are derived from activities that generate pollutant load to the lake, and therefore increase with pollutant load, and from the ecosystem services provided by the lake which decrease if high pollutant inputs shift the lake into the low-value state. The 'optimal' pollutant load, which maximizes the net benefits from polluting activities and ecosystem services, can be calculated as shown in Appendix 2.

As the probability increases that the low-value state exists, the optimal pollutant loading rate declines (Fig. 17.9a). For this set of parameters, the optimal input rate reaches zero when the probability of the low-value state is about 0.3. The expected utility (net value) derived from the lake also declines as uncertainty increases (Fig. 17.9b), because the relatively profitable activities that generate pollutant load must be curtailed as the probability of irreversible change grows. Thus, the net value can be increased if scientific evidence accumulates against the existence of the low-value stable state. On the other hand, if scientific research shows that the low-value state is more likely to exist, then policy choice should be even more cautious and utility will decline.

How does scientific research affect the decision process? If the research decreases the probability of the low-value state, then managers will increase pollutant inputs and the utility extracted from the system. The marginal utility of scientific precision (rate at which utility increases per unit increase in precision, Appendix 2) is greatest when uncertainty is relatively high (Fig. 17.9c, d). On the other hand, if the research finds that the probability of the low-value state is greater than previously believed, then pollutant loads will be reduced. Utility will decline because of diminished pollutant loads, but ecosystem services will be sustained and society will be spared the burden of an irreversibly degraded ecosystem. Before the research is done we cannot know whether it will increase or decrease our estimate of the probability of the irreversible low-value state. Regardless of the true state of nature, however, scientific research on alternate states has the greatest potential benefit in situations where the probability of a low-value attractor is equivocal.

The specific results shown here are sensitive to the particular parameter values used in the calculations (Appendix 2). However, for a wide range of parameters: (i) even a modest probability of alternate states may have significant impact on policy choice; and (ii) new information has the greatest potential impact in the region where policy choice is most sensitive to changes in the probability of alternate states. Where data are equivocal yet decisions must be made, it may be possible to choose actions that decrease future uncertainty, i.e. manage adaptively (Walters 1986). For discussions of the feedbacks between scientific learning and decision

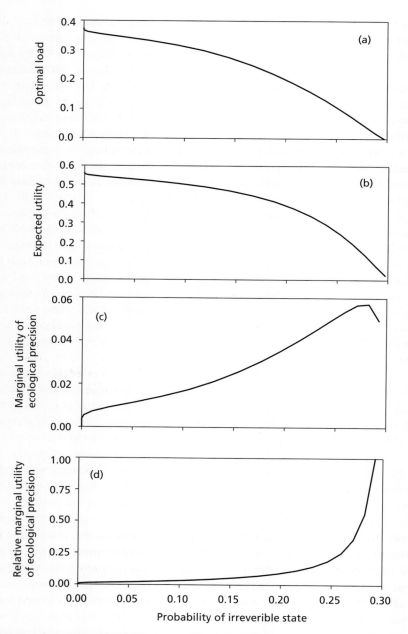

Figure 17.9 Dependence of policy choice on the probability that a low-value stable state exists, in the economic analysis of lake eutrophication. (From Carpenter *et al.* 1999b.) (a) Optimal load of phosphorus; (b) expected utility; (c) marginal utility of ecological precision; (d) relative marginal utility of ecological precision. See Appendix 2 for definitions of terms and an explanation of the calculation.

analyses, see Lindley (1985), Walters (1986), Easley and Kiefer (1988) and Carpenter et al. (1999b).

In many important cases where the stakes are high and the questions are difficult, new research will not yield clear answers before decisions must be made. In such cases, decision making will have to be based on pre-existing data, which may be inconclusive. We have seen that equivocal probabilities of alternate states (in the range of 0.2–0.8) can have substantial effects on policy choice. Even small risks of shift to undesirable states may favour caution in environmental decisions.

A conventional approach to ecological inference could have disastrous consequences. Suppose we state the null hypothesis that alternate states do not occur in a particular system, and the data fail to reject this hypothesis. If, on the basis of this study, a manager assumed that the null hypothesis was true, the optimal loading would be set at the level appropriate for zero probability of shift to the low-value state. But the analysis has given no evidence for this probability. Because the uncertainty is stated inaccurately, the optimal policy is calculated incorrectly. A disastrous collapse to the low value attractor may occur. At least two mistakes have occurred in this chain of inference.

1 Failure to reject the null hypothesis does not imply that one should accept the null hypothesis, unless the power of the statistical analysis is acceptably high. This point is made in nearly all statistics texts and often repeated in the ecological literature, yet often ignored.

2 The null hypothesis is irrelevant to the information needed by the manager, who requires instead an estimate of the probability of shift to the low-value state. The type of analysis necessary to estimate this probability is different from the analysis used to evaluate the null hypothesis (Lindley 1985; Walters 1986; Ludwig 1995; Hilborn & Mangel 1997). Bayesian statistics are generally used to calculate the probabilities used in decision analyses, whereas null hypothesis tests are more commonly used for scientific inference (Dennis 1996; Ellison 1996).

It is a more costly error to conclude that alternate states are absent when they are in fact present, than to falsely conclude that they are present. Therefore the burden of proof for alternate states should be reversed: the analysis should be required to prove that the probability of alternate states is negligibly low. So far, no ecological study that has rejected the hypothesis of alternate states has met this standard of inference. Such studies are equivocal, and may or may not provide evidence against the existence of alternate states. The presence or absence of alternate states has substantial effects on ecologists' ability to predict ecosystem change, and the pursuit of more rigorous and thorough evidence for or against alternate states should therefore be a research priority. The need is even more urgent for ecosystem management. Decision analyses are very sensitive to assumptions about the reversibility of ecosystem response to change in the variable being managed (such as pollutant loading rate or resource harvest rate). Generally, the assumption of smooth and reversible ecosystem dynamics leads to relatively high pollutant loads or harvest rates, because if a mistake is made its effects can be quickly reversed. In contrast, the assumption of alternate states will lead to more cautious loading or harvest rates, to account for the probability of shift to a low-value state followed by a costly period of recovery. In the absence of substantial and convincing evidence, it is dangerous to assume that this probability is negligible.

Acknowledgements

I have learned a great deal about multiple stable states from W. A. Brock, C. S. Holling, A. R. Ives, S. A. Levin, D. Ludwig, R. T. Paine, M. Scheffer, B. Walker, and C. Walters. T. Havlicek provided valuable help in searching the literature for case studies. W. A. Brock, A. R. Ives, T. Havlicek and D. Ludwig wrote helpful reviews of the manuscript. This work was supported by the Andrew W. Mellon Foundation, the North Temperate Lakes Long-Term Ecological Research site, and the National Science Foundation. No thanks would be too much for Ralf Yorque and the Resilience Network.

Appendix 1

The probability of detecting alternate states in time series observations of an ecosystem variate X was calculated by repeated simulations with the difference equation

$$X_{t+1} = X_t - \beta_0(X_t - \beta_1)(X_t^2 - \beta_2) + v_t \tag{1}$$

The βs are parameters and v is a normally distributed process error with mean zero and standard deviation s_v. If $\beta_2 < 0$, there is one stable equilibrium at $X = \beta_1$. If $\beta_2 > 0$, there are three equilibria corresponding to β_1, $\sqrt{\beta_2}$, and $-\sqrt{\beta_2}$; the largest and smallest of these three are attractors, and the intermediate one is a repeller. Observations of X were simulated as

$$X_{obs,t} = X_t + w_t \tag{2}$$

where w_t is a normally distributed observation error with mean zero and standard deviation s_w.

For simulations shown in this paper, parameter values were $\beta_0 = 0.1$, $\beta_1 = 1$, $\beta_2 = 0.16$, $s_w = 0.01$, and $s_v = 0.05, 0.1, 0.2$ as specified in the presentation of results. Thus, there were two stable points at $X = -0.4$ and $X = 1$, with an unstable separatrix at $X = 0.4$.

To generate non-experimental data, simulations of a given number of time steps were calculated from a randomly chosen starting point on the interval between the two stable points. To generate experimental data, simulations were restarted at a new random starting point every 12 time steps. This represents regular manipulations that reset the initial conditions of the ecosystem. For both non-experimental and experimental scenarios, 1000 independent data sets were simulated for a given number of time steps. Time series were fitted to a cubic polynomial

$$X_{t+1} = b_0 + (1 + b_1)X_t + b_2 X_t^2 + b_3 X_t^3 \tag{3}$$

by least squares, and the number of stable steady states of difference equation 3 were determined. The probability of detecting alternate states was calculated as the proportion of simulated data sets that yielded two stable states.

Appendix 2

Effect of uncertainty about the occurrence of irreversible alternate states was calculated for the model of lake eutrophication introduced by Carpenter et al. (1999):

$$X_{t+1} = X_t + L_t - bX_t + [X_t q/(1 + X_t q)] \qquad (1)$$

In this dimensionless form of the model, X is the amount of phosphorus in the lake water during summer, L is the annual loading of phosphorus to the lake, b is a parameter that controls the removal of phosphorus by sedimentation or flushing, and the term in square brackets represents recycling of phosphorus from sediments to the overlying water. Because of this recycling, the lake has alternate stable states for certain values of the parameters (Carpenter et al. 1999b). Here we assume that b is an uncertain parameter of the ecosystem and that L is a control variable that may be manipulated by a decision-maker.

Suppose a decision-maker is uncertain about the value of b. How should this uncertainty affect the choice of annual phosphorus loading, L? One possible solution to this problem is to calculate the value of L that makes the probability of a shift to the high phosphorus state acceptably low. Such a calculation bases management choices entirely upon the state of the ecosystem. However, in most situations there will be economic value derived from activities that create phosphorus input, as well as economic benefits derived from the lake such as fish production, water for irrigation, industry, drinking, and so forth. Thus, the choice of L will be based on economic as well as ecological criteria. Economists examine the costs and benefits of phosphorus input using a utility function such as

$$U_t = c_1 L_t + c_2 X_t - c_3 X_t^2 \qquad (2)$$

(Carpenter et al. 1999b). Utility U increases with loading. It increases with X at low values, then reaches a maximum and declines at higher values of X. This hump-shape accounts for the fact that utility increases at low levels of phosphorus, due to increases in amenities such as fish production, but decreases at high levels of phosphorus due to the damaging consequences of eutrophication.

To balance the costs and benefits of loading, one can calculate the schedule of loadings that maximize the expected value of utility over infinite time. This expected value is defined as

$$V = \sum \delta^t E(U_t) \qquad (3)$$

where the summation is over infinite time, E denotes mathematical expectation, and the discount factor δ reflects the preference for short-term utility vs. long-term utility (Heal 1997). The expectation operator is necessary because U is a random variable due to the uncertainty about b. The problem of maximizing expected utility for this model of lake eutrophication has been addressed by Carpenter et al. (1999b).

We define irreversible values of b as those for which a shift from the low-phosphorus state to the high-phosphorus state that cannot be reversed even if L is reduced to zero. Such irreversible shifts are possible when $b < 0.686$ (given $q = 8$, Carpenter et al. 1999b). The probability of irreversible values of b is defined as the probability that $b < 0.686$. To compare policies for different values of this probability, b was modelled as a normally distributed random variable with mean 1 and a range of values of the standard deviation s_b. Thus, for a given value of s_b, one can compute the probability that $b < 0.686$. For each value of s_b, the load that maximized V was calculated as described by Carpenter et al. (1999b), starting each time from a low value of X. This calculation yields a stable value L^* corresponding to annual returns U^*. The increment in annual utility that could be derived by reducing uncertainty is $-dU^*/ds_b$, which was calculated by finite difference. This is referred to as the marginal utility of reducing uncertainty. This quantity can be divided by utility to calculate the relative marginal utility of reducing uncertainty, $-(dU^*/ds_b)/U^*$. Parameter values used in the calculations shown here are $q = 8$, $c_1 = 1, c_2 = 2, c_3 = 4, 0 < s_b < 0.6, \delta = 0.99$.

References

Austin, M.P. & Cook, B.G. (1974) Ecosystem stability: a result from an abstract simulation. *Journal of Theoretical Biology* **45**, 435–458.

Barkai, A. & McQuaid, C. (1988) Predator–prey role reversal in a marine benthic ecosystem. *Science* **242**, 62–64.

Blindow, I., Anderson, G., Hargeby, A. & Johansson, S. (1993) Long-term pattern of alternative stable states in two shallow eutrophic lakes. *Freshwater Biology* **30**, 159–167.

Braconnot, P., Joussaume, S. & Marti, O. (1999) Synergistic feedbacks from ocean and vegetation on the African monsoon response to mid-Holocene insolation. *Geophysics Research Letters* **26**, 2481–2484.

Brock, W.A., Dechert, W.D., LeBaron, B. & Scheinkman, J.A. (1996) A test for independence based on the correlation dimension. *Econometric Reviews* **15**, 197–235.

Brock, W.A. & Durlauf, S. (2001) Interactions-based models. In: *Handbook of Econometrics*, Vol 5 (eds J. Heckman & E. Leamer), North-Holland, Amsterdam, in press.

Broecker, W.S. (1997) Thermohaline circulation, the Achilles heel of our climate system: Will man-made CO_2 upset the balance? *Science* **278**, 1592–1598.

Bromley, D.W. (1990) The ideology of efficiency: searching for a theory of policy analysis. *Journal of Environmental Economics and Management* **19**, 86–107.

Broström, A., Coe, M., Harrison, S.P. et al. (1998) Land surface feedbacks and paleomonsoons in Northern Africa. *Geophysics Research Letters* **25**, 3615–3618.

Brown, P. (1991) Why climate change is not a cost-benefit problem. In: *Global Climate Change* (ed. J.C. White), pp. 33–44. Elsevier, New York.

Burnham, K.P. & Anderson, D.R. (1998) *Model Selection and Inference*. Springer-Verlag, New York.

Carpenter, S.R. (1988) Transmission of variance through lake food webs. In: *Complex Interactions in Lake Communities* (ed. S.R. Carpenter), pp. 119–135. Springer-Verlag, New York.

Carpenter, S.R., Brock, W.A. & Hanson, P.C. (1999a) Ecological and social dynamics in simple models of ecosystem management. *Conservation Ecology* **3**. Available on the Internet. URL: www.consecol.org/Journal/vol3/iss2.

Carpenter, S.R. & Kitchell, J.F. (1993) *The Trophic Cascade in Lakes*. Cambridge University Press, Cambridge.

Carpenter, S.R. & Leavitt, P.R. (1991) Temporal variation in a paleolimnological record arising from a trophic cascade. *Ecology* **72**, 277–285.

Carpenter, S.R., Ludwig, D. & Brock, W.A. (1999b) Management of eutrophication for lakes subject to potentially irreversible change. *Ecological Applications* **9**, 751–771.

Carpenter, S.R. & Pace, M.L. (1997) Dystrophy and eutrophy in lake ecosystems: implications of fluctuating inputs. *Oikos* **78**, 3–14.

Carpenter, S.R., Kraft, C.E., Wright, R., He, X., Soranno, P.A. & Hodgson, J.R. (1992) Resilience and resistance of a lake phosphorus cycle before and after food web manipulations. *American Naturalist* **140**, 781–798.

Connell, J.H. & Sousa, W. (1983) On the evidence needed to judge ecological stability or persistence. *American Naturalist* **121**, 789–242.

Costanza, R. (2000) Social goals and the valuation of ecosystem services. *Ecosystems* **3**, 4–10.

Dechert, W.D. & Hommes, C.H. (2000) Complex nonlinear dynamics and computational methods. *Journal of Economic Dynamics and Control* **24**, 651–662.

DeMenocal, P., Ortiz, J. & Guilderson, T. (2000) Abrupt onset and termination of the African Humid Period: Rapid climate responses to gradual insolation forcing. *Quaternary Science Review* **19**, 347–361.

Dennis, B. (1996) Discussion: Should ecologists become Bayesians? *Ecological Applications* **6**, 1095–1103.

Dennis, B., Desharnais, R.A. & Cushing, J.M. (1995) Nonlinear demographic dynamics: mathematical models, statistical methods, and biological experiments. *Ecological Monographs* **65**, 261–281.

Dublin, H.T., Sinclair, A.R.E. & McGlade, J. (1990) Elephants and fire as causes of multiple stable states in the Serengeti-Mara woodlands. *Journal of Animal Ecology* **59**, 1147–1164.

Easley, D. & Kiefer, N. (1988) Controlling a stochastic process with unknown parameters. *Econometrica* **56**, 1045–1064.

Ellison, A.M. (1996) An introduction to Bayesian inference for ecological research and environmental decision-making. *Ecological Applications* **6**, 1036–1046.

Ellner, S. & Turchin, P. (1995) Chaos in a 'noisy' world: New methods and evidence from time series analysis. *American Naturalist* **145**, 343–375.

Frelich, L.E. & Reich, P.B. (1999) Neighborhood effects, disturbance severity and community stability in forests. *Ecosystems* **2**, 151–166.

Guckenheimer, J. & Holmes, P. (1983) *Nonlinear Oscillations, Dynamical Systems and Bifurcations of Vector Fields*. Springer-Verlag, New York.

Hansson, L.A., Jeppesen, E., Søndergaard, M. *et al.* (1998) Biomanipulation as an application of food-chain theory: Constraints, synthesis and recommendations for temperate lakes. *Ecosystems* **1**, 558–574.

Heal, G.M. (1997) Discounting and climate change. *Climatic Change* **37**, 335–343.

Heal, G.M. (2000) Valuing ecosystem services. *Ecosystems* **3**, 24–30.

Hilborn, R. & Mangel, M. (1997) *The Ecological Detective*. Princeton University Press, Princeton, NJ.

Hilborn, R. & Walters, C. (1992) *Quantitative Fisheries Stock Assessment: Choice, Dynamics and Uncertainty*. Chapman & Hall, New York.

Holling, C.S. (1973) Resilience and stability of ecological systems. *Annual Review of Ecology and Systematics* **4**, 1–23.

Ives, A.R. (1995) Measuring resilience in stochastic systems. *Ecological Monographs* **65**, 217–233.

Jeppesen, E., Ma. Søndergaard, Mo. Søndergaard & Christoffersen, K. (eds) (1998) *The Structuring Role of Submerged Macrophytes in Lakes*. Springer-Verlag, New York.

Knowlton, N. (1992) Thresholds and multiple stable states in coral reef community dynamics. *American Zoologist* **32**, 674–682.

Kuznetsov, Y.A. (1995) *Elements of Applied Bifurcation Theory*. Springer-Verlag, New York.

Laycock, W.A. (1991) Stable states and thresholds of range condition on North American grasslands — a viewpoint. *Journal of Range Management* **44**, 427–433.

Lewontin, R.C. (1969) The meaning of stability. In: *Diversity and Stability in Ecological Systems*, pp. 13–24. Brookhaven Symposia in Biology, 22. Brookhaven, New York.

Lierman, M. & Hilborn, R. (1997) Depensation in fish stocks: a hierarchic Bayesian meta-analysis. *Canadian Journal of Fisheries and Aquatic Science* **54**, 1976–1984.

Lindley, D.V. (1985). *Making Decisions*. John Wiley & Sons, New York.

Ludwig, D. (1995) A theory of sustainable harvesting. *SIAM Journal of Applied Mathematics* **55**, 564–575.

Ludwig, D. (2000) Limitations of economic valuation of ecosystems. *Ecosystems* **3**, 31–35.

Ludwig, D., Walker, B. & Holling, C.S. (1997) Sustainability, stability and resilience. *Conservation Ecology* [online] **1**, 7. Available from the Internet at URL: http://www.consecol.org/vol1/iss1/art7.

May, R.M. (1977) Thresholds and breakpoints in ecosystems with a multiplicity of stable states. *Nature* **269**, 471–477.

Noy-Meir, I. (1975) Stability of grazing systems: an application of predator-prey graphs. *Journal of Ecology* **63**, 459–481.

Nürnberg, G. (1984) Prediction of internal phosphorus load in lakes with anoxic hypolimnia. *Limnology and Oceanography* **29**, 135–145.

Pace, M.L., Cole, J.J., Carpenter, S.R. & Kitchell, J.F. (1999) Trophic cascades revealed in diverse ecosystems. *Trends in Ecology and Evolution* **14**, 483–488.

Persson, L. (1999) Trophic cascades: abiding heterogeneity and the trophic level concept at the end of the road. *Oikos* **85**, 385–397.

Peterman, R.M. (1977) A simple mechanism that causes collapsing stability regions in exploited salmonid populations. *Journal of the Fisheries Research Board of Canada* **34**, 1130–1142.

Peterson, C.H. (1984) Does a rigorous criterion for environmental identity preclude the existence of multiple stable points? *American Naturalist* **124**, 127–133.

Petraitis, P.S. & Latham, R.E. (1999) The importance of scale in testing the origins of alternative community states. *Ecology* **80**, 429–442.

Post, D.M., Carpenter, S.R., Christensen, D.L. et al. (1997) Seasonal effects of variable recruitment of a dominant piscivore on pelagic food web structure. *Limnology and Oceanography* **42**, 722–729.

Pritchard, L., Folke, C. & Gunderson, L. (2000) Valuation of ecosystem services in institutional context. *Ecosystems* **3**, 36–40.

Rahmstorf, S. (1997) Bifurcations of the Atlantic thermohaline circulation in response to changes in the hydrological cycle. *Nature* **378**, 165–167.

Ricker, W.F. (1963) Big effects from small causes: two examples from fish population dynamics. *Journal of the Fisheries Research Board of Canada* **20**, 257–264.

Sagoff, M. (1988). *The Economy of the Earth*. Cambridge University Press, Cambridge.

Sanderson, B.L. (1998) *Factors regulating water clarity in northern Wisconsin lakes*. PhD Thesis, University of Wisconsin-Madison.

Sanderson, B.L., Hrabik, T.R., Magnuson, J.J. & Post, D.M. (1999) Cyclic dynamics of a yellow perch population in an oligotrophic lake: evidence for the role of interspecific interactions. *Canadian Journal of Fisheries and Aquatic Sciences* **56**, 1534–1542.

Scheffer, M. (1998) *Ecology of Shallow Lakes*. Chapman & Hall, New York.

Smith, V.H. (1998) Cultural eutrophication of inland, estuarine and coastal waters. In: *Successes, Limitations and Frontiers in Ecosystem Science* (eds M. Pace & P. Groffman), pp. 7–49, Springer-Verlag, New York.

Sousa, W.P. & Connell, J.H. (1985) Further comments on the evidence for multiple stable points in natural communities. *American Naturalist* **125**, 612–615.

Spencer, P.D. & Collie, J.S. (1997) Effect of nonlinear predation rates on rebuilding the Georges Bank haddock (*Melanogramus aeglefinus*) stock. *Canadian Journal of Fisheries and Aquatic Science* **54**, 2920–2929.

Steele, J. & Henderson, E. (1984) Modeling long-term fluctuations in fish stocks. *Science* **224**, 985–987.

Sutherland, J.P. (1990) Perturbations, resistance and alternative views of the existence of multiple stable points in nature. *American Naturalist* **136**, 270–275.

Taylor, K. (1999) Rapid climate change. *American Scientist* **87**, 320–327.

Van Donk, E. & Gulati, R.D. (1995) Transition of a lake to turbid state six years after biomanipulation: mechanisms and pathways. *Water Science and Technology* **32**, 197–206.

Vandermeer, J. & Yodzis, P. (1999) Basin boundary collision as a model of discontinuous change in ecosystems. *Ecology* **80**, 1817–1827.

Walters, C.J. (1986) *Adaptive Management of Renewable Resources*. Macmillan, New York.

Weitzman, M. (1998) Gamma discounting. *Harvard Institute of Economic Research*, Discussion Paper 1843, Harvard University, Cambridge, MA.

Whang, Y. & Linton, O. (1999) The asymptotic distribution of nonparametric estimates of the Lyapunov exponent for stochastic time series. *Journal of Econometrics* **91**, 1–42.

Wilson, J.B. & Agnew, A.D.Q. (1992) Positive-feedback switches in plant communities. *Advances in Ecological Research* **23**, 263–336.

Zimmer, C. (1999) Life after chaos. *Science* **284**, 83–86.

Zimov, S.A., Chuprynin, V.I., Oreshko, A.P., Chapin, F.S. III, Reynolds, J.F. & Chapin, M.C. (1995) Steppe-tundra transition: a herbivore-driven biome shift at the end of the Pleistocene. *American Naturalist* **146**, 765–794.

Part 5
Concluding remarks

Chapter 18
Concluding remarks

J. H. Brown

To me falls the honour of trying to summarize in a few brief remarks this special meeting and the resulting symposium volume. This was an unprecedented occasion: the first joint meeting of the British Ecological Society and the Ecological Society of America. It celebrated our science on the occasion of the millennium. It presented an opportunity to look back on our history and to look forward to our future.

I will cast my remarks in the context of the two themes of the meeting: achievement and challenge.

Achievement

Phenomenal is the one word that I can find to best describe the accomplishments of ecology during the last two or three decades. Even though I have been a practising ecologist throughout this period, I did not really appreciate the breadth and magnitude of our accomplishments until this meeting. It is as if I were an undergraduate again. The plenary papers and the resulting chapters in this volume are a refresher course in contemporary ecology: a review of important recent research by some of the world's most eminent ecologists. I suspect that I am not alone in having been so wrapped up working in my own specialized niche that I have failed to appreciate many important advances in other subdisciplines of ecology.

It would be redundant to try to highlight particular research achievements or refer to individual chapters or authors here. I hope that all ecologists will read this entire volume. Instead, I will try to point out a few themes that are common to many chapters, cut across specialized subdisciplines, unify modern ecology, and provide a common basis for recent accomplishments.

One theme is the ability to make important connections across the enormous breadth of the discipline. Ecologists study organism–environment interactions at levels of organization from individual organisms, to populations and communities, to ecosystems and landscapes, to the entire globe. They study how the biological processes of anatomy, physiology, genetics, development and behaviour affect and are affected by biological, geological, chemical and physical characteristics of the environment. In recent years ecological research has become increasingly

Department of Biology, University of New Mexico, Albuquerque, NM 87131, USA

broad and integrative. Physiological ecologists study the genetic and developmental basis of tolerance and acclimation. Population ecologists study how differences among individuals in morphology, physiology and behaviour affect demography and population dynamics. Plant ecologists study how animal herbivores and mutualists and microbial symbionts affect plant performance. Community and ecosystem ecologists study the influence of individual species on biogeochemical processes and vice versa. Landscape ecologists investigate patterns and processes that characterize spatial relationships between the biota and the abiotic template of geology, topography, soil and climate. Global ecologists ask how the physiology of plants and the biogeochemical processes of particular habitats affect and are affected by changes in atmospheric gasses and climate. The old specialized subdivisions—plant or animal; terrestrial, freshwater or marine; physiological, population, community or ecosystem—are disappearing as ecologists pursue questions about processes across different taxa of organisms, levels of organization, kinds of environments, and spatial and temporal scales.

Modern ecology has embraced, adopted and applied the latest technological advances. With some justification, ecologists were once depicted as doing their science wearing a pair of scuffed and muddy boots, carrying a net or pair of binoculars, and recording data by hand in Higgins Eternal Ink in a field notebook. No longer. Now they use molecular genetic markers to monitor the fate of individuals and populations, stable isotopes to identify sources of critical resources for plant or animal populations, satellites to track migratory animals and to monitor disturbance events and vegetation change, state-of-the-art computer hardware and software to simulate global patterns of climate and gas fluxes, and chaos, cellular automata, and partial differential equations to model population and landscape dynamics. The result of using these emerging technologies has been new and improved ecological science—increased success in addressing big, important questions, and increased respect from scientists in other disciplines.

Ecology has recognized the importance of scale, and found innovative, integrative ways to study it. Invaluable insights into contemporary ecological relationships have come from expanding the temporal scale of study. New long-term studies, such as the international network of Long-Term Ecological Research (LTER) sites, have been initiated. Palaeo-ecologists have made unique contributions by analysing the fossil record for information on past environments and the organisms that lived in them. There has been enormous progress in studying space and spatial scale. It has been facilitated by conceptual and technological advances: Geographic Information Systems, spatial statistics, satellite imagery, landscape perspectives and spatially explicit models. Ecology has matured from debating which is the 'right scale' at which to study a particular phenomenon to studying the phenomenon of scaling itself, making comparisons and integrating across wide ranges of relevant scales.

Contemporary ecology has become increasingly pluralistic and tolerant of diversity. A few decades ago there were rancorous debates between zealous advocates of particular approaches: field experiments vs. non-manipulative observations, null models vs. mechanistic hypotheses, and analytical mathematical models vs. computer simulations. Evolutionary ecologists complained that ecosystem ecologists did not understand natural selection, and ecosystem ecologists replied that evolutionary ecologists lacked training in basic physics and chemistry. Today those arguments have largely faded away. There is widespread recognition that nearly all areas of ecology are benefiting from diverse approaches: mathematical theory, simulation modelling, highly controlled laboratory measurements, manipulative field experiments, comparisons of unmanipulated systems, and studies that span a wide range of temporal and spatial scales. Rather than debating the relative importance of biotic interactions and physical-chemical factors, ecologists are trying to understand how biotic and abiotic components of the environment interact to affect all levels of ecological organization, from individual organisms to the biosphere.

Ecology has become increasingly interdisciplinary and collaborative. Only a few decades ago it was easy for a single professional ecologist, typically trained as a biologist and perhaps assisted by a small number of undergraduate or graduate students, to do state-of-the-art research. It may still be possible. But increasingly the exciting research frontiers lie at the interfaces of the traditional subdisciplines of ecology and at interfaces between ecology and other sciences. The background and skills required to do such interdisciplinary research often exceed the expertise of any one scientist. As a result, we are increasingly seeing collaborative research teams composed of multiple investigators with different backgrounds and skills. Ecologists collaborate with climatologists and atmospheric chemists to study global gas fluxes, with economists to devise conservation and resource-harvesting policies, with physical oceanographers to understand the composition of plankton, with sociologists to study the ecology of humans and human-modified ecosystems, with geologists and hydrologists to understand relationships between streams and their drainage basins, and with molecular biologists to catalogue and map the diversity of microbes. As a result of these collaborative interdisciplinary endeavours, ecology has become more respected and better integrated into the fabric of contemporary science.

Finally, ecology has become recognized by society as a respected science that has important things to say about our own species. There is increasing recognition that humanity's most pressing problems—malnutrition, disease, pollution, poverty, economic disparity, social strife and political instability—are largely the result of a growing human population placing increasing demands on a finite Earth. This is probably the first time in the 4.5-billion-year history of life that a single species has become so powerful as to affect the entire Earth. Humans have changed climate, transformed habitats, caused the extinction of native species and the spread of exotic ones, and altered the chemical composition of air, water and soil. If ecologists can claim special understanding of current human and environmental problems, they must also assume special responsibility for solving them. The distinction between 'basic' and 'applied' research is breaking down. Ecologists are studying the ecology of *Homo sapiens* and human-dominated ecosystems, and applying their expertise to practical issues of policy and management. In 1950 few people knew the meaning of the word 'ecology'. Now most people have heard a great deal about ecology, and they increasingly look to ecologists for solutions to humanity's problems.

Challenge

Complexity is the one word that I can best choose to describe the future challenge to ecology. It is trite to say that ecological systems are complex. They are composed of many different parts of many different kinds, both living organisms and inanimate materials, which interact with each other in many different ways and on different temporal and spatial scales. The papers at this meeting and the resulting chapters in this symposium volume describe many spectacular recent achievements, but they only hint at the longstanding challenges posed by complexity. Darwin (1859; p. 373) eloquently described this challenge in his famous passage on the 'tangled bank.'

> It is interesting to contemplate a tangled bank, clothed with many plants of many kinds, with birds singing on the bushes, with various insects flitting about, and with worms crawling through the damp earth, and to reflect that these elaborately constructed forms, so different from each other, and dependent upon each other in so complex a manner, have all been produced by laws acting around us.

What follows is my assessment of the problems and prospects, 150 years after Darwin, for discovering those laws that have produced the complexity of the 'tangled bank'.

The great achievements in late 20th-century science were primarily reductionist. Scientists took nature apart, identified its fundamental components, and learned how they worked in isolation or in interaction with a few other components. The much-heralded accomplishments of particle physics and molecular biology are perhaps the two most spectacular examples. Recent advances in ecology, highlighted in this volume, are also primarily reductionist. Ecologists have learned a great deal about the molecular basis of physiological adaptation, the genetic and social structure of populations, the interactions between pairs of species, the sources of pollutants in air and water, and the processes of the global carbon cycle.

20th-century science was much less successful at understanding the structure and dynamics of entire, complex systems. Molecular biologists have nearly sequenced the genomes of humans and other organisms, but they have made far less progress in understanding how the genetic code directs development and, in interaction with the environment, produces a phenotype. Similarly, ecologists have been able to identify and map the distribution of trees in a forest, but we do not understand how hundreds of species are able to coexist in just 1 hectare of Amazonian rain forest.

I will suggest that there are two kinds of ecological complexity, or at least two extremes in the spectrum of complex structures and dynamics exhibited by ecological systems. The first class I will call inherently complex phenomena. They exhibit complicated, non-linear behaviours because they represent the outcomes of many different kinds of components interacting on different temporal and spatial scales. I assert that it will be nearly impossible and always impractical to 'predict' the behaviour of such systems very far in advance, although it may be possible to 'understand' their behaviour after it has occurred. To illustrate this class of complex systems I will draw first on an analogy. Despite the enormous advances of modern medicine, it is impossible to predict with any precision the health crises that one of us may face a decade from now, although it may be fairly straightforward to understand the course of events after they have occurred. The health history of an individual is a reflection of the behaviour of a complex system, a human body. I believe that many ecological phenomena fall into a similar category of complexity.

As a real ecological example, consider the long-term dynamics of five species of rodents at my study site in the Chihuahuan Desert (Fig. 18.1). When we began our work in the 1970s, we intended to test the prevailing hypotheses that seed-eating rodent populations fluctuated in response to the precipitation regime: increasing in response to food availability after rainfall events and declining in response to food shortage during subsequent droughts. Among other predictions, we expected population increases following El Niño events, which were reportedly associated with unusually heavy precipitation in the south-western United States. The first 10 years of data provided tentative support, which we published (Brown & Heske 1990). With another decade of data we expected to understand the causes well enough to predict the rodent fluctuations. But now, after more than 20 years, we are no closer. The reason is the complexity of population dynamics shown in Fig. 18.1. Most species were not present continuously; they colonized and went locally extinct, sometimes repeatedly, during the 22-year period. Increases were not predictably associated with El Niño events, in part because El Niño did not reliably bring exceptionally heavy precipitation to our part of the south-western United States. Decreases were not predictably associated with droughts; in fact, catastrophic population crashes were caused by two isolated, exceptionally large precipitation events.

So there appears to be a class of ecological complexity that is inherently idiosyncratic and unpredictable. At least some examples of this class, such as a single individual or a population of a single species, are complex systems embedded within other, larger complex systems. The behaviour of the smaller systems are influenced by many state variables and mechanistic processes of the larger system that have varying strengths, operate on varying temporal and spatial scales, and interact with each other in various ways. The net result is that the smaller systems exhibit complex structures and non-linear dynamics. Rodent populations are affected by relatively short-term fluctuations in food supply and predation risk, but also by longer-term changes in vegetation and climate, and by infrequent, episodic disease epidemics and extreme weather events. If we could measure all the relevant

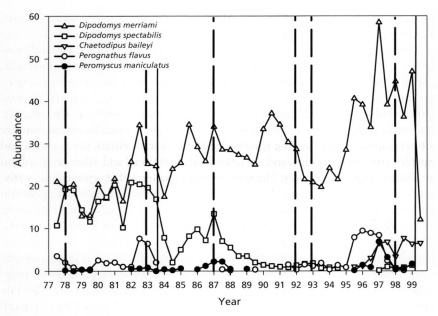

Figure 18.1 Fluctuations of five rodent species at Portal, Arizona, over a 22-year period. Note that each species exhibited distinctive, idiosyncratic population dynamics. The two continuous vertical lines indicate two extreme short-term rainfall events: a 3-day tropical storm that completely saturated the soil in October 1983 and a 1-h thunderstorm that caused surface flooding in August 1999. The former caused the gradual decline of *Dipodomys spectabilis* and *Perognathus flavus*. The latter caused the dramatic crash of *D. merriami*. The dashed vertical lines indicate the El Niño events, which were not consistently followed by increases in abundance of any rodent species, including *Peromyscus maniculatus*, the reservoir for the Sin Nombre strain of hantavirus.

variables and calibrate the rodent responses, prediction of the population dynamics might be possible in principle. But it will never be practical. Even if it were accomplished for one rodent species, the result would have almost no generality: it would not allow us to predict the population dynamics of a different species at the same site or of the same species at a different site.

There is a practical message here. Many of the systems that ecologists have traditionally studied fall into this class of complexity. That is why they are so wonderfully diverse and so frustratingly idiosyncratic. For these kinds of systems, we will have to be satisfied with achieving understanding rather than predictability. Recognizing the kinds of ecological systems that fall into this class of complexity should help to recognize the limits of our science and to explain those limitations realistically to the public, managers and policy-makers. For example, appreciation for this kind of complexity should give us pause in forecasting outbreaks of certain human diseases, such as hantavirus and Lyme disease. Note that the deer mouse *Peromyscus maniculatus*, the primary reservoir of the Sin Nombre strain of hantavirus in the south-western United States, is one of the rodent species that exhibited complex population dynamics at my long-term study site (Fig. 18.1).

I will end on an optimistic note by considering the other kind of ecological complexity. It should probably be called emergent simplicity rather than complexity, but it arises from the diverse structures and non-linear dynamics of complex ecological systems. Unlike the first class, this one is manifested in patterns and processes that are simple and general rather than complicated and idiosyncratic. Examples include the well-known distributions of abundance, body size, and size of geographical range among species, food-web characteristics, species–area and species–time relationships, and the latitudinal, elevational and other geographical gradients of species diversity. These are empirical patterns that represent the statistical outcomes of large numbers of individuals and species interacting within systems governed by powerful physical and biological constraints. They are so general that they appear to hold clues to the operation of underlying mechanistic laws (Brown 1999; Lawton 1999).

I will present just one example of emergent simplicity. The late plant ecologist and conservationist A. H. Gentry surveyed more than 200 forests world-wide, recording the identity and diameter of all trees within standardized 0.1-hectare plots. Analysis of Gentry's data by Brian Enquist, Karl Niklas and John Haskell reveals an amazing result. A single power–law relationship with a slope of −2 describes the relationship between number and stems and their diameter for all trees, irrespective of species, in nearly every forest from high latitudes to the tropics (Fig. 18.2). Furthermore, a simple model of resource-based thinning can predict this relationship (Enquist *et al.* 1998; B. J. Enquist, G. B. West, and J. H. Brown, in progress). The model is based on the assumptions that the total resources used by all trees are constant, set by the variables limiting productivity of the forest, and the division of these resources among individuals reflects powerful constraints of plant size on resource use.

This second class of ecological complexity as emergent simplicity should encourage those of us seeking generality, predictability and universal principles. There is great promise for developing truly novel and very general ecological theories. We now have a model that purports to explain the distribution of stem sizes in a tropical forest. We still await a theory to explain how those stems are apportioned among species, or how the number of species changes with spatial or temporal scale. What could offer greater excitement than the possibility of discovering new ecological laws?

In my opinion, then, the greatest challenge for ecology in the 21st century will be to grapple with complexity. I have suggested that most ecological systems fall along a spectrum from inherent complexity to emergent simplicity. If this is correct, it raises challenging questions. Where in the spectrum does a particular system lie, and how can it most profitably be studied? When should we be satisfied with mechanistic understanding and limited, small-scale, short-term predictability, and when should we seek general laws and near-universal predictability? I suspect that we will soon have a general theory that will predict the distributions of abundances

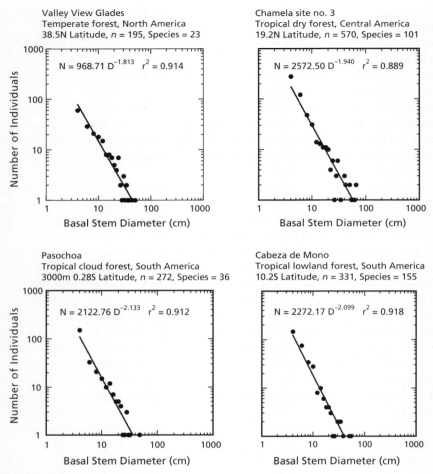

Figure 18.2 Relationship on logarithmic scales between number of stems and stem diameter for all of the trees in four 0.1-hectare plots of mature forest: temperate deciduous forest in Missouri, tropical dry forest in Mexico, tropical cloud forest in Ecuador, and tropical rain forest in Peru. Note that while the number of individuals (n) varies from 195 to 570 and the number of species from 23 to 155, the data for each forest fit a power function (straight line) with an exponent (slope) very close to the theoretically predicted value of -2.0. Data: Gentry database (http://www.mbot.org); analysis: B. J. Enquist, K. Niklas and J. P. Haskell.

among the hundreds of tree species in a tropical forest and the flows of energy and materials through these basal components of the food web. This is because we see hints of the operation of universal ecological laws in the emergent patterns of tree size, abundance and species diversity. I suspect that it will be nearly impossible to forecast well in advance the next epidemic of Sin Nombre hantavirus. This is because many contingencies and non-linearities affect the chain of causation running from El Niño to precipitation to seed and insect availability to deer mouse population growth to increased incidence of hantavirus in the mice to contact between infected mice and humans. When should an epidemic be expected and the public be alerted? Presumably the answer lies somewhere between two extremes: (i) an El Niño signal in the Southern Oscillation Index, which may be a long term, but highly inaccurate predictor, leading to many false alarms; and (ii) high populations of hantavirus-infected mice, which may allow accurate, but undesirably short-term forecast. Clearly, the complexities of ecological structures and dynamics pose worthy challenges for the coming generations of ecologists.

Summary

Nearly all attendees at the first joint meeting of the British Ecological Society and the Ecological Society of America accorded it a great success. Those who did not attend missed out on the excellent poster presentations and stimulating discussions. They do, however, have the opportunity to read the published versions of the plenary talks in this volume. These chapters highlight the spectacular recent achievements of ecological science. We should celebrate these advances. We should temper our enthusiasm, however, by recognizing that they, like other accomplishments of 20th-century science, were primarily reductionist. Ecologists have done a great job of taking ecological systems apart, describing the components, and figuring out how they work in isolation or very simple subsystems. But the greatest challenge lies ahead. The challenge of the 21st century is to untangle ecological complexity—to discover the laws that govern the structures and dynamics of Darwin's tangled bank.

Acknowledgements

The ideas expressed here, especially regarding ecological complexity, have benefited from many discussions with the students, postdocs and faculty who comprise the informal 'ecological complexity group' at the University of New Mexico and Santa Fe Institute. The National Science Foundation has supported my field research in the Chihuahuan Desert (most recent grant DEB-9707406); a Packard Interdisciplinary Science Grant, the Thaw Charitable Trust and the Santa Fe Institute have supported my more theoretical work on allometric scaling and ecological complexity. Morgan Ernest and Brian Enquist kindly provided Figs 18.1 and 18.2, respectively.

References

Brown, J.H. (1999) Macroecology: progress and prospect. *Oikos* **87**, 3–14.

Brown, J.H. & Heske, E.J. (1990) Temporal changes in a Chihuahuan Desert rodent community. *Oikos* **59**, 290–302.

Darwin, C.R. (1859) *The Origin of Species by Means*

of Natural Selection, or the Preservation of Favored Races in the Struggle for Life. John Murray, London.

Enquist, B.J., Brown, J.H. & West, G.B. (1998) Allometric scaling of plant energetics and population density. Nature **395**, 163–165.

Lawton, J.H. (1999) Are there general laws in ecology? Oikos **84**, 177–192.

Index

Note: page numbers in *italics* refer to figures; those in **bold** refer to tables.

abundance, 393–4
    genetic aspects, 25, 37–8
Acaulospora, 104, 106
Acaulospora longula, 104
Accipiter nisus see sparrowhawk
Acer saccharum, 212
achlorophyllous obligate mycorrhizal plants, 107
Aegolius funereus, 84, *85*
Africa
    human–mammal conflicts, **320**, *320*
        mammalian defaunation, 321
    savannah, 365
African elephants, 59–60, 321, 324, 326
African wild dogs, *52*, 52
age-related breeding/survival (performance)
    banded bird studies, 67, 68–71, *69*, *70*
        breeding site fidelity, 81, *82*
        deterioration with senescence, 71
    genetic aspects, 8, 9, *10*
ageing
    definitions, 7
    evolutionary aspects, 3, 7–11
    mutations accumulation, 7, 8, 9
    natural populations, 7, 86–7
    pleiotropy (trade-offs), 7, 8–10
    study approaches, 8–9, 86
Allee effect, 50, *51*, 51
alligator juniper (*Juniperus deppeana*), 131
Ammophila arenaria, 98
amphibians
    small stream habitats, 311
    threatened species, 325
Andropogon gerardi, 96
Andropogon scoparius, 96
Angiosperm evolution, 284–5
ant lion (*Myrmeleon immaculatus*), 11
Antonovics' tenets, 25–43
Apis melifera, 11
arbuscular mycorrhiza, 95–111
    diversity, 98–107
        environmental factors, 105–6
        surveys, 100, **101**
    experimental approaches
        additive treatment with community reconstructions, 98, *99*
        biocide application, 96, *97*
        sampling effect, 98
    genetic aspects, 106–7, 239
    host specificity, 102, *103*, 103–4, 107

impact on plant community structure, 96–8, 109–11, *110*
laboratory culture problems, 104
linkages, 107–11, *110*
    carbon transport, 108–9
multiple root infections, 104
    ineffective associations ('cheats'), 104–5
    pathogenic fungi protective function, 105, *106*
phosphorus transport, 104, *105*
spatial/temporal distribution patterns, 104, *105*
Arctic skua (*Stercorarius parasiticus*), 68
Arenaria serpyllifolia, 96
arid-land ecosystem, 116, 129–33
    precipitation variation, 129
        seasonal distribution, 129–31, *130*
    primary productivity, 129
    vegetation precipitation exploitation
        near-surface water, 131, *132*, 133
        small event utilization, 132
        trade-offs, 133
Asio flammeus, 84
Asio otus, 84
Atlantic silverside (*Menidia menidia*), 11
atmospheric CO_2, 117, 249–65
    arid-land ecosystem water use efficiency, 131
    C_3/C_4 photosynthetic pathway evolution, 117, *118*
    global carbon cycle studies, 121, 123
    human activities impact, 116, 249
    Phanerozoic era, 272, *273*
        global surface temperature trend simulations, 285–6
        plant biodiversity relationship, 284–6, *285*, *286*
        stomatal density relationship, 280–1, *281*, *282*
    terrestrial sequestration *see* carbon dioxide sequestration
    Triassic–Jurassic boundary, 281, *282*, *283*, 283
    see also carbon dioxide enrichment studies

bacterial competition, 26–30
    see also competitive fitness
barn swallow (*Hirundo rustica*), *70*
basswood (*Tilia americana*), 214, 216
beaver (*Castor canadensis*), 328
    cottonwood interaction, 328–9
beech (*Fagus grandifolia*), 212, 219
beech (*Fagus sylvatica*), *236*, 236
benomyl application studies, 96, *97*
Betula alleghaniensis, 213–14, 216
Betula papyrifera, 108
BIODEPTH, 143–6, 185

397

INDEX

BIODEPTH (cont.)
  plant productivity, 144, *145*
    local determinants, 144
biodiversity *see* diversity
bird leg bands, 67
birds
  age-related breeding/survival (performance), 67, 68–71, *70*, 86
    initial improvements with age, 68, *69*
    residual reproductive value, 71
  banding studies, 67–87
    data analysis developments, 87
  dispersal, 76–86
    nomadic species, 82, 84–6, *85*
  inverse density-dependent effects, 62
  lifetime reproductive success, 72–8, *73*, *74*
    proportion of productive/unproductive individuals, 72, 74, **75**
black rhino, 321
black-capped chickadee (*Parus atricapillus*), *70*
black-legged kittiwake (*Rissa tridactyla*), *70*
blue tit (*Parus caeruleus*), 72, 77–8
bluebell (*Hyacinthoides non-scripta*), 104
body size
  evolution with polygyny, 55, *57*
  laboratory thermal selection, 13, *14*
  latitudinal clines
    Drosophila melanogaster, 11, *12*, 12–13
    ectotherms, 11
    environmental temperature relationship, 13
Brandt's cormorant (*Phalacrocorax pencillatus*), 71
breeding behaviour, 47
breeding site fidelity, banded bird studies, 78, **79**, 80–1
  migratory species, 82
brocket deer, 323
brook trout (*Salvelinus fontinalis*), 311
Burnupena papyracea, 366

C_3 photosynthesis
  atmospheric CO_2 sequestration, 228
  carbon isotopes ratios, 123–4, *125*, *126*, *127*
  physiological constraints, *125*, 125, 126
C_3/C_4 pathways, 116–21
  atmospheric CO_2 enrichment responses, 256
  evolutionary aspects, 117, *118*
  palaeoecology
    carbon isotope ratios, 118–19, *120*, *121*
    glacial periods, 118
    global faunal changes, 119, *122*
  quantum-yield cross-over model, 117, *118*, 119
Calidris temminckii, *70*
CarboEurope, 260, 265
carbon budget, 124–7
  eddy covariance technique, 127
  flux partitioning technique, 127–8, *128*
  Keeling plot analysis, 124–6, *125*, *126*
  net exchange–biodiversity relationship, 287–8, *288*
  plant leaf wax isotope analysis, 128
  wet/dry period comparisons, *126*, 126
carbon certificate trading, 231
carbon cycle, 121, 123–9, *250*, 251

biopshere carbon exchange
  global warming simulation, 275–6, *276*
  turnover time, 274
carbon sinks
  northern hemisphere terrestrial, 123
  oceanic, 274–5
Phanerozoic era, 272
role of marine organisms, 251–2
seasonal variations, 123, 129
terretrial ecosystems, 254
vegetation–climate feedback simulations, 276, *277*, *278*, *279*
carbon dioxide (CO_2) enrichment studies, 228–30, 256
  complexity levels, 234, 239–40
  ecological relevance criteria, 240, 241–3
  light availability influence, *235*, 235
  mycorrhizal influence, 239
  plant developmental phase effects, 232
  precision, 239, 240
  research tools, 234
  soil effects
    moisture, 237
    substrate quality, *236*, 236–7
  species-specific responses, 236, 237–8
  stomatal aperture responses, 237, *238*
  time of sampling influence, 232, *233*
carbon dioxide (CO_2) limitation, 230
carbon dioxide (CO_2) sequestration, 228, 229
  carbon pool size, 231
  plant developmental phase effects, 232
  terrestrial
    climate impact, 273–4, *274*
    crop plants, 233
    forest uptake/storage, 228, 229, 231–2, *232*
    'missing carbon', 228, 229
    northern hemisphere ecosystems, 123
    silicate bed rock weathering, 272
    vegetation–climate feedback simulations, 276, *277*, *278*, *279*
  transient responses to resource supply, 232, *233*
  see also carbon sinks
carbon flux tracking methods, 263, **264**
carbon isotope ratios, 116
  C_3 plant tissues, 123–4
  C_3–C_4 ecosystem palaeoecology, 118–19, *120*, *121*
  carbon cycle studies, 121, 123–9
    northern hemisphere terrestrial carbon sink, 123
    seasonal variations, 123
  ecosystem carbon balance over time, 124–7, *125*, *126*
  plant leaf wax, 128–9
  Triassic–Jurassic boundary leaf material, 281
carbon sinks, 249, *250*, 250–62
  control, 256–8
  location, 258–60, *259*
  ocean, 250, 251–2
    carbon turnover time, 274
    global warming response, 257–8
  terrestrial, 252–5, **253**, 260–1, 276, *277*, 287
    global warming simulation, 275, *276*
    increased temperature response, 257
    Kyoto Protocol mechanisms, 262–3

398

net ecosystem productivity measurement, 253, *254*
Phanerozoic era, 272, *273*, 273
see also carbon dioxide (CO$_2$) sequestration
carbon transport, arbuscular mycorrhizal linkage, 108–9
Carnivora, 324
Castor canadensis, 328
Cebus apella, 322
Cedar Creek diversity experiment, 202
Ceratodon purpureus, 96
Chrysomella confluens, 328
Cladonia rangiformis, 96
climate change, 139–56
 biodiversity impact, 139
  local vs. regional, 151–3, *152*
 ecosystem processes impact, 139
 experimental approaches
  controlled environmental facilities (CEFs), 154–5
  reductionist experiments, 147–8
 hardwood forest species range extensions, 219–21
 individualistic nature of species responses, 141, 146–7
 rapid shifts, 362
 species distribution changes, 148–50
  prediction from climate mapping ('climate envelopes'), 149, *150*
climate–plant interactions, 271–90
 feedback simulations, 276, *277*, *278*, *279*
 modelled vegetation changes over next century, *284*, 284
 Phanerozoic era, 272–3, *273*
 time scales, 274
 vegetation cover–precipitation interaction, 288, *289*, 289
climatic simulations, 275
 Triassic–Jurassic boundary extinction event, 281, *282*, *283*, 283
coexistence
 arbuscular mycorrhiza multiple root infections, 104–5
 bacterial ecotypes, 29–30, *30*, 39
 biodiversity–productivity relationship, 193, 194, 199
 patchy resource distribution, 194
 sensitive and resistant bacteria, T4 virus interaction, 35–6, **36**, 39
 trade-offs, 39
commercial exploitation of animals, 322
common crossbill (*Loxia curvirostria*), 84, 86
common gull (*Larus canus*), 70
community structure, 38–9, 183
 bacteria
  ecotypes development, 29–30, *30*
  T4 virus interaction, 35–6, **36**, 39
 climate change impact, 139, 147
  reductionist experimental approaches, 147–8
  species distribution changes, 148–50
 human activities impact, 201
 plant–arbuscular mycorrhiza interactions, 96–8, 109–11, *110*
 prediction of change from plant functional types, *166*, 167
competition
 bacteria, 26–30
 density-dependent processes, 49

diversity–productivity relationship, 190–1, 193, 199–200
diversity–temporal ecosystem stability relationship, 187, *189*, 189
ecosystem productivity, 177, *178*
interspecific differentiation, 194
inverce density-dependent processes (Allee effect), 50, *51*, 51
niche differentiation, 201
plant functional types *see* competitors
sampling effect, 201
 antagonistic efficiency, 191, 193
 resource exploitation, 190–1
sedge phosphorus uptake, 176
social vertebrates, 47
 between groups, 47, 50–6
 within groups, 49–50
spatial scale of biodiversity experiments, 201
competitive fitness, bacterial adaptive evolution, 26–43
 cell size changes, 27, *28*, 38
 demographic parameter changes, 28–9, *29*
 ecotypes development, 29–30, *30*, 41
 life history change, 28–9, *29*
 resource limitation, 26–7, *27*, 40
 thermal selection, 30–4, *31*, *32*, *33*, 38, 40, 41
 thermotolerant mutants emergence, 33–4
 time scale, 41–2
 trade-offs, 32, 33, 36–7, 40–1
 viral partial resistance, 37–8, *38*
 viral predator interactions, *34*, 34–7, 40, 41–2
 virus extinction, 36, 37
 virus-mediated coexistence of sensitive and resistant genotypes, 35–6, **36**, 39
competitors, plant functional types, 165, 167, 168
 ecosystem resilience relationship, 169, 170, 171
 mineral nutrient recycling, 173
complex systems, 390–3
conservation, social vertebrates, 60–1
controlled environmental facilities (CEFs), 154–5
 climate change impact investigation, 140, *141*
cooperative breeders, 47, 61
 group extinction rates, 52–3
 inverse density-dependent processes (Allee effect), 50–1
 obligate, 52
 sex ratio of offspring, 52
copepods, 11
cost–benefit analysis, 374
cotolerance of climatic stresses, 170
Crepis capillaris, 96
critical depensation, 358
crop plants
 CO$_2$ sequestration, 233
 stomatal aperture, CO$_2$ enrichment response, 237
culling rates, 60, 61

dams, 300, 321
Danthonia spicata, 98
decision analysis, 374–5, 377
deforestation, 231, 249, 253, 319, 320
density-dependent prices, 342, *343*

density-dependent survival/breeding success, 49
Dermoptera, 324
detrital food webs, headwater streams, 296–7, *297*
dispersal, 47, 67, 140
  banded bird studies, 76–86
    breeding, 76, 78–82, **79**, *80*
    fluctuating habitats/food supplies, 82, 84
    natal, 76, *77*
    nomadic species, 82, 84–6, *85*
    non-breeding, 76, 82–6, **83**, *85*
  climate change response, 146–7
  group size influence, 53
  sex differences in frequency, 53–6, *56*
  sociality effects, 53–6, *54*
distribution
  arbuscular mycorrhiza, spatial/temporal patterns, 104, 105
  climate change response, 148–50
    prediction from climate mapping ('climate envelopes'), 149, *150*
  genetic aspects, 25, 37
diurnal raptor dispersal, 84
diversity, 25, 38, 39
  carbon exchange relationship, 287–8, *288*
    Phanerozoic era, 284–5, *285*, *286*
  climate change impact, 139, 146
    local vs. regional diversity, 151–3, *152*
    species distribution changes, 148–50
  CO_2 enrichment responses, 237–9
  confounding variables, 184–5
  definitions, 184–5, 200, 203
  ecological economics, 350–1, *351*
  ecosystem processes impact, 139, 141–6, 183, 184
  ecosystem stability relationship, 183–4, 186–90, *188*
    resistance/resilience, 169, 171, 172
    resource competition, 187, *189*, 189
    statistical averaging effect, 187, 188, 189
  energy input relationship, 153–4
  human activities impact, 201
  mammalian defaunation consequences, 326
  tropical forest understorey, 329–30, *330*
  plant communities, 95
    perennial, genetic basis, 163–4, *164*
  productivity relationship, 173–6, 177, 183, 190–9, 287
    experimental approaches, 194–9, *196*, 200–1
    microcosm experiments, 174–5, *175*
    niche differentiation, 190, 193–4, *197*, 198
    sampling effect, 174, 190–3, *192*, 197, 198
  small stream ecology, 310–11
  spatial scale of experiments, 201
  species–area relationship, 202–3
  temperature increase response, 286–7, *287*
DNA analysis
  bird extra-pair fertilization, 74, 76
  mycorrhizal fungi identification, 99
Drosophila
  body size
    altitude-related increase, 13
    fitness correlations, 14–15
    latitudinal clines, 11
    thermal selection, 13, 14–16
  growth efficiency, *14*, 15–16, 17, 18
  temperature cline microcosm experiments, 150
Drosophila melanogaster
  ageing, 8–10
  body size
    latitudinal clines, 11, *12*, 12–13
    thermal selection response, 16–17, *17*
Drosophila subobscura, 11
drought, 186
  ecosystem resistance
    biodiversity relationship, 186–7
    plant functional type responses, 170–1
  temperate hardwood forest disturbance, 214

economics, ecological, 337–54, 374
  consumer preferences, 353
  control variables, 340, 344
  ecosystem services, 338–9
    harvests, 338
    in situ services, 338–9
  investment returns, 351–3, *352*
  multispecies models, 350–3
    weeding activities, 350, *351*
  optimal zonation, 344–5, *345*
  production functions, 339, 340, *341*, 344
  single-species models, 338–49
  targets, 339–44, *342*, *343*
  see also harvesting
ecosystem processes, 183
  biodiversity influence, 183, 184
    necessary species number estimates, 202–3
  mammalian defaunation consequences, 326
  plant diversity reduction impact, 141
    across-site effects, 142–3, *143*
    BIODEPTH experiment, 143–6
    primary productivity, 141, *142*
    within-site effects, 142–3, *143*
ecosystem stability
  biodiversity relationship, 183–4, 186–90, *188*
  plant functional groups, 187
  see also resilience, ecosystem; resistance, ecosystem
ECOTRON, 154, 185, 195
ectomycorrhiza
  host dependence, 102
  reciprocal carbon transport, 108
ectotherm body-size
  latitudinal clines, 11
  thermal evolution, 3, 11–18
eddy covariance technique, 127
effort quotas, 340
El Niño, 129, 131, 132, 251, 391, 395
elephant, 59–60, 321, 324, 326
  African grassland creation from woodland, 365
emergent simplicity, 393
energy input, biodiversity relationship, 153–4
environmental economics, 338
equids, palaeoecological dietary data, 119, *120*, *121*
ericoid mycorrhizas, 107
Erodium cicutarium, 96
Escherichia coli, 26–30, *27*, *28*
  T2 virus interaction, 37, *38*

T4 virus interaction, *34*, 34–7
T5 virus interaction, 36
Etheostoma boschungi, 311
Etheostoma trisella, 311
Eucalyptus, 96
EUROFLUX, 260
experimental approaches, 7
  biodiversity
    definitions, 185
    spatial scale, 201
  biodiversity–productivity relationship, 194–9, *196*, 200–1
    sampling effect, 174
    synthesized ecosystems, 173–4, *176*
  climate change impact, 140, *141*, 147–8
  ecosystem resistance/resilience, climate manipulation responses, 171, *172*–3
  multiple stable states detection, 369
  mycorrhizal reciprocal carbon transport, 108
  precision, 239, *240*
  soil microbe–plant interactions, impact on plant community structure, 96, *97*, 98, *99*
  survey of use by ecologists, 3, 4, **5**, **6**
  see also carbon dioxide enrichment studies; experimental plant ecology
experimental plant ecology, 227–45
  carbon sequestration studies, 228–30
  data handling, 227
  ecological relevance criteria, 240, 241–3
    atmospheric environment, 242
    below-ground environment, 242–3
    experimental setting, 242
    plant community, 243
    plant species, 243
  methodological drivers, 233–4, *244*
  need for large/complex experiments, 234, 244
  resource limitation, 227, 230, 244
extinction, 42, 140
  fossil plant record, 271
  harvesting targets, *346*, 346, 349
  human activities impact, 201
  Phanerozoic era, 278, *279*
    plants as climatic indicators, 278, *279*, *280*, 280–4
    Triassic–Jurassic boundary, 278
  plant functional types, 167–8
  recovery of diversity, 283
  social vertebrate home-range size, 60
extra-pair fertilization, 74, 76

FACE (free-air CO_2 enrichment), 234
Fagus grandifolia, 212, 219
Fagus sylvatica, CO_2 enrichment responses, *236*, 236
Falco naumanni, 78
fecundity, genetic aspects, 25
Festuca ovina, 98
Ficedula hypoleuca, 70, 78
fire
  African grassland creation from woodland, 365
  temperate hardwood forest disturbance, 214, *215*, 221
fish
  body size latitudinal clines, 11–12
  inverse density-dependent effects, 62
  recruitment cycles, 366–7
  shallow lake phytoplankton vs. macrophyte dominance, 365
  small stream habitats, 311
fitness, 25, 39–40
  ecosystem productivity, 177
  lifetime reproductive success as measure, 72, 74
fluvial earth-moving activity, 298–300, *299*
flux partitioning technique, 127–8, *128*
FLUXNET, 253, *254*
forest
  atmospheric CO_2 enrichment response, 256, *257*
    stomatal aperture, 237, *238*
  carbon flow data, 254, *255*
  carbon sinks, 228, 229, 231–2, *232*, 253, 254–5
    Kyoto process, 231
    Kyoto Protocol mechanisms, 262–3
  FACE (free-air CO_2 enrichment) studies, 234
  fragmentation, 321–2
  logging, 231
  old forest preservation, 231
  spatial scale of biodiversity experiments, 202
  stem number–diameter relationship, 393, *394*
  understorey, mammalian herbivory impact, 329–34, *330*, *332*, *333*
  see also hemlock–hardwood forest
Formica propinqua, 328
fossil plant record, 271
fossil pollen studies, hemlock–hardwood forest, 212, 217–19, *218*, 220
Fulmarus glacialis, 70
Fundulus heteroclitus, 11
Fusarium oxysporum, 105

genetic aspects, 3–18
  Antonovics' tenets, 25–43
  arbuscular mycorrhiza, 106–7
  ecosystem resistance/resilience, 169
  plant perennial community biodiversity, 163–4, *164*
  stress tolerant plant functional types, 169–70
  study approaches, 7
    survey of use by ecologists, 3, 4, **6**, 6–7
genetic diversity, 25, 38, 39
geographic information system (GIS), 217
geographical range
  climate change impact, 140, *141*, 146–7, *149*, 219–20
    prediction from climate mapping ('climate envelopes'), 149, *150*
  mammalalian contraction, 321
Gigaspora, 106
global change, 115–34
  atmospheric CO_2, 116
global warming, 249, 250
  rapid climate shifts, 362
  terrestrial biosphere carbon exchange, 275, *276*
  Triassic–Jurassic extinction event, 280, 281
Glomalean mycorrhiza *see* arbuscular mycorrhiza
Glomus, *102*, 104, 106
Glomus etunicatum, 107
Glomus mosseae, 100, *102*, 107

401

INDEX

grasses
  atmospheric CO_2 enrichment response, 256–7
  soil microbe interactions, 96, 98
grassland, 183–203
  CO_2 enrichment response, soil moisture dependence, 237
  ecosystem resistance/resilience, 170–2, *172*
  fire-related creation, 365
  spatial scale of biodiversity experiments, 201–2
grazers, 57
  C_3/C_4 plant diet, palaeoecological data, 119, *120*, *121*, *122*
grazing pressure, 95
great grey owl (*Strix nebulosa*), 84
great skua (*Stercorarius skua*), 68, 71
great tit (*Parus major*), 70

HadCM2, 275
haddock stocks, 367, *368*
Haematopus ostralegus, 68
hantavirus, 392, 394–5
harpy eagle, 321
harvesting, 338
  alternate state dynamics, 358, 360
  decision analysis, 374, 377
  production function, 339, *341*
  quotas, 340, 341, *342*
  targets, 339–42, *342*
    density-dependent prices, 342, *343*
    quasi-stationary distribution approach, *346*, 346–9, *349*
    traditional optimum, 340
hemlock (*Tsuga canadensis*), 212
hemlock–hardwood forest, 211–22
  climate-related range extensions
    beech, 219–20
    hemlock, 220–1
  composition changes, 211–12
  disturbance, 213–15
    drought, 214
    fire, 214, *215*
    related succession, *364*, 364
    wind, 213, 214
  fossil pollen studies, 212, 217–19, *218*, 220
  long-term observations, 211
  patch dynamics, 215–19
    hemlock invasion of maple stands, 216–17
    origin of hemlock stands, 217
    sugar maple/hemlock-dominated patch persistence, 215–16, *216*
  upper Great Lakes Region, 212, *213*
herbivory
  African grassland creation from woodland, 365
  see also mammalian herbivory; phytophagous insects
herring gull (*Larus argentatus*), 71
Heteromys, 326
Hirundo rustica, 70
honey bee (*Apis melifera*), 11
house fly (*Musca domestica*), 11
human activities impact, 319, 337, 389–90
  biodiversity, 201

earth-moving, 298–300, *299*
human–mammal conflicts, 320–1
  Africa, **320**, 320
  see also mammalian defaunation
stream ecology, 295, 296, 298–301, 304, *305*, 305, *306*, 307, 312
human disease outbreaks, 392
human population density, plant functional type correlations, 167, *168*
hunting
  elephants, African grassland–woodland balance, 365
  subsistence, 322–3
Hyacinthoides non-scripta, 104
Hymenoscyphus ericae, 107

Icteria virens, 78
in situ ecosystem services, 338–9
  rent targets, 342, *343*, 344
  weeding activities, 350
inbreeding, 47
  avoidance in social vertebrates, 53–5
inbreeding depression, 8
individual animal studies, 67–87
  markers, 67
infanticide, 49
insects
  inverse density-dependent effects, 62
  see also phytophagous insects
integrative approaches, 387–8, 389
interdisciplinary research, 389
inverse density-dependent processes (Allee effect), 50, *51*, 51–2
  group extinction rates, 52–3
  group size-dependent dispersal rates, 53
IRONEX, 258
ironwood (*Ostrya virginiana*), 216
IUCN Red List of Threatened Animals, *324*, 324–5, *325*

jaguar, 321, 322
Jasus lalandii, 366
Juniperus deppeana, 131
Juniperus osteosperma, 131

keystone predators, 39
Kyoto process, 231
Kyoto Protocol, 249, 262
  forest carbon sequestration, 262–3

La Niña, 129
Lagomorphs, 324
Lake Guri, 321
lakes
  eutrophication
    alternate states, decision analysis, 375, *376*, 379–80
    phosphorus availability, 362–4, *363*
    phytoplankton vs. macrophyte dominance, 365–6
Large Scale Biosphere–Atmosphere Experiment in Amazonia (LBA), 260, 265
large-scale ecological processes, 161
Larus argentatus, 71
Larus canus, 70

402

Larus occidentalis, 68, 71
leaf wax carbon isotope analysis, 128
lemurs, 324
lesser kestrel (*Falco naumanni*), 78
life history, individual animal studies, 67
life history traits, 7
    bacterial competitive fitness, 28–9, *29*
    genetic aspects, 25
    trade-offs, 8, 10
lifespan
    artificial selection experiments, 8
    individual animal studies, 67
lifetime reproductive success, 67
    banded bird studies, 72–8
    definition, 72
    DNA analysis/extra-pair fertilization, 74, 76
    proportion of productive/unproductive individuals, 72, 74, **75**
    sparrowhawk (*Accipiter nisus*), 72–6, *73*, 74
long-eared owl (*Asio otus*), 84
Long-Term Ecological Research (LTER) sites, 388
Los Tuxtlas, Mexico, 329, 330, 331
Loxia curvirostria, 84, 86

macrophytes, shallow lakes, 365–6
mammalian defaunation, 319, 321–34
    commercial exploitation, 322
    direct causes, 322–3
    ecological/biodiversity consequences, 326
    forest fragmentation, 321–2
    global level, *324*, 324–6, *325*
    habitat destruction, 321
    indirect causes, 321–2
    subsistence hunting, 322–3
mammalian herbivory, 326–34
    beaver–cottonwood interaction, 328–9
    phytophagous insect interactions, *327*, 327, 328, 331–4, *332*, *333*
    trophic interactions, 326–7, *327*
    tropical forest understorey interactions, 329–34, *330*, *332*, *333*
marine benthos predation, *366*, 366
MARK, 87
mate retention, banded bird studies, 81, 82
mating systems, 55, 57–60
    DNA analysis/extra-pair fertilization, 74, 76
mazama deer, 322
meerkats, 53, *54*
Menidia menidia, 11
Mexico, mammalian defaunation, 321
microcosm experiments
    ecosystem productivity, 174–5, *175*
    plant perennial community biodiversity, genetic basis, 163–4, *164*
migration
    birds, return to breeding/wintering localities, 82, **83**
    response to climate change, 146–7
mineral nutrient recycling, 173
molecular analysis, 161
    arbuscular mycorrhiza root colonization, 104
    bird extra-pair fertilization, 74, 76

mycorrhizal fungi identification, 99
Montes Azules, 329, 330
Morone saxatilis, 11
mortality, genetic aspects, 25
MOSAIC, 216
multiple stable states, 357–80
    African grassland–woodland, 365
    detection, 367–73
        experimental design, 369
    evidence, 361–2
    fish recruitment cycles, 366–7
    haddock stocks, 367, *368*
    hemlock–hardwood forest succession, *364*, 364
    hysteretic shifts, 360–1, 367, *368*
    management issues/policy choice, 374–8, *376*
        decision analysis, 374–5, 377
    marine benthos predator dominance, *366*, 366
    phosphorus flux during lake eutrophication, 362–4, *363*
    rapid climate change, 362
    shallow lake phytoplankton vs. macrophyte dominance, 365–6
    theoretical aspects, 358–61, *359*
    time series data, 369, *370*, 370–1, *372*, *373*, 378–9
    trophic cascades in lakes, 366–7
Mummichog (*Fundulus heteroclitus*), 11
Musca domestica, 11
mutualistic interactions, 141
    see also arbuscular mycorrhiza
mycorrhizal symbiosis *see* arbuscular mycorrhiza
Myrmeleon immaculatus, 11
myxomatosis, 95

natural resource economics, 338
net ecosystem productivity
    measurement, 253, *254*
    terrestrial carbon sinks, 252–3
niche differentiation, 141, 190, 193–4, 195, 197, 198
    ecosystem processes impact, 140–1
    productivity differences, 141
nitrogen fertilization, 227, 228–9
    carbon sink influence, 258, 262
    small streams, 300, 309, 310
northern fulmar (*Fulmarus glacialis*), *70*
Norway spruce (*Picea abies*), 84
nutrient uptake length, 309

observational studies, 4, **5, 6**
ocean carbon sinks, 250, 251–2
    carbon turnover time, 274
    global warming response, 257–8
ocean circulation shifts, 362
Omphalea oleifera, 331
Ostrya virginiana, 216
Ovis aries, 60, 74
oystercatcher (*Haematopus ostralegus*), 68

palaeoclimate reconstructions, 362
Palmer Drought Severity Index (PDSI), 214
Panicum sphaerocarpon, 98
Park Grass Experiment, 186

Parus atricapillus, 70
Parus caeruleus, 72, 77–8
Parus major, 70
Passer montanus, 78
patch dynamics, hemlock–hardwood forest, 215–19
patchy resource distribution, 194
pathogens
    establishment in social groups, 49
    human activity-related introduction, 201
peccary, 322, 323, 329
pelagic trophic cascades, 366–7
Peromyscus maniculatus, 392
Perryssodactyla, 324
Phalacrocorax pencillatus, 71
Phanerozoic era
    atmospheric CO_2, 272, *273*
        global surface temperature trend simulations, 285–6
        plant biodiversity relationship, 284–6, *285*, *286*
        stomatal density relationship, 280–1, *281*, *282*
    carbon cycle, 272
    climate–plant interactions, 272–3, *273*
    extinction, 278, *279*
        plants as climatic indicators, 278, *279*, *280*, 280–4
        Triassic–Jurassic boundary, 278
    terrestrial carbon sinks, 272, *273*, 273
phosphorus availability
    arbuscular mycorrhiza transport, 104, 105
    carbon sink influence, 262
    human activity impact, 300
    lake eutrophication, 362–4, *363*
    legume CO_2 enrichment responses, 236–7
    sedges, 175, 176
    small stream ecology, 309
photosynthesis, 117
    atmospheric CO_2 enrichment response, 256, 257
    carbon sinks, 252, 253
    marine organisms, 258, 274
    nutrient supply limitation, 258, 262
    ocean, 252, 258
    see also C_3 photosynthesis; C_3/C_4 pathways
physiological ecological approaches, 115, 116
phytophagous insects, impact of mammalian herbivory, *327*, 327
    beaver–cottonwood interaction, 328
    tropical forest understorey, 331
    Urania fulgens–Omphalea oleifera interaction, 331–4, *332*, *333*
Phytophthora cinnamomi, 96
phytoplankton, shallow lakes, 365–6
Picea abies, 84, 107
pied flycatcher (*Ficedula hypoleuca*), 70, 78
Pinus edulis, 131
Pinus monophylla, 131
pinyon pine (*Pinus edulis*), 131
plant functional types, 147, 161–78
    architecture–climate relationships, 165
    competitor strategy *see* competitors
    ecosystem resistance/resilience, 168–73, 187
        drought responses, 170–1
        experimental climate manipulation, *172*, 172–3

principal component analysis, *172*, 172
protocol for testing predictions, *166*, 166–7
rarifaction/extinction, 167–8
global scale, 162
hierarchy, *162*, 162–3
human population density correlations, 167, *168*
local scale, 163
perennial community diversity, microcosm experiment, 163–4, *164*
productivity relationship, 173–8, 195
    habitat disturbance, 165
regenerative strategies, 170
regional scale, 162–3
ruderal strategy *see* ruderals
stress-tolerator strategy *see* stress-tolerators
plant–climate interactions *see* climate–plant interactions
plant–mammal interactions, 319–34
plethodontid salamanders, 311
Poephila guttata, 67
pollutant discharge, decision analysis, 374–5, 377
polygyny, 47, 55, 57–8, 62
    grazing behaviour, 57
    sexual size dimorphism evolution, 55, 57
population dynamics, 7
    adult sex ratio impact, 59–60
    mating systems, 55, 57–60
    sex differences in dispersal, 55
    sociality effects, 61–2
predation
    alternate state dynamics, 358
    marine benthos, *366*, 366
primary productivity
    atmospheric CO_2 enrichment response, 256
    biodiversity relationship, 190–1
        BIODEPTH experiment, 144, *145*
        plant diversity reduction effect, 141, *142*
    terrestrial carbon sinks, 252, **253**
Primates, 324
Proboscideans, 324
production functions, 339, 340, *341*
productivity
    biodiversity relationship, 173–6, 177, 183, 190–9, 287
        experimental studies, 194–9, *196*
        microcosm experiments, 174–5, *175*
        niche differentiation, 190, 193–4, 197, 198
        sampling effect, 174, 190–3, *192*, 195, 197, 198
    determinants, 173
    net ecosystem productivity
        measurement, 253, *254*
        terrestrial carbon sinks, 252–3
    plant functional types, 173–8
    terrestrial carbon sinks, 252–3, *254*
    see also primary productivity
Prunus serotina, 98
Pseudotsuga menziesii, 108
Puffinus tenuirostris, 70
Pythium, 98

Quercus gambelii, 131
Quercus turbinella, 131

rain forest
  carbon sinks, 253, 254
  vegetation cover–precipitation interaction, 288, *289*, 289
Rana sylvatica, 13
raptor dispersal, 84
reafforestation, 231
red deer
  adult sex ratio bias, 58–9, 60
  breeding success
    group size relationship, 49, *50*
    male body size, 55, 57
  culling rates, 61
  sex differences
    dispersal, 55, *56*
    survival, 57–8
reductionist methods, 161
  climate change experiments, 147–8
regenerative strategies, plant functional types, 170
reproductive suppression, 49–50
residual reproductive value, 71
resilience, ecosystem, 169
  plant functional types, 168–73
    competitors, 169, 170, 171
    ruderals, 169, 170, 171
  plant regenerative strategies, 170
  prediction, 169
  see also ecosystem stability
resistance, ecosystem, 168–9
  biodiversity relationship, 186–7
  plant functional types, 168–73
    stress-tolerators, 169–70, 171
  plant regenerative strategies, 170
  prediction, 169
  see also ecosystem stability
resource limitation
  bacterial adaptive evolution, 26–7, *27*, 40
  experimental plant ecology, 227, 230, 244
    definitions, 230
    relation to biomass, 230
rhizosphere population diversity, 100
Rissa tridactyla, *70*
rock lobster (*Jasus lalandii*), 366
rodent-eating owl dispersal, 84, *85*
rodents, 324
  population fluctuations, 391, *392*
ruderals, plant functional types, 165, 167, 168
  ecosystem resilience relationship, 169, 170, 171
    regenerative strategies, 170
    mineral nutrient recycling, 173
Rumex acetosella, 96

Salvelinus fontinalis, 311
sampling effect, 141, 195
  competition, 201
    antagonistic efficiency, 191, 193
    resource exploitation, 190–1
  diversity–productivity relationship, 174, 190–3, *192*, 197, 198
  soil microbe–plant interactions, 98

Scottolana canadensis, 11
Scutellospora, 104
Scutellospora dipurpurescens, 106
sedges
  dauciform roots, 175
  phosphorus capture, 175, 176
seeds
  ecosystem resilience, 170
  synthesized ecosystem experiments, 174, 175–6
sex differences, 47
  dispersal, 53–5, *56*, 62
  size dimorphism, 55, 57, 62
    energetic requirements, 57
    impact on grazing behaviour, 57
    survival influence, 57–8
sex ratio
  at birth, 57–8
  polygyny-related adult bias, 58–9, 62
Seychelles' warblers, 52
short-eared owl (*Asio flammeus*), 84
short-tailed shearwater (*Puffinus tenuirostris*), *70*
silicate bed rock weathering, 272
Sitta europaea, 78
slackwater darter (*Etheostoma boschungi*), 311
Soay sheep, 60, 74
sociality, 47–63
  competition
    between groups, 50–6
    within groups, 49–50, 61
  conservation implications, 60–1
  cooperative breeders, 50–1, 61
  culling rates, 60, 61
  density-dependent survival/breeding success, 49
  disease/parasite prevalence, 49
  dispersal effects, 53–5, *54*
  inbreeding avoidance, 53–5
  inverse density-dependent processes (Allee effect), 50, *51*, 51, 52–3
  mating systems, 55, 57–60, 62
  range size, 60, *61*
  reproductive suppression, 49–50
  sex ratio at birth, 57–8
soil
  bacterial diversity, 99
  carbon balance
    carbon stores, 252, 255
    climate warming response prediction, 260–2
    respiratory flux, 253–4, 257, 260, 261
    temperature increase response, 257
  microbe–plant interactions, 95
  see also arbuscular mycorrhiza
Soldago missouriensis, 104
Sorghastrum nutans, 96
sparrowhawk (*Accipiter nisus*), 67
  age-related breeding/survival, 68, *69*, *70*, 71
  dispersal
    breeding, *80*, 80–2
    mate retention relationship, 81, *82*
    natal, 76–8, *77*
  lifetime reproductive success, 72–6, *73*, *74*

405

INDEX

spatial variation, 53
species richness *see* diversity
species–area relationship, 202–3
spider monkey, 321
steppe–tundra transition, 365
Stercorarius parasiticus, 68
Stercorarius skua, 68, 71
stomatal aperture, CO_2 enrichment response, 237
stomatal density
    atmospheric CO_2 enrichment response, 256
    Phanerozoic atmospheric CO_2 trend, 280–1, *281*
    Triassic–Jurassic boundary, 281, *282*
stream ecology, 295–313
    biodiversity, 310–11
    cataloguing small streams, 301–2
    detrital food webs, 296–7, *297*
    headwater stream loss, 302–7, *303*
        ecosystem processes/biotic impact, 307–11, **308**
    headwater streams, 296–8
    impact of human activities, 295, 296, 298–301, 304, *305*, 305, *306*, 307, 312
    nutrient dynamics, 296–8
    nutrient uptake length, 309
    river continuum concept, 298
stress-tolerators, plant functional types, 165, 167, 168
    ecosystem resistance relationship, 169, 171, 172
    mechanical damage, 170
    mineral nutrient recycling, 173
    spectrum of resistance, 170
striped bass (*Morone saxatilis*), 11
Strix nebulosa, 84
subsistence hunting, 322–3
sugar maple (*Acer saccharum*), 212
    see also hemlock–hardwood forest
SURGE, 87

T2 virus, *Escherichia coli* dynamic interaction, 37–8, *38*, 39
T4 virus, *Escherichia coli* dynamic interaction, *34*, 34–7, 40
    prevention of competitive exclusion, 35–6, **36**, 39
    resistant mutants emergence, 34–5, *35*, 37
T5 virus, *Escherichia coli* dynamic interaction, 36, 37, 40
    viral extinction, 36, 37, 39
tapir, 321, 322, 323, 326, 329
Temminck's stint (*Calidris temminckii*), *70*
Tengmalam's owl (*Aegolius funereus*), 84, *85*
territorality, 47–8
    breeding site fidelity, banded bird studies, 78, **79**, 80–1
Thuja plicata, 108
Tilia americana, 214, 216
time scale of genetic/ecological change, 26, 41–2
time series data, alternate states detection, 369, *370*, 370–1, *372*, 373, 378–9
trade-offs, 3, 7
    ageing evolution, 7, 8–10
    arid-land ecosystem precipitation utilization, 133

bacterial competitive fitness, 32, 33, 40–1
    viral predator interactions, 36–7, 39
C_3/C_4 photosynthesis quantum-yield model, 117, *118*
coexistence, 39, 194
    plant functional types, 165
    resistance/resilience, 169
tree sparrow (*Passer montanus*), 78
tree-seed-eating bird dispersal, 84, 86
Triassic–Jurassic boundary
    extinction event, 278, *279*, 283
    leaf material carbon isotope ratios, 281
    simulated atmospheric CO_2/temperature changes, 281, *282*, *283*, 283
    stomatal density, 281, *282*
Trifolium, 236
Trifolium pratense, 144
trispot darter (*Etheostoma trisella*), 311
trophic cascades, pelagic ecosystems, 366–7
tropical forest
    deforestation, 319, 320
    understorey, mammalian herbivory impact, 329–34, *330*, *332*, *333*
Tsuga canadensis, 212
    see also hemlock–hardwood forest
turbinella live oak (*Quercus turbinella*), 131

ungulates
    adult sex ratio bias, 58–9, 60
    economic yield, 60–1
    polygyny, 57
Urania fulgens, 331
Urania fulgens–*Omphalea oleifera* interaction, 331–4
    herbivory, 331–2, *332*
    larval density, 332, *333*
    larval parasitism, 332–3, *333*
    mammalian herbivory influence, 333–4

vertebrate sociality *see* sociality
vesicular–arbuscular mycorrhiza *see* arbuscular mycorrhiza
virtual rent, 339
Vulpia ciliata, 105

Web-FACE, 234
weeding activities, 350
western gull (*Larus occidentalis*), 68, 71
whelk (*Burnupena papyracea*), 366
windthrow damage, forest disturbance, 213, 214
wintering site fidelity, migratory birds, 82, **83**
wood nuthatch (*Sitta europaea*), 78

yellow birch (*Betula alleghaniensis*), 213–14, 216
yellow-breasted chat (*Icteria virens*), 78

zebra finch (*Poephila guttata*), 67